# North American
# Reference
# Encyclopedia of  Ecology And Pollution

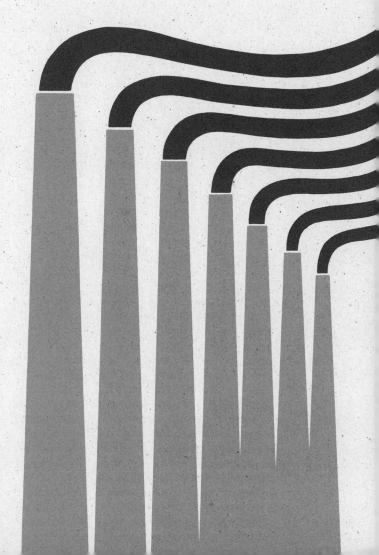

# North American Reference Encyclopedia of **Ecology And Pollution**

EDITORS
William White Jr., Ph.D.
and
Frank J. Little Jr., Ph.D.

North American Publishing Company
Philadelphia, Pennsylvania

*International Standard Book Number*
0-912920-06-8
*Library of Congress Catalogue Card Number*
72-84736
*Order Number*
7211


Staff:
Editorial Director:    Dr. William White Jr.
Editor: Dr. Frank J. Little Jr.
Research Editor: John M. Mankelwicz.
Production Assistance: Bernadette McGauley.
Illustration: Robert Richards and Anthony Hillman.

Irvin J. Borowsky, Publisher
North American Publishing Co.
134 North Thirteenth Street
Philadelphia, Pa. 19107

# CONTENTS

# Preface

This third volume of the North American Reference Library continues the innovative method of organizing and presenting material for reference and study. The articles are all written with the aim of presenting "the state of the art" of ecology to readers of all levels. The individual articles are cross-referenced in terms of both concepts and scientific principles, while the exact terminology of each special approach is repeated in the index.

The distinguished scholars whose opinions and facts are included in the following pages have all published papers in their areas of expertise. In many cases they are the persons who coined or assigned the meanings to the terms they employ. Following the format and purpose of the two previous volumes in the series, the editors have sought clarity and timeliness first of all. As in any project of this size a large number of people have contributed to its success. We would like to acknowledge the invaluable aid of some of them here. Special thanks must be extended to Dr. Howard T. Odum, Dr. Jay M. Savage, Dr. Robert J. Byers, Dr. M. Taghi Farvar, Dr. Ariel G. Lowey, Dr. Vivian Nahemias, and Dr. James Bolke. Materials including photographs were kindly contributed by Dr. John J. Craighead, Dr. A. B. Graf, Dr. S. I. Auerbach, Mr. Daniel Saults, Dr. Robert N. Colwell, and a large number of institutions including The Oak Ridge National Laboratory, The National Forestry Service, Bio Sciences Information Service, Argonne National Laboratory, The National Aeronautics and Space Administration, The Department of the Interior, and the New York Scientists' Committee for Public Information.

The libraries and collections found to be most useful and which graciously extended facilities to the project were The Academy of Natural Sciences, The University of Pennsylvania, Princeton University, Columbia University, Drexel Institute of Technology, Johns Hopkins University, The Free Library of Philadelphia, The Philosophical Society, The Thomas Jefferson University, Temple University, Haverford College, and The University of Toronto.

Through inter-library and other loans, material was drawn from the repositories of the University of California, Berkeley; University of Florida; University of Michigan; University of Montana; University of Georgia and many others, all of whom deserve our appreciation.

# A Preface On Paper

In producing a major reference compilation of original and primary statements on ecology it is necessary to practice what we preach. Few technical processes are as old or as basic to civilization as the making of paper. Unfortunately few processes are as messy, requiring large amounts of water and producing massive volumes of waste and sludge. Since all paper is made from some naturally derived fiber papermaking is a process which lends itself to recycling.

Throughout ancient and early medieval times civilization's chief writing material was probably papyrus, a reed that grows widely in the Eastern Mediterranean coastlands. It was so strong and durable when made into sheets that it could be used several times over to conserve expense. The old ink was scraped off, the fiber pumiced or rechalked and the new writing set down on top. This sort of recycling led to the formation of one of the strangest literary items, the palimpsest, a manuscript with two or more totally different texts written on top of each other. A much later example of this is a quaint little book once owned by the Anglo-American actress, Laura Keene (1826-1873) which contains popular French stage songs printed over *assignats*, the devalued currency of the French Revolution.*

After the fall of Rome and the triumph of Judeo-Christian thought over paganism, many thousands of Greek and Latin manuscripts containing classical texts were summarily recycled by being carried from the library to the privy. This direct connection between the academy and the outhouse has continued through the centuries.

By the advent of the thirteenth century, the copying of books by hand had reached a high point of art and skilled copyists would labor for years over a single issue of the Latin Bible, a prayer book, or a courtly tale. With the general dawning of something like culture after the Renaissance the Chinese art of paper making made its way to Europe. In time it became a highly skilled trade and its chief staple of raw fiber was the linen and cotton rag. The invention of printing by an obscure German in the Fourteenth Century, brought with it the innovation of many other sublime arts. However one of the less than heavenly aspects of the trade was the recycling of precious medieval manuscripts and incunabula (books printed before 1500 A.D.) to produce the gray colored heavy paper stock used to wrap better grade writing papers.

This wrap was called in Old Dutch, *"monnikegrauw,"* which meant "monk's gray." The name had reference to the practice of the Calvinist craftsmen of Amsterdam by which they recycled the books written by the monks of the Catholic church. In 1636 a law was passed by the burghers which levied a tax upon the consumption of old books by the paper mills. The argument against the tax stated by the paper makers' guild maintained that since the makers, publishers, and owners of the originals were dead, the materials were "sent from the other world" and thus could not be taxed in this one.

This sort of recycling was celebrated in a Polish poem from the district of Kalish dating from the same Seventeenth Century. It points up the art and thrift of papermakers who were able to retrieve rags from the dung hills from which they made paper for various noble purposes which was still later used for wrapping and finally for the purpose which returned it to the dung hill, a complete and economical recyclement.

In 1695 a little known Dane, one Georg Balthasar Illy, wrote to the king expostulating on his new found techniques for turning old printed paper into new white writing stock. Just how he deinked the paper is unknown, probably by a series of soaks and washes, possibly with a natural digester. However some were not really impressed by such progress and attempted to find other sources of natural or raw fiber. One such experimenter, Dr. Christian Schaeffer of Regensburg, Germany tried all manner of natural materials for papermaking, among these were weeds, grasses, hornet's nests, potato skins, straw, and barks from a wide variety of trees. He began publishing the results along with sample pages of his researches in 1765. This redoubtable scientist also suggested that wood itself might be used.

---

* This and other specimens mentioned in this preface have been shown to the editors by Mr. Henry Morris, proprietor and printer of the Bird and Bull Press and an expert on the history of paper and printing. Details and original documents have been published in his books such as: *Five On Paper, Omnibus, Old Ream Wrappers,* and *The Bird & Bull Commonplace Book.*

It was the English papermaker, Mathias Koops, who continued to experiment until he discovered a process to manufacture paper from wood fiber. His first success was noted in 1801. From that moment on the need for recycling became much greater. Throughout the Nineteenth Century rag paper was still the standard. The demand for rags was met from every conceivable source. The Italian circus strongman Belzoni had popularized the great caches of mummies to be found in Egypt. Since the Arabs had used ground mummy dust as a medicine and British had fired up the boilers of their trains with them, there must have been a goodly supply. An enterprising businessman from central New York imported them by the boatload and used them (at least the wrappings) for paper pulp. Not a few Victorian novels were published upon the last will and toga of some decayed Egyptian.

Although the methods existed for removing writing inks by a series of soaks and washes, it was as yet impossible to consistently remove the heavy oil-based inks employed by printers. The ultimate solution was to use chlorine bleaches which formed very toxic pollutants when released into the local water supply.

With the vast tonnage of paper made and used yearly in the United States, which requires the removal of millions of trees and the production of noxious chemicals and millions of pounds of waste, it is absolutely necessary that paper recycling be practiced. Unfortunately there is a good deal of confusion in the public mind as to precisely what is involved in recycling.

Paper makers have always practiced some method or technique of recycling ever since paper was invented. Usually this involved taking the odd scraps and cuttings of paper around the establishment and rerefining them along with the new fiber. However this yields no real saving of trees or linen or whatever, and does nothing to relieve the vast piles of waste paper already consumed by the population.

The real crux of the matter is whether or not the recycling took place before or after the paper was used by the consumer. This is currently defined as pre- and post-consumer waste. Again even within the handling of post-consumer wastes there is a recycling which involves bleaching out the inks and resins and which adds a great deal of fouled water and strong chemicals to the already over-burdened eco-system. The ecological purist should thus seek the utilization of recycled, non-deinked, post-consumer waste fiber.

That is precisely what these words you are reading are printed upon.

*To our knowledge this is the first reference book in the biological sciences, and the only one in ecology printed on completely recycled material. If the color or texture is new to you then perhaps you should take note of this "first" in the long history of papermaking and printing. The book you are holding not only presents ecology and champions conservation, it is ecological and it does conserve.*

# Chapter 1

# Introduction to Reference Encyclopedia of Ecology and Pollution

Inventing possible futures was a glamorous intellectual sport in the 1950's and 1960's. It was then fashionable to assume that most of man's ancient problems would soon be solved by science, and that the year 2000 would mark the dawn of a technological utopia—physical work would be entirely taken over by robots; dreams would be programmed at will; disease and pain would be eliminated by magic drugs; there would be permanent stations on the moon and on the floors of the oceans.

This euphoric mood had almost completely disappeared by 1970. Indeed, the pessimists now believe that environmental pollution and the depletion of natural resources are taking industrial societies on a suicidal course and that the year 2000 might well be the beginning, not of an era of comfort and plenty, but rather of a gloomy sunset for the human race. Even those who do not take such a tragic view of the present environmental crisis are inclined to believe that there will be little chance for the development of esoteric new technologies in the near future, because we must devote our energies to the urgent task of correcting the damage which has been done to earth by thoughtless industrial growth. Many students of social systems have recently emphasized that the inescapable constraints on the size of the world population, the limited supply of natural resources, and the rapid accumulation of pollutants, have brought us close to the limits of industrial growth. The widespread awareness of the limitations of the spaceship earth makes it likely that in the years to come, technological glamor will have to take second place to pollution control.

A striking illustration of the change of attitude with regard to the near future was provided recently by the results of a survey conducted among some 3,000 Japanese scientists—who were asked in 1971 to state what they regarded as the most likely scientific and technologic development for the year 2000. It would have been expected that in view of Japan's phenomenal successes in electronics, this field would have headed the list of further advances. But in fact, the development of the TV-telephone was almost at the bottom of the list, whereas pollution control was at the top.

In the section of the survey dealing with the social aspects of science and technology, 98 percent of the Japanese scientists suggested that the most important achievements over the next 25 years would be "the determination of the level of tolerance to pollutants by human beings." Furthermore, 90 percent of them stated that it was of the greatest importance to establish "a network of regional monitoring and warning systems" for pollutants. At about the same time, one member of the Japanese government expressed in the strongest possible terms the view that technological prosperity had been achieved at the cost of an intolerable deterioration of the Japanese environment: "The government . . . has done little to regulate the activities of polluting businesses except when the harm was apparent. Environmental disruption has proceeded in a creeping way without drawing much attention at first, but then, with the quickening of technical progress . . . it has developed into a *monstrosity*." (italics mine) It is unfortunate that few American and European statesmen have dared be as explicit with regard to environmental degradation in their respective countries.

The massive production of environmental pollutants and the accumulation of solid wastes constitute of course disgraceful characteristics of all industrial societies. But contrary to general belief, these problems do not result from the fact that modern man has suddenly acquired wasteful and careless habits. Waste accumulation antedates industrialization and urbanization. Its origins can be traced far back into prehistory. Gorillas, chimpanzees, and other great apes living in the wild are notoriously wasteful and careless. And so was early man.

Caves occupied by Neanderthal people some 100,000 years ago have been found to be littered with animal bones and stone artifacts. More recent archeological sites likewise contain immense amounts of bones, tools, and fragments of pottery. In fact, archeologists thrive on the accumulation of solid waste abandoned by early man. The artifacts made of stone, ivory, or pottery which litter prehistoric sites are the equivalent of the gadgets, plastic containers, and aluminum cans with which we litter our landscapes. The gross pollution of water in many primitive settlements and John Evelyn's famous account of London's air pollution in the seventeenth century make it clear that polluted environments have been the rule rather than the exception, long before the Industrial era. As to the biological, chemical, and waste pollution during the early phases of the Industrial Revolution, one needs only read the testimony of novelists like Dickens or scientists like Faraday to realize that the good old days were even more shocking to sensitive people than our slums are today.

One century ago, industrial cities all over the Western world were indeed vastly more polluted than they are now, and they were especially much dirtier. The only saving grace was that most solid wastes were readily biodegradable and were commonly eaten by animals. As late as the mid-19th century, it was taken as a matter of course that pigs roaming on Broadway should act as scavengers for the garbage dumped on the street pavement. A clean environment, free of pollutants, is an ideal which emerged slowly from the late 19th century Sanitary Revolution.

There is no evidence that we are more wasteful or careless than our prehistoric ancestors or than our 19th century predecessors. But what is certain is that our wastes and pollutants are chemically different from those of the past and that we produce much more of them—both collectively and per capita. Even more important perhaps is the fact that because many of the wastes and pollutants of modern industrial civilizations have been produced on a huge scale only during recent decades and because most of them are now on earth neither nature nor man have had time to develop adaptive mechanisms to deal with them.

In the course of evolution, there have appeared an immense variety of bacteria, fungi, protozoa, and insects, and other living things capable of decomposing the wastes of animals and plants. Under normal conditions, leather, wood, cotton, paper and all the common organic materials of natural origin used in human artifacts are rapidly decomposed in soil or water by biological and chemical action. Even the old-fashioned tin can will break down in 5–10 years. Since all organic materials were biodegradable until a few decades ago, they had little chance to accumulate in nature. But the natural processes of decomposition fail to operate effectively when the environment is grossly unnatural, as is commonly the case in modern urban industrial areas. Furthermore, and most importantly, nature does not know how to deal with

aluminum, steel, plastics, and a host of other industrial man-made products which are new and with which it has had no experience in the evolutionary past.

Like the other organisms in nature, human beings are ill-equipped biologically to deal with many of the man-made substances produced by modern technology and which have become inescapable parts of daily life. We still operate with the same genetic equipment as our Stone Age ancestors, and therefore have the same biological limitations as well as the same physiological needs. We have no instinct or organ to warn us of the dangers that lurk in an invisible beam of radiation or in toxic substances which are tasteless and odorless in the concentrations commonly encountered in the so-called "normal" environment of technological societies.

Some of the most important health problems in the modern world arise from environmental threats which are not perceived by the senses and which do not produce acute toxic effects; as a consequence, these cannot be readily linked to their causes. More than ever before, the health of modern societies is thus threatened by a "pestilence that steals in the darkness." For example, the increase in the incidence of chronic pulmonary disease, of certain forms of cancer, and of some genetic defects can almost certainly be traced to pollutants which are ubiquitous in urban industrial areas, but which are commonly present in concentrations low enough not to be readily detected. We absorb lead, asbestos, pesticides, and probably thousands of other potentially toxic and mutagenic chemicals without even being conscious of their presence in our environment.

Everyone is now aware of the gross aspects of pollution of air, water, and food, but the more subtle and probably more dangerous pollutants still escape notice. While Pittsburgh boasts of having gotten rid of soot, the clear air of that city is still contaminated with invisible chemicals which have long-range toxic effects. Similarly, water which has been purified by the ordinary processes of treatment can be clear yet dangerous; dirt is removed but poisons remain.

The detection of potentially deleterious substances is rendered even more difficult by the fact that the effect of many of them are extremely slow in developing. Since years may elapse before the damage becomes manifest, elaborate epidemiological studies are usually required to establish cause-effect relationships. Such is the case for radiations, cigarette smoking, inhalation of asbestos, and of mutagenic substances, and probably indeed for the great majority of noxious agents.

A further complication arises from the fact that many substances which exert no apparent ill effect in normal organisms may become dangerous under conditions of physiological stress. It has long been known for instance that lead stored in the bones is suddenly released into the general circulation during episodes of pneumonia. Many toxic or carcinogenic compounds are stored in a harmless state in the liver by conjunction with glucoruonic acid. The process of detoxification, however, can be impaired by liver disease; it can be swamped by overload with other ligands and it can be reversed by glucuronidase in the bladder.

Recent studies have established that DDT, dieldrin, and other halogenated pesticides which do not cause obvious damage when stored away in body fats can generate gross and even fatal pathologies when they are suddenly released into the general circulation. Release can occur in various forms of physiological stress such as pregnancy, lactation, food deprivation, intoxication or acute infection, or even as a result of social disturbances associated with crowding. Synergistic effects have been described in several other systems—for example in systems involving a polychlorinated biphenyl hydrocarbon (PCB) and the duck hepatitis virus, or the enzymes used in laundry products and bacterial pathogens.

Thus, many different kinds of physiological or infectious stresses can act as triggers to reveal potential toxicities that remain unmanifested under conditions of health.

The results of the Japanese survey mentioned at the beginning of this essay placed great emphasis on the need to determine the "level of tolerance to pollutants by human beings." But the determination of levels of tolerance is extremely complex for the reasons mentioned in the preceding paragraphs. Since toxic potentialities may not become manifest until after long periods of time following initial exposure, or after they have been triggered into activity by unrelated physiological stresses, the very concept "level of tolerance" involves factors which do not lend themselves to definition, let alone measurement.

There is reason to fear that in practice the phrase "level of tolerance" will be equated with concentrations of pollutants which are sufficiently low not to cause obvious immediate disturbances. Yet such low levels will in many cases be nevertheless capable of causing harm to a large percentage of the population in the long run. Few

persons are seriously handicapped in their daily activities by the concentrations of air pollutants presently considered acceptable, yet these concentrations are almost certainly capable of increasing the incidence and severity of chronic pulmonary disease.

New kinds of toxicological and epidemiological studies will certainly be required to determine the long-range and indirect effects of environmental pollutants. These studies will be costly, but much less than advertising. In any case, they should be regarded as legitimate parts of the cost of research and development of new technological innovations. No product coming under control of the Food and Drug Administration can be released for general public use until it has been proven to be effective and safe. It does not seem unreasonable to hope that a similar social philosophy will be formulated for all other technological innovations.

Fears have been expressed that rigid safety controls over new procedures and new products would retard and even prevent technological progress. But this need not be so. A sophisticated scientific society could certainly learn to devise administrative methods for differentiating between innovations that have real potential usefulness and are therefore worth a gamble, and other innovations which are promoted chiefly or even exclusively for commercial advantages. There was no doubt from the beginning of the biological tests that DDT had great potential usefulness in medicine and agriculture, a fact which provided justification for distributing it widely before completion of long-range toxicological studies. In contrast, certain food additives have no real usefulness and should not be released to the general public until thoroughly tested for safety under practical conditions.

As recommended in the Japanese survey, monitoring and warning systems should be established as essential components of pollution control systems. The limitation of monitoring, however, is that many agents which are potentially dangerous have not yet been recognized as such. Asbestos and PCB were put on the danger list only a very few years ago and new kinds of substances constantly come under suspicion, for example, monosodium glutumate and saccharin. Thousands of new products are put into circulation every year without any significant toxicological studies. Furthermore, a very large percentage of the microscopic particulate matter in the air of cities is still chemically unidentified; yet it is probable that some of the colloidal components which are small enough (below 5 micra) to reach deep into the lungs can exert deleterious effects—were it only by adsorbing toxic substances, mutagens, and carcinogens.

In the present state of ignorance concerning the biological activities of most substances and procedures used in everyday life and in view of the fact that testing facilities will long remain inadequate, it is unavoidable that many technological innovations will be marketed and widely used before their potential dangers have been recognized. Tragedies, analogous to the production of tumors by ionizing radiations, of deformities by thalidomide, or by emphysema and mesothelioma by asbestos will continue to happen, not due to negligence, but to lack of information.

The spread of disasters caused by new toxic agents might perhaps be kept to a minimum by an adequate warning system. One can envisage a kind of prospective epidemiology, so designed as to alert public health authorities concerning new trends in pathological conditions. For example, the systematic testing of population samples for abnormalities in certain physiological and hematological characteristics might help in the early detection of biological disturbances, and in the identification of their origin, before the process had reached epidemic proportion. But while such a science of prospective epidemiology is readily imagined in theory, its practical instrumentation would be costly and difficult. The testing of adequate population samples is a very complex affair. And the interpretation of the findings will be greatly complicated by the fact that many new substances are constantly being introduced into daily life. There will be many candidates among them for each new physiological abnormality. One needs only remember the vast numbers of air pollutants which are suspected of creating pulmonary disorders, or the nutritional habits which have been suspected of playing a role in the increase in cancer of the colon.

It has been argued that the human and ecological problems created by technological societies can best be solved by imitating as closely as possible natural processes— on the assumption that "nature knows best." But in reality "nature" knows only how to deal with situations it has experienced repeatedly in the evolutionary past. It is usually impotent in face of the new situations created by man because these rarely have precedents in natural processes.

In the phrase "nature knows best," the word nature denotes the self-sustaining, self-reproducing systems which have come into being by evolutionary adaptation. Natural processes imply the operation of integrated mechanisms through which such systems maintain themselves in dynamic equilibrium. In technological societies, however, human interventions commonly introduce new and powerful forces which disturb the equilibrium. The disturbance cannot be corrected by the system itself because nature, i.e., the system in question, has had no experience with these forces. Man who is the creator of the disturbances must accept responsibility for the management of their consequences. While "closing the circle," to use Barry Commoner's felicitous phrase, is as essential in a human controlled system as it is in natural economy, the ways to close the circle differ from case to case. Nature knows how to do it in the situations which are the result of undisturbed evolutionary processes, but man must devise new ways for the new situations he creates.

In particular, as already mentioned, we have no instinct to warn us of the dangers posed by modern pollutants, and most of them indeed cannot possibly be detected by our senses. Neither can we in many cases determine a priori how certain technological innovations will affect ecological systems. The more innovative a society is, the more it will have to depend therefore on scientific knowledge as a substitute for the deficiencies of our senses and our commonsense, as well as for the deficiencies of nature.

The effects of ecological disturbances and pollutants on human beings usually take into account only physical health. But other criteria should also be considered. Environmental pollutants cause damage to buildings, plants, and animals—in cities, along highways, and on the farm. Environmental pollutants, furthermore, lower the quality of life by spoiling the fragrance of the air and the taste of the water, and by decreasing the luminosity of the sky during the day and masking the Milky Way at night.

It is impossible to evaluate the social importance of the multifarious effects of environmental degradation, in part because of their ill-defined complexity and even more because practically all of them involve values which transcend economics. How can one evaluate the contribution to the quality of life of a luminous, fragrant Spring day, or of a sparkling river which invites little boys to go swimming, adults to go fishing, and lovers to go dreaming. But measurements are hardly needed to show that economists and technologists are taking a narrow view of the problem when they emphasize exclusively— as many of them are prone to do—the cost of pollution control. The more important question probably is "How much does it cost society *not* to control pollution?" or even more importantly "Can a society long remain desirable if it does not control pollution?"

Technological societies are only now beginning to concern themselves with the creation of desirable environments. So far they have been satisfied with attempts to prevent environmental degradation from going beyond socially acceptable limits. The phrase "socially acceptable" is loaded with ethical implications. Human life is an adventure and any form of progress implies risks, including risks involving environmental quality. Man in general, Western man in particular, is a hunter after the unknown, wherever the hunt may lead him. But technological societies cannot afford to ignore that they will experience human and ecological disasters if they do not learn to give as much attention to the quality of the environment, as they do to the production of goods and the accumulation of wealth.

René Dubos, Ph.D.
Professor Emeritus,
The Rockefeller University,
New York City.

# Chapter 2

# Introduction to a History of Ecology

## Definition

The term "ecology" was probably coined during a casual conversation sometime in the early 1860's. From its obscure beginnings it has developed over the intervening century into a moral alarm and a political watchword. The term was synthetically formed from two roots common in classical Greek, οικος *oikos*, "house" and λογος *logos*, "word" or "study" to yield a new combination meaning, "the science of surroundings or environment." The original German form of the term, *"oecologie-sche"* was either coined or introduced by the biologist Ernst Haeckel (1834–1919) or one of his friends or associates in the Jena circle of intellectuals. It was many decades later that the term found a specific application and received precise definition. Haeckel first employed the term in a public address in 1869 in reference to the body of knowledge then being gathered about the relationships of organisms to their organic and inorganic environment.

Haeckel was one of the more romantic figures in an age when romanticism was the popular mode of science. He was widely read and admired among the Darwinians of England and so the term soon crossed the channel and took the English form, "ecology." However it was not regularly used in the biological literature until the very end of the century. Haeckel and his students extended the ecological concept to both biotic (living) and abiotic (non-living) components of environment, subject model, figure 1.

In the century or more since its appearance the term "ecology" has taken on two different although related meanings.
  a. Ecology = the body of knowledge and the study of the interactions and interrelationships between living organisms and their environments.
  b. Ecology = the study of the economics, the gain and loss of energy by organisms, the dynamics of organisms in their environment, the total effect of all organisms in a locality.
Since the leading biological researchers of the mid-nineteenth century were concerned with systematic classification (Taxonomy) and morphology (Anatomy and Physiology) these two methods of experimentation and research dominated early ecology. The term came to be applied primarily to the taxonomic and anatomical/physiological characteristics of the interaction of an organism with its environment.

Ecology developed as a subordinate application of Darwinian organic evolution. Haeckel himself defined ecology in relation to Darwin's notion of the development of organisms. For nearly fifty years after its appearance ecology was interpreted by Darwinian concepts. When their veracity began to be questioned, ecology suffered a period of reduced emphasis.

In the twentieth century molecular biology and biophysics developed from the chemical and electronic technology which was produced after World War I. The functions, behavior, and morphogenesis of organisms are subjects of ongoing research. The increased awareness of the interdependence of living things has been reinforced by the extraterrestrial vantage point of space travel. This emerging consciousness will no doubt produce another phase in the understanding of ecology. This will be an embracing, functional concept of the whole unified earth, a biosphere. Mutual interdependence will be seen as the basis and condition for life. Such views are developed in the pages of this book. The ecology of the near future will be subordinate to a total systems insight.

The growth of biological science has caused the multiplication of specialties and a number of newer fields built upon ecology as the basic unit. Among them are ecological geography, micro-ecology, paleoecology, and ecological evolution. It may easily be said that no major modern biological field is totally divorced from the impact of ecology. Two different applications of the term have developed in spoken and literate usage:

### • The Scientific Meaning
Ecology is used as a scientific term in most of the world's written languages as the theoretical and applied study of the interdependence of the

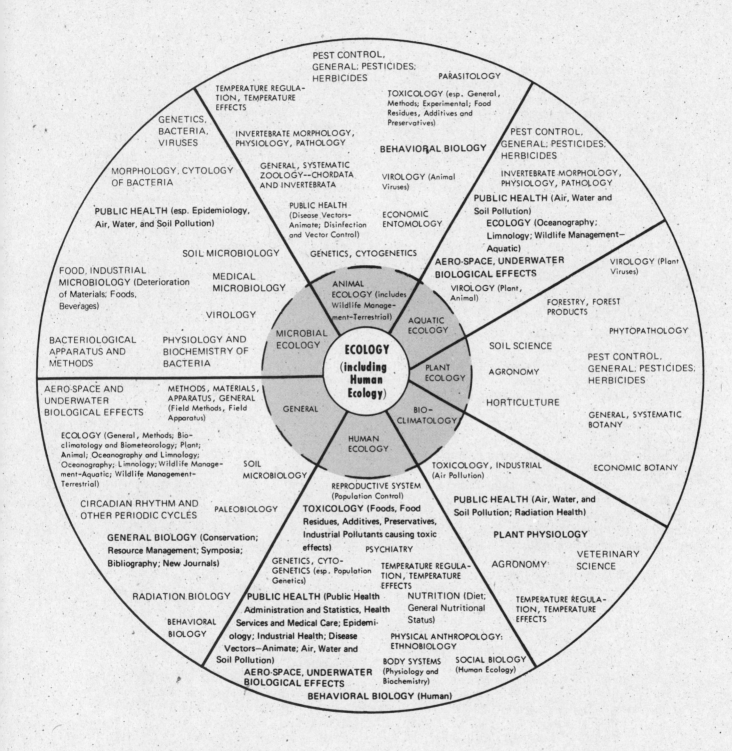

**Figure 1.** A subject-model of the relationship of Ecology to the other special sciences reprinted courtesy of Bio Sciences Information Service, Philadelphia, Pennsylvania.

biosphere, the impact of environment on organisms and the impact of organisms on the environment. It is a universally recognized branch of biological science, a research field supported by government and private resources, and a curricular subject at all levels of education.

• **Popular Usage**

The common everyday speech and mass media employ the term in reference to the present environmental crisis. This involves a negative judgment against pollution and a positive desire to improve the physical conditions of life. Ecology represents in the minds of most citizens the effort to obtain pure water, air and food and the restriction by law and prohibition by authority of activities and operations which would distribute organic and inorganic wastes and pollutants. It also assumes the context of natural resources.

**The Rise and Development of Ecological Science**

The inquiry into the relations of organisms to environment began in the earliest days of biology, although the term denoting this area of research developed late. The peoples of the early river-valley civilizations of the ancient Near East developed the breeding of domestic animals and the knowledge of agriculture. Some few treatises on these subjects were set down in their complex writing systems. Both the Greeks and the Romans were fascinated by travel stories and wondered at animals and plants brought from distant places. A number of adventurers wrote about the fauna and flora that they had seen including no less a teller of tales than Herodotus of Halicarnassus (Fifth Cent. B.C.). A century and a half later Aristotle sailed among the Greek islands in search of marine life which he examined and dissected. He attempted to write a catalogue of animals only as it lent evidence for his cosmological theories. All told he makes reference to over five hundred species now recognizable. However his essential classification rested upon four characteristics the first of which was the animals' mode of life. His student, Theophrastus, (c. 287 B.C.) wrote extensively on botany and improved the technique of classification.

When Alexander's Empire dissolved into petty kingdoms in 322 B.C., Greek intellectual dominance was lost over the whole of the Mediterranean. The Roman civilization took up the challenge of culture but without the philosophical gift of Greece. The Roman scholar Pliny the Elder (A.D. 23–79) employed an environmental scheme in classifying animals. This was the framework on which the collected wisdom of the ages concerning animals and plants was hung. Although trial and error experimentation was attempted in his day, Pliny was satisfied to rest upon the past. In the next seven hundred years the dead hand of classical tradition was to weigh heavily upon the mind of Christian Europe. The first dawning of a new scientific spirit was signaled by Islam. Through Muslim and Jewish scholars the amassed wisdom of classical antiquity passed through Byzantium and traveled to Italy. Out of the confusion of Italian duchies and mercantile cities medicine and biology along with art and literature were reborn. With the Renaissance of the Fourteenth Century, the pursuit of knowledge by scientific inquiry spread over the Alps and through all of Europe. In its wake a new consciousness of the world and its variety of peoples and places swept over all levels of society. In another century the voyages of discovery and colonizing expeditions were sending back to Europe a vast array of life forms never before seen by Western Man. In specimen or anecdote the organisms of the earth were collected, studied, and classified. The ancient study of "natural history" blossomed anew.

The social and political upheavals of the Protestant Reformation and the wars of religion raised up a new social entity, the middle class, which had been quietly evolving in the womb of medieval feudalism. Overnight the wisdom of the ages became nonsense and heresy became fashionable. New physics, new anatomy, new biology, and new astronomy followed the new religion. Otto Brunfels of Mainz (148?–1534) produced a massive herbal or catalogue of medicinal plants. This included not only those habitat details known to the author but quotations from classical sources. Soon afterward Konrad Gesner of Zurich (1516–1565) produced his history of animals in 3,500 pages with woodcuts of each form. His was a work halfway between two worlds, its plan and intent were medieval but the execution was amazingly modern.

The early modern states arose in the Seventeenth Century and their newly enfranchised middle classes sought to make their livelihood through private farming, commerce, and industry. These were the inventors and innovators of Europe's new wealth. They practiced printing, building, animal husbandry, weaving, and every day chemistry. It was their well-to-do offspring, thoroughly convinced of the values of science, which founded the scientific societies and tended the botanical gardens. Throughout this popularizing of science a vast amount of new information was gained and published in the journals and annals of the societies. The culmination of this effort and interest was the formulation of the Newtonian synthesis in the physical and mathematical sciences, and the work of the Swedish biologist, Carl Linnaeus (1707–1778) in the life sciences. He not only set the systematics of classification on a firm foundation of dissection and anatomy but made astute judgments on habitat. He is to be credited as near the beginning of serious ecological observation and analysis. However he did not deviate from the encyclopedic fashion then in vogue but continued to add more material up until the end of his life. Other of his contemporaries made many investigations into the theory of population which was stated in mathematical terms by the English clergyman, Thomas Malthus (1766–1834). The rationalistic philosophy then in its heyday, combined with deistic theology to motivate the quest for "natural laws." The biologists although inclined toward qualitative rather than quantitative research desired to isolate one group of mechanistic principles which would explain the system of living organisms. From 1790 on, environmental studies and observations on habitat and life cycle began to appear more frequently in the literature.

On the continent the analytical philosophy of Kant triumphed in the early years of the Nineteenth Century and romantic idealism reigned supreme in the sciences. Poets joined biologists in extolling the glories of nature and the artists gained inspiration from the physical environment. Goethe, Beethoven, Hölderlin and many others fled to the rural villages and celebrated the joys of untrammeled nature. This same spirit fired the enthusiasm of the transcendentalists of New England and motivated the American school of nature painters and conservationists. Their counterparts in England were already protesting the excesses of the industrial revolution.

The triumph with which Newton's synthesis explained the cosmos in mathematical perfection was the envy and model of the life sciences. The advancing notions of organisms and environment challenged many biologists to try to formulate a new mechanistic explanation for the differentiation and succession of organisms. The French nobleman, Georges Buffon (1707–1788) applied the current advances in physics and geology to his conception of biology. He properly assessed the significance of environment upon organisms and tried to find in it a solution to animal speciation. His was a dynamic biology and it contributed much to the underlying structure of later ecology. At the same time advances in the exact sciences, the

perfection of measurement, purification of reagents. and the standardizing of quantitative methods, provided new tools for investigating physiology, embryology, and inheritance. A countryman, Jean Lamarck (1744–1829) brought the work of his predecessors to the conclusion that the environment caused changes in organisms and that their offspring inherited the altered or acquired characteristics. But as grand and simple as this solution appeared to be, the experimental facts did not validate it.

The apparent and accepted solution was presented by Charles Darwin (1809–1882) as the dynamic result of organic evolution. Darwin was well aware of the incredible complexity of the environment where myriad organisms in quality and quantity may impinge upon each other. The unraveling of the succession of the coral reel and the earthworm community only uncovered the superficial complexities and revealed a glimpse of the mind-shocking immensity which lay beneath. Even before the controversy over his views had died down, the mathematicization of physiology had taken place and yielded accurate and precise measurements of the life functions. The tools had been forged and the theories generated for full scale ecology to advance.

From 1870 to the end of the century, ecology was moving toward the forefront of biological interest. Data about habitat and consumption, statistics of population and reproduction were gathered in all parts of the world. As the more rapacious aspects of colonialism gave way to the technical exploitation of raw materials "rules of animal and plant" distribution were forthcoming. An added interest in animal geography began and a positive interest in forestry and conservation arose. At the turn of the century many public figures around the globe began to be concerned with the preservation of the ecological balance. Unfortunately much of the effort was directed only to the conservation of primeval wilderness areas while the effluents of industry and urbanization were left uncontrolled. When the study of plant ecology started earlier in the Nineteenth Century, it was quickly equaled by the growth of animal ecology emphasizing the special areas of physiology and community. Oceanographic expeditions and dredging operations had spanned the Nineteenth Century and increased thereafter. Slower in developing was the investigation of freshwater bodies and rivers, termed limnology. The French scholar F. A. Forel published his pioneering study on Lake Geneva and based the new science on a good foundation. "Response physiology" or "response ecology" was the major point of investigation during the pre-World War I period. During that time a new school of American ecologists began to form. The British Vegetation Committee seeded the British Ecological Society in 1913 with A. G. Tansley as first president, two years later the Ecological Society of America was founded with V. E. Shelford as its first president. The botanical journal, *Plant World* was turned over to the new association which renamed it *Ecology*.

The group which met to form the new society was largely the product of the teaching and research begun in the state and university of Illinois by S. A. Forbes. A number of younger colleagues, including H. C. Cowles and F. E. Clements also influenced by the publications of European ecologists, began to train generations of students in the methods of ecological investigation. From the labors of these and other American scholars the "Chicago school of ecology" came into existence. A survey of the first twenty years of the journal *Ecology* reveals that this group of researchers set the tone for American ecology until the Second World War. One of the prime movers was C. C. Adams, for whom the Center for Ecological Studies is named. Unfortunately the teaching of ecology became a somewhat dogmatic proposition and the field declined in support and interest in the decades between the two world conflicts. This does not mean that men of significant calibre and work of the highest quality were not being devoted to ecology, it simply means that newer and more pressing problems were catching the popular scientific fancy.

The development of ecology in the Twentieth Century has been rapid and encompassing. From a minor biological specialty of the decade 1900 to 1910 it has now become a major concern of mankind. As more was discovered about human and animal physiology, and biochemistry, closer investigation of the tissue level of organisms was the natural outcome. Histology, the study of tissues with the light microscope, had been well founded in the preceding century in many parts of the world after Pasteur's work on the germ theory of disease gained universal acceptance.

The study of organs and their functions and the invention of other specializations of optical instruments, such as the phase contrast condenser and monochromatic light microscope furthered the study of cells as they lived and reproduced. The invention of submicroquantitative chemical analysis beginning with refinements of spectrographic analysis and moving rapidly to chroma-

tography pushed biological and chemical investigation down to the molecular level. The success in unraveling the constituents of proteins and genes has led to a still further penetration, submolecular investigation. The physical investigation of the life processes divided into two broad fields—those of energy and control.

All of this mass of information brought forth some telling concepts. 1. The cell obeys some very simple thermodynamic principles. 2. Energetic equilibrium is one of the central principles of the living cell. 3. Energy transformations at the cellular level are isothermic and guided by enzymatic catalysis. These catalytic agents might range down to the finite quantity of only two or three strands of protein in a single celled organism.

Although these principles were suspected as early as 1910 the actual proof of them was far beyond the chemical technology and instrumentation of the time. Sad to say, it was the pressure of conflict brought about by two world wars which hastened the breakthroughs in electronics, optics, and chemistry which rendered molecular biology and cellular physiology possible. And it was the economic surplus produced by wartime economy which gave support to basic research and encouraged many younger investigators to enter the field.

The early Twentieth Century was a time of increasing prosperity and reckless optimism founded on the abiding faith in the ultimate triumph of the machine. In the 1850's Henry David Thoreau had faint misgivings about the gospel of progress which he stated in indirect transcendental fashion in his only two works, *A Week on the Concord and Merrimack Rivers* (1849), and *Walden* (1854). There is no doubt that he was deeply influenced by the German natural philosophers who shared his Kantian point of view. At the same time natural historians and writers such as John Muir, the Scotch-American traveler, and Robert Underwood Johnson, the journalist, were campaigning for preservation of the great Western wilderness. They finally prevailed upon Theodore Roosevelt who established the conservation commission under Gifford Pinchot in 1906. In the minds of most North Americans however, the frontier was endless and its resources were being constantly replenished. Clearly this was not an age when the delicate balances of nature were apparent to the majority of beholders.

Already by 1900 the human migration from Europe was turning to a flood and smog laden, slum slicked urbanization was growing out of control. "Response Ecology," as it was called became the major focus of attention within the field and followed the growing interest in behavior and psychological experimentation excited by Pavlov and Freud and their hosts of European disciples. Although developmental embryology and physiology were active areas in which great strides were being made, their ultimate impact on the classical biology was still in the future and was known to few biologists in the field. It was from these studies that the first notions of external and internal environment were to come. Although the principles of thermodynamics had been introduced by Sadi Carnot as early as 1824 they were not to make their way felt in biology until after the First World War and not to be comprehended in Ecology until many years later. It is a principle of any historical period that it always sees its own time in terms of some past culture or previously popular point of view. Although the first years of the Nineteenth century were times of revolutionary thinking in the mathematical and physical-chemical disciplines, this was not understood until nearly a generation afterward. The early years of the century basked in the romantic ideal that "man was the measure of all things," and while industry and commerce hummed, scientific investigation and technological application moved rapidly forward. However the first few hints that some problems might come to light were voiced by physicians involved in public health. England had faced the problems of the industrial revolution and most of the industrialized nations had followed her example by ameliorating the grossest aspects of human exploitation with wide ranging labor laws. But the actual technologies themselves were coming under suspicion. The more violent poisons had been traced to their sources by the early part of the Eighteenth century. The progress in the development of the light microscope and the principles and equipment for qualitative and quantitative chemical analysis gave the study of human pathology new motivation from the middle of the Nineteenth century until the turn of the Twentieth. A French Physician, M. Auribault noticed the effects of asbestos on workers in the mills where it was spun in 1906, and American researchers traced respiratory maladies in the cities to the discharge of sulphur dioxide ($SO_2$) as early as 1910. But the full realization of the environmental damage, the negative aspect of ecology, was to wait for decades.

A growing interest in natural history prevaded basic biology and a number of field scientists and talented amateurs wrote on their special subjects in the decade up until 1910. However the over-whelming majority of physically inclined biologists were chemists by training and inclination. Their material was of considerable value as the raw observation from which selective data could be drawn. Scientific ecology was just being developed but still the lack of exact scientific quantification made much of the work resemble a technique more than a science. The search for new species and new biologic associations would run its course until no more frontiers were left and all of the major genera had been described. Few biologists of note did not have one or more obscure species named after them. But this desire to fill up the catalogue of creatures, although legitimate in the light of the opinions of the time also tended to retard the quantitative aspects of ecology in developing.

These trends continued until 1930. As the new sciences of biochemistry and cell physiology began to rise in the number of investigators and the extent of their findings, ecology seemed to be the least likely to benefit. The rapid advances in atomic physics and the increased understanding of chemical dynamics caught the attention of more and more biologists before the interruption of the Second World War. The growth of air travel and international communications technology did as much to turn the popular attention to the environment as any scientific discovery made since 1945. Slowly the systematic nature of the major diseases, cancer, heart disease and the like, and the unity of all life processes became recognized by more and more people.

When electronically controlled and infinitessimally precise instruments were turned upon the living cell, the incredible complexity of life began to unfold. The study of fixed, stained cells, called histology, had been developed to a high art by the classical microscopists of the Nineteenth Century. The study of living tissues had to await the invention of biochemical techniques and radioactive tracers which could quantify their functions. Molecular biology was first applied to ecology in attempts to explain the effects of antimetabolites such as insecticides upon populations of organisms. The full measure of this method is yet to be taken. One result is the development of two newer specialties, the energetics of ecological systems, and the theory of ecological evolution.

Ecology proceeds upon two fronts, the quantitative understanding of organisms in their environment, the economics of organisms, and the more traditional qualitative study of the interrelationships of organisms with their environment. There is no doubt that they reinforce each other. But the more mathematical economics is still not widely understood. However ecology is now much more than a biological curiosity. It has become the scientific explanation for the degradation of the biosphere. Daily news media around the globe display the latest results of investigation. The current insight into the subject is all pervasive with deep seated philosophical ramifications and embraces the whole content of human life. In the second half of the twentieth century ecology enforces an alternate life style. The materialistic pragmatism which has powered the growth of the American economic colossus is being challenged face to face. In the wake of the materialist society has come insatiable want for material benefits not available to all. This produces a differentiation, a gap between what is desirable and what is affordable. The history of the recent past demonstrates that this led to many quick schemes to denude the land of raw materials for instant profit. In the society of the last century this was condoned and lauded if it brought obvious riches and success. The automation and mechanization of the means to exploit has denuded more area and dug to greater depths than had been accomplished in all of the preceding one million years.

Ecology has been interpreted to mean not only living with the land and its resources but living for the land. The hylozoistic universe, the ancient notion of the great living cosmic being, has been reborn in our time. Such pantheistic notions are not confined to the young and the radical, many sober scientists have been disgusted by the ravishing of the biosphere and pleaded for reform. Their imagery is often similar to the Greeks of the pre-Socratic age, wishing for some primordial love of the land and a religious reverence for all of life. Whether or not this adulation and respect is an integral part of the scientific nature of ecology will be seen in the years which lie ahead. It is a feature which has swept on beyond the careful, incisive growth of ecology as a body of verifiable scientific knowledge and turned into a crusade, a holy war against those who would defile the earth. The philosophical motive for this reawakening of Western man's sense of biotic responsibility has been influenced by two scientist-humanists. The writings of the Alsatian physician Albert Schweitzer, particularly his *Out Of My Life And Thought*, have expressed his notion of "reverence for life." While the French Jesuit-anthropologist, Pierre Teilhard De Chardin, has expressed his motif of cosmic evolution in his, *The Phenomenon of Man*. Both have had great impact on the interpretation and direction of the life sciences. The notion of

"science" changed considerably from the time of Newton to the age of Einstein as the certainties of a closed, lawful, and perfect systems gave way to the universe of probability and relativity in which truth itself was ever changing and the nature of science was no longer discovery, but quest. In the Nineteenth Century another change came about, the slow effacement of academic science with its rigorous mental discipline and all encompassing passion for knowledge. In its place the Twentieth Century continued to contribute to the peculiarly American notion of pragmatic-industrialized science. This latter science seeks few ultimate answers and is advanced by the practical questions it solves and the marketable products it produces. Like the capitalistic philosophy of economics it functions solely for the production and continuation of profit. While the system worked very well in raising the standard of living for a vast host of human societies in North America and Western Europe, its chief feature was environmental exploitation.

The gouging out of enormous scars by strip mining, the deforestation of millions of acres, and the utter annihilation of millions of organisms as in the case of the passenger pigeon and the American bison was obvious enough evidence that something of the primal wilderness had to be preserved or posterity would live on a bald desert. Not only was conservation advanced but the initial attempts to limit the complete and total destruction of the environment by exploitation was begun. This was possible in 1890 and may have been successful without the increase in population and without increasingly more gigantic methods of production. In all too many cases however conservation meant the preservation of a small area as a natural terrarium without glass. The power industry came into its own in the 1920's and early 30's and the chemical industry boomed after the second world conflict. Both are staples and necessities of the contemporary population explosion, to some extent they have aided and abetted the explosion through reduction in disease and increase in leisure. However they also stripped millions of acres of raw material and spewed out massive effluents.

The result in environmental study has been to turn a little-known pedantic enterprise of the 1920's into a worldwide necessity in the 1960's and a crusade in the 1970's. The advance of ecologic and environmental studies is presently so rapid as to be beyond the scope of any one investigator. It has not only restructered the science of biology, its teaching and investigations, but is now in the process of restructuring all of science and especially those aspects which were the unwitting tools of environmental exploitation and destruction. There is every possibility that in the last quarter of the Twentieth Century—physics will become ecophysics, and engineering become eco-engineering. In all of these disciplines common-sense considerations of human values will be given equal weight with the quantitative proofs of the problems concerned.

Although it is impossible to deal with all of the many different types of research carried on in modern ecology it is necessary to single out a few of the major specializations and the individuals of note who have advanced the science. Certainly the first and largest group are the plant ecologists. Stemming from the double investigations of agriculture and conservation the plant communities received a great deal of attention at the close of the last century and for most of the present one. Frederic C. Clements (1874–1945) was primarily active in the study of soil and the ecology of plants. His observations embraced the concepts of succession and vegetation analysis and his large publication written with C. L. Sheer on the *Genera of Fungi* and published in 1931, it is still a standard reference. His entire researching and teaching career concentrated on the Middle Western United States. W. D. Billings has contributed substantially to the areas of both vegetational description and statistical study. His major work, *Plants, Man and Ecosystems* (1969) gives a fine overview of his interests. Pierre Dansereau has also been involved in vegetational description and statistical study but with the further application of these principles to biogeography. Chief botanist at the New York Botanical Garden for many years he has written two widely regarded works: *Biogeography: An Ecological Perspective* (1957) and *Challenge for Survival* (1970). In a parallel field has been the efforts of Henry J. Oosting presented in his work, *Study of Plant Communities* (1956). C. Raunkier has developed the notion of plants as life forms while John E. Weaver has advanced the fundamentally important study of grassland ecology, his major work being, *Prairie Plants and Their Environment* (1967). Needless to say there are a great many other investigators who are making innovative contributions but especially important is the work of Robert H. Whittaker whose work on communities and ecosystems has been proposed in a book of that title published in 1970. His researches also expand from an ecosystems insight to important statements on the animal component of such systems.

The general and mathematical theories of ecology have taken up the interest of a large number of biologists, biostatisticians and biometricists. Among the leaders have been such Europeans as Vito Volterra who independently worked out a basic model of the predator-prey relationship. He has also contributed to the understanding of competition exclusion and of other basic principles. One of the major American ecologists who worked in this field was V. E. Shelford who wrote a great number of original papers which are widely distributed through the literature from 1900 on. He taught at the University of Illinois and was interested in life histories and their effect on ecology. He was a pioneer student of community and population ecology, he is credited with many of the basic maxims of the science including "Shelford's Law of Tolerance," and similar statements. His work is found year by year in the first few decades of the journal, *Ecology*.

One of the leaders of the school of researchers contemporary with Shelford was Warder C. Allee (1885–1955) whose principle researches were in the areas of aggregations and densities of organisms. He discovered that both low and high densities in a given area may slow or inhibit population growth. He was a coeditor and contributor along with A. E. Emerson, P. Park, T. Park, and K. P. Schmidt to the first major modern text in ecology, *Principles of Animal Ecology* (1949) and a prime contributor to the best recognized American work on animal geography up to its time, Richard Hesse's *Ecological Animal Geography* (1951). His own researches were presented in his two books, *Animal Aggregations* (1931), and *Social Life of Animals* (1938). Thomas Park, another faculty member of the University of Chicago innovated experimental ecological studies by using a microcosm constructed in the laboratory. He employed the flour bettle *Tribolium* and published a number of papers on the biology of populations.

Other areas including theory have been advanced by the work of George Evelyn Hutchinson who is highly regarded as an authority on many topics including competition and systems ecology. Primarily a limnologist he has been a faculty member of Yale University for a number of years. His broad concern for the environment has been expressed in two of his more popular books, *Enchanted Voyage and other studies* (1962) and *Ecological Theater and the Evolutionary Play* (1965). A Yale colleague of Hutchinson, Edward S. Deevy has devoted much time to life table studies and pleistocene ecology generally, like Hutchinson he is also a limnologist and studies the ecological aspects of rivers and other freshwater bodies. The Cornell University ecologists, Lamont C. Cole has published on many topics but especially well regarded are his studies of cyclic phenomena and life histories. He is especially interested in woodlands and the phenomena of autecology.

Herbert G. Andrewartha, a leading theorist in his own right has worked closely with L. C. Birch, an advocate of density independence to produce two valuable works, *Distribution and Abundance of Animals* (1945) and *Introduction to the Study of Animal Populations* (1961). In the more quantitative aspects of species diversity and abundance and advanced "niche" theory, Robert H. MacArthur of Princeton has made a number of contributions. His *Biology of Populations* (1966) employs much of his investigation of birds and advances theoretical work in ecology. A colleague of MacArthur, Joseph H. Connell coauthored the work with him and has taught at Santa Barbara, worked extensively in marine and tropical distribution. Less mathematically oriented but still involved with modeling and theory are A. J. Nicholson an investigator of population regulation and ecological evolution, and Richard Levins a model builder and fitness theorist. More specialized studies utilizing one class of organism have led to many general theoretical insights.

Among the more specific investigators must be mentioned, Ronald A. Fisher who is best known as a population geneticist, but who has made decided contributions to the areas of modeling, mimicry, and populations. With F. Yates he wrote *Statistical Tables for Biological, Agricultural and Medical Research* (1964), and his own *Theory of Inbreeding* (1965) is an accepted work in the field. David Lack has done important researches on bird ecology, the theory of bird reproduction rate and clutch size regulation. Among his books are: *Natural Regulation of Animal Numbers* (1954); *Population Studies of Birds* (1966); *Ecological Adaptations for Breeding in Birds* (1968); and *Ecological Isolation in Birds* (1971) he is also interested in the history of biology and ecology and has published the major work on *Darwin's Finches*. Alfred J. Lotka independently worked out a basic model of predator-prey interaction and proposed influential viewpoints on competition exclusion. His general efforts are explained in his *Elements of Physical Biology* (1957). Jane Brower has carried out many field studies on mimicry from her laboratory at Amherst, while Manfred D. Engelmann has pursued research into soil and energetics systems utilizing arthropods.

Two of the best known biologists have become involved with ecology and added their intense interest in the humanities to popularize the biological and ecological sciences. W. C. Wynne-Edwards proposed a hotly controversial theory on animal distribution, social behavior and population regulation. His most widely circulated writing has been his book *Animal Dispersion in Relation to Social Behavior* (1962). Ernst Mayr of Harvard has been a leader in the investigation of geographical speciation in plants and animals and the results of his research is voluminous. However he is also concerned about the overall history of the biosphere and the place of man in nature and his two books *Systematics and the Origin of Species* (1942) and *Animal Species and Evolution* (1970) have been widely read outside the field of strict biology.

Another field of modern ecological inquiry has been community description which goes back to the early part of this century. F. W. Preston was a pioneer thinker in the problems of species abundance and H. A. Gleason of the New York Botanical Garden attempted to devise models and trace descriptions of communities. One of the best names in the early popularization of ecology was Charles Elton who wrote a premier text *Animal Ecology* in 1927. His early attempts at constructing a theory of community and his descriptions of communities are recounted in numerous articles. He is ascribed the concept of "Elton's Pyramids." The description of communities in rivers and streams was pioneered by the Philadelphia scientist, Ruth Patrick and her group at the Academy of Natural Sciences. They have written many reports and studies on limnology for both government and private organizations. Among the Europeans engaged in community investigation, J. Braun-Blanquet introduced the notion of "faithful" rather than dominant species as criterion of community description. He was a leader of the phyto-sociological school of European ecologists and his *Plant Sociology* (1932) stated his views concisely.

Less controversial but equally important has been the work of four Americans, Rexford F. Daubenmire of Washington State University who has studied life zonation and vegetation and written *Plant Communities* (1968); Nelson G. Hairston of the University of Michigan at Ann Arbor for his models and descriptions of communities, particularly centering on invertebrates; Ramon Margalef who pioneered the use of information theory in descriptive models in his *Perspectives in Ecological Theory* (1968); and Frederick E. Smith who has studied models and descriptions of communities but also written very extensively books such as *Killing for the Hawks* (1968).

Behavioral ecology has been popular and gained more attention as psychology and behavioral science has generally advanced in the recent decade. Three leaders in this field have been N. Tinbergen, whose theories of aggression and territoriality have found expression in his *Herring Gull's World* (1961) and *Animal Behavior* (1968) he is one of the founders of the newer science of ethology; Konrad Lorenz has been involved with many of the same areas of study as Tinbergen but he has also been interested in human behavior, his books *Evolution and Modification of Behavior* (1965) and *Studies in Animal and Human Behavior* (2 vols. 1970–71) have been very influential. Arthur D. Hasler of the University of Wisconsin has done pioneering observations in the area of homing. He is a limnologist and his *Underwater Guideposts: Homing Instincts of Salmon* (1966), and other writings have been widely read and encouraged other similar studies.

The difficult study of Ecological genetics has been advanced by Eugene B. Ford in both the theoretical and practical realm. H. B. D. Kettlewell's work on British moths and the tracing of certain factors as morph ratios, predation, and coloration lent a new facet to the study of ecological genetics. The stature of Theodosius Dobzhansky as a geneticist and expert on microevolution has given great impetus to his work in the genetics of species and their ecological aspects. From his position at the Rockefeller Institute, now the Rockefeller University have come two very strong treatises: *Evolution, Genetics,*

*and Man* (1955) and *Biological Basis of Human Freedom* (1956).

Biophysics and bioenergetics have been interdependent from their earliest days. Raymond L. Lindeman has been the founder of the ecological school which emphasizes this aspect. The relationship of energetics and metabolism has been investigated and discussed by Oliver P. Pearson, while Lawrence B. Slobodkin linked population theory with energetics. His book *Growth and Regulation of Animal Populations* (1961) and his articles stress this correlation. Of all the researchers into this area of ecology the brothers Howard T. Odum and Eugene P. Odum have been especially important in the development of the ecosystem concept and other conceptual schemes. See their contributions to this reference work. Three other biologists have made a broad appeal to the general population to inform the citizen of the problems of pollution and the biosphere. They are Paul R. Ehrlich with his best selling *The Population Bomb* (1968) which incited sufficient interest to force the federal government to begin to study the problem of population expansion and control; Barry Commoner whose group at Washington University in St. Louis has been influential at every level through their books and magazine *Environment*, and René Dubos whose medical knowledge and ecological interest have been shown in many efforts such as *Man, Medicine and Environment*. This attempt to popularize their findings immediately has given ecology a new and dynamic image which has been lacking in most of the life sciences. It remains to be seen if their efforts and their warnings have been in time.

William White Jr., Ph.D., Editor
North American Publishing Co.,
Philadelphia, Pennsylvania

# Chapter 3

# Ecological Energetics

The ultimate source of energy in biological systems is the sun. As this energy enters the earth's atmosphere, about half of it is reflected and about ten percent of it is scattered or absorbed by atmospheric gases or small particles in the air.

Energy may be said to be in various forms. Solar radiation is energy in radiant form. Some of it will later be transformed by green plants into chemical energy in the process of photosynthesis. Still later it may be transformed into mechanical energy and heat in the processes of cell metabolism.

The incoming solar radiation is in the form of infra-red, ultra-violet, visible light, X-rays, gamma rays, and radio waves. Only about half of it is in the spectrum of visible light.

Radiant energy absorbed in the tropospheric layer of the atmosphere is then scattered in all directions in the form of infra-red. Some of this new infra-red goes downward to the earth's surface, adding to the total amount of insolation there. A heavy cloud cover seriously reduces the total amount of radiation reaching the earth.

That radiation which does penetrate the atmosphere acts to heat the lower atmosphere, land, water, and all the plants and animals on earth. Shiny portions of the earth's surface reflect a tremendous amount of the radiation back into the atmosphere. Ecologists are very much interested in the utilization, movement, distribution, form, path of transfer, and pattern of loss of the energy from solar radiation.

## Basic Factors of Ecological Energetics

The study of these aspects of energy as it relates to biological systems is called ecological energetics. The earth can be considered as a huge closed system, i.e., a system where the amount of matter remains constant but in which energy flows in and out and all through the system. Ecological energetics is concerned primarily with the fate of the energy while it is in the system. It describes the movement of this energy in terms of physics. The *First Law of Thermodynamics* states that energy may undergo transformations from one form to another, but the amount of energy remains constant; energy is never destroyed and new energy is never created. The *Second Law of Thermodynamics* states that energy transformations take place only when there is some increase in the randomness or nonusability of the energy. Heat is energy in the form of random movement of matter. It is to heat that the solar radiation eventually proceeds in the processes, physical and biological, of the earth. And dissipation of heat is the end product.

Systems proceed to a state of equilibrium in which entering energy equals energy that is leaving. Thus the earth-system is at an equilibrium where the energy from insolation matches the energy escaping the system as heat. A more precise statement must, of course, include definite units of energy in the different forms and a quantification in terms of time. There is other energy reaching the earth's surface, usually as heat, from geological and geophysical forces operating in the earth itself. These energy sources, however, are of relatively little interest to ecologists presently specializing in energetic topics as they are not usable by organisms.

Studies of the total energy utilized by plants, the total that is actually incorporated in plants, and the total that is available to animals from a food source are all of great interest. The manufacture of plant material by plants is called primary productivity. (Animal matter production is secondary productivity) The term *gross primary productivity* refers to the total amount of energy the plants use per unit area per unit time. Plants have respiration and maintenance needs, and thus not all of the gross primary productivity is incorporated into plant material by the growth process. The overall relation can be stated thus:

$$\left(\begin{array}{c}\text{Gross primary}\\\text{productivity}\end{array}\right) = \left(\begin{array}{c}\text{Net primary}\\\text{productivity}\end{array}\right) + \left(\begin{array}{c}\text{Energy used}\\\text{in respiration}\end{array}\right)$$

Note also that the three terms are each stated in different units: gross primary productivity is given in terms of solar radiation, energy in radiant form; net primary productivity is given in terms of chemical energy; respiration energy is in terms of released heat. A calorie is defined as the amount of energy (in the form of heat) required to raise the temperature of one gram of water from 14.5°C to 15.5°C, a rise of one degree. One thousand calories make up one kilogram-calorie (kcal.).

The actual energy that plants can use varies in amount at different parts of the earth. The degree of chronic cloud cover and the elevation are only two of the variables. Some annual amounts (gross averages) are: Great Britain, $2.5 \times 10^8$ cal/m²/yr.; Georgia, $6 \times 10^8$ cal/m²/yr.

*Photosynthesis* is the process by which green plants harness the solar radiation as a source of energy. Only the red and blue hues of visible light, however, are of use in the process, and non-visible light is not used. Chlorophyll, the photosynthetic pigment, is green or on some less frequent occasions blue. Thus it is a superior absorber of the red and blue hues, while tending to reflect the green hues. The basic process can be summarized in the equation:

$$6CO_2 + 12H_2O \xrightarrow[\text{chlorophyll}]{673\ Kcal} C_6H_{12}O_6 + 6O_2 + 6H_2O$$

carbon    water    within green plants    sugar  oxygen  water
dioxide

Plants in turn are the food energy source for animals, which are incapable of photosynthesis. Animals which eat plants only are called *herbivores.* Those which eat herbivores in turn are called *carnivores.* An animal that eats both plants and other animals is called an *omnivore. But ultimately* whether a herbivore or a carnivore the source of energy is the stored sugar of plants.

A given series of one plant eaten by one herbivore, eaten by one carnivore is called a *food chain.* The chain may be extended, that is, there may be animals that eat carnivores; such an animal could be called a *second order carnivore.* An animal eating a second order carnivore would be *third order carnivore,* and so on. A good example is the cartoon illustration of fish each being eaten by a successively larger one, Figure 2. In practice, there are rarely more than a few links in a food chain. Most food chains weave together with other chains to form a *food web.* Energy flows along the chain or web from organism to organism. However much of it is lost at each step.

## Energy capture efficiencies

In any area plants actually use only a small portion of the available light for primary productivity, probably no more than one to five percent of the available energy. Such a percentage value reading is called an *efficiency of energy capture.* In aquatic ecosystems, the amounts of energy available are even lower, because of the absorption of light as it passes through water. Correcting calculations of aquatic energy capture efficiencies accordingly, the new values are still below five percent in most cases.

It can be safely inferred that light is not usually the limiting factor to primary productivities, since plants use so little of it. This is not always the case, however. In aquatic systems, for instance, there is a point below which so little light penetrates that no photosynthesis can take place. Further, red hues are absorbed more rapidly than blue and often disappear in the first few meters of water. The predominant algae here are green and blue-green algae that are superior in the absorption of red hues. Red algae, superior absorbers of blue hues, tend to stratify in the lower waters.

Under a heavy forest canopy, the shading may deplete the light reaching the ground. Some forest plants are adapted to lower light levels. Other bloom before the appearance of thick foliage. Shading seems to also be a key factor in forest succession, since the progeny of shade-tolerant species have superior survival and growth on the

**Figure 2.** The simplest form of food chain.

forest floor. In the Eastern United States this usually means the replacement of sun-favoring conifers by shade tolerant hardwoods.

### Biomass and energy

Living material is called *biomass*, whether it be plant, animal, or microbial in nature. Estimations of the total biomass in an ecological system and of the energy stored in the biomass are considered very important ecological data. Biomass measurements are measurements of mass, i.e., weight. The organisms may be weighed either living or in a dehydrated form. The latter is the preferred way. The amount of stored chemical energy can be calculated directly by burning the biomass in a calorimeter, a device for measuring the caloric content of an object. The typical procedure is to calculate the rise in temperature of a surrounding water bath as the matter burns. This method is quite applicable to energetic studies on animals, on prepared specimens, and on single organisms. It is of some value in green plant studies. More general and less tiresome procedures are used in estimating primary productivity. Most of these employ the fixed nature of the photosynthetic equation to calculate the remaining terms from what data is available.

There are variations in the energy content of biomass on a gram for gram basis, depending upon the species sampled and the tissue type used. It has been estimated that most plant biomass has stored in it about four to five kcal per gram. While energy is used in the system and is lost to it in the form of heat, biomass remains. It is broken down upon the death of the organism. Any ecological system requires *decomposers* (usually microorganisms) for this purpose if it is to avoid a build up of dead biomass, also called *organic detritus*.

The term *standing crop* refers to the total amount of biomass present in an ecological system or in a given area at any one time. This is a most useful concept. Reasoning from raw standing crop data alone, however, can lead to grave confusion. In a lake for instance, there would be times when the actual standing crop of fish would far exceed that of the microscopic algae which form the base of the food chain. Calculating the total energy utilization in the lake over an interval of a season, however, one would quickly realize that the reproduction and turnover rate are so great that there are astronomical numbers of algae relative to the fish and that the total biomass of algae produced in a season's time far exceeds that of the fish. It is because of limitations on the use of any single concept or principle that ecologists have devised

the indexes, ideas, and models discussed below.

### Pyramids and efficiencies

It was previously mentioned that energy travels the path of the food chain or food web, with much being lost at every point. The term *ecological efficiencies* is used to describe the ratios between the energy flow at different points, expressed in percentages. Energy flow at a given point in a food chain can here be taken as simply the sum of the biomass production and the respiratory needs of the organisms at that point.

Let us consider a few of the conceptual terms used to describe and analyze energy flow in the food chain; there are at least two more to be considered before proceeding. An animal is rarely able to utilize all of the food eaten, using it either for biomass production or for respiration. Rather there is typically much food material whose energy is never tapped. The total amount of ingested food is called the *energy intake.* The term *assimilation* refers to the total ingested energy that is actually used by transformation within the cells. It also can be said to equal the sum of production and respiration. An example of an often unutilized source of energy is cellulose which is made up of bonded sugar molecules, the constituent of plant cell walls. Relatively few animals can assimilate its energy directly. Thus the final decomposition of the cellulose itself must be left to specialized organisms, often microbes. Chitin is a chemical relative of cellulose, and it is the tough structural material of arthropod external skeletons such as bugs and lobsters and the cell walls of some fungi. It also often passes through the gut of animals undigested.

Using ratios between the above terms, ecologists have constructed a series of efficiency measures. These are usually combined with models and other theoretical tools in at least three ways in constructing further ecological theory and presenting descriptive data on specific species, food webs, or environments. First, *pyramid models*, starting with plants as producers at the base and rising through the different levels of animals as consumers. The quantitative data in the pyramids may be given in terms of numbers of organisms, biomass, or energy flow. Analyses using all of the above are the most meaningful from the standpoint of understanding ecological systems. Each level in the pyramid is designated by a name or number: plants are *level one*, primary consumers (herbivores) constitute *level two*. Secondary consumers (carnivores) are *level three*. Second order or higher order carnivores belong to the upper levels. Note that the formation of a food web in nature involves the intertwining of food chains, with most species preying upon and preyed upon by more than one other species. Thus an individual species may actually belong to more than one level in the pyramid, it may prey upon many smaller organisms and be the prey of organisms larger than itself.

These pyramid levels are called *trophic levels*. Note that such a broad approach tells us little of the actual detail of specific species relationships and of the functioning of the total community of organisms in terms of nutrient or mineral cycling, or even water cycling. There is also the further difficulty in this and the other two common approaches of just where to place the decomposers in the overall scheme of things. The view of the importance of these microorganisms is that they have been too long denied consideration as a food source as well as an agent in organic recycling.

Sometimes a detailed, diagrammatic approach is used to describe the entire food web, with the efficiencies and other data inserted or given in tables. Systems flow diagrams, using the language and mathematics of theoretical model building are often employed. An entire system or a portion of it may be modelled; single species data are often lumped together. This approach is especially applicable to computer studies and simulations. The key ecological data become movement parameters or storage quantities.

### Estimating productivity

The production of plant material may be estimated in various ways. Most, as previously mentioned, are variations using the form of the photosynthetic equation, in which the quantities of one or more terms are measured and the others calculated by rearranging the equation.

The most direct method is, of course, to harvest the plant material and to weigh it with corrections where needed.

Easier methods are usually employed, however, which measure the use of carbon dioxide or the production of oxygen. Carbon dioxide production or use in land systems can be measured by the technique of infrared gas analysis in a closed container; in aquatic systems other methods are used.

Oxygen production methods, employing the newer electronic meters and probes are very much in use. The use of radio-isotope marked chemical tracers has also provided the opportunity for interesting studies. Other methods are sometimes employed in various situations.

In most of the methods, both the use or production of a material in photosynthesis and its contrary process in respiration must be determined in the calculations.

There is some evidence that there may be some degree of consistency in the efficiency in which energy is transferred from one trophic level to another: an efficiency in the neighborhood of ten percent. Other studies do not support this conclusion, showing no universal pattern between levels or within and between systems.

As population increases and the world's store of fossil fuels diminishes the study of ecological energetics must be advanced so that an efficient and equitable distribution of the energy sources available for all life, plant, animal, and human can be maintained.

## REFERENCES

Eckardt, F. E., (ed.), *Functioning of Terrestrial Ecosystems at the Primary Production Level*, (symposium) Paris: Unesco. (1968).

Engelmann, M. D., "Energetics, terrestrial field studies, and animal productivity," *Advances in Ecological Research.* 3:73-115. (1966).

Goldman, C. R., ed., *Primary Productivity in Aquatic Environments.* (symposium) Berkeley and Los Angeles: University of California Press. (1966).

Kormondy, E. J., *Readings in Ecology.* Englewood Cliffs, New Jersey. Prentice Hall, Inc. (1965).

Kormondy, E. J., *Concepts of Ecology.* Englewood Cliffs, New Jersey. Prentice Hall, Inc. (1969).

Lindeman, R. L., "The trophic-dynamic aspect of ecology," *Ecology*, 23:399-418. (1942).

Odum, E. P., *Fundamentals of Ecology.* third ed., W. B. Saunders Co., Philadelphia, Pa. (1971).

Odum, H. T., "Primary production in flowing waters," *Limnology and Oceanography.* 1:102-117. (1956).

Phillipson, J., *Ecological Energetics*, London. Edward Arnold Publishers, Ltd. (1966).

Slobodkin, L. B., "Energy in Animal Ecology," *Advances in Ecological Research.* 1:69-101. (1962).

Southwood, T. R. E., *Ecological Methods.* London. Methuen and Co., Ltd. (1966).

Turner, F. B., editor, "Energy flow and ecological systems," (symposium) *American Zoologist.* 8:10-69. (1968).

Whittaker, R. H., *Communities and Ecosystems.* London. The MacMillan Co. (1970).

Whittaker, R. H., "Forest dimensions and production in the Great Smoky Mountain," *Ecology.* 47:103-121. (1966).

Westlake, D. F., Comparisons of plant productivity. *Biological Reviews.* 38:385-425. (1963).

John Michael Mankelwicz, M.Sc.
Willow Grove, Pennsylvania

# Chapter 4

# The Chemistry of Life

Rapid advances in the field of modern biochemistry have placed increasing demands upon the seasoned reader in this area. His knowledge of the terminology must be ever expanding to include the new terms that are being added each day. Therefore, the beginning reader, or one who has limited experience in biochemistry must have some facility with the basic terminology which has become everyday language within the scientific community. This brief introduction to the chemistry of life is a small effort to define some of the basic concepts and terms that the reader will encounter throughout many of the subsequent articles. It is not meant to be a complete compendium, and the reader is advised to keep in mind that as new terms are introduced in the text, many of them will be defined in the context of the article in which they appear.

*Biochemistry* is a study of chemical and molecular processes essential for life. It consists of two major approaches:

*Descriptive biochemistry* which involves the precise chemical structure of living matter and;

*Dynamic biochemistry* which deals with the chemical changes or metabolism associated with living systems.

An analysis of living substance reveals three major types of organic molecules:

*Carbohydrates* which contain carbon, hydrogen and oxygen;

*Proteins* which contain nitrogen, sulfur and small amounts of other elements in addition to carbon, hydrogen and oxygen; and

*Lipids* which are also made up of carbon, hydrogen and oxygen. These substances, either alone or in combination, serve as the building blocks of living matter.

Carbohydrates are the primary source of energy which is used in maintaining *anabolic* (synthetic or building) processes in living organisms. Proteins serve both structural and dynamic functional roles in the life process. Most chemical reactions, either *catabolic* (degradative or breaking down) or anabolic require rather extreme conditions to occur when they are conducted in a test-tube. Temperatures, *pH* (hydrogen ion concentration) and pressures that are far from physiological are required to produce the breakdown or synthesis of many of the simplest organic molecules. Thus, one can readily see that some unique mechanism must exist within living organisms to enable these reactions to occur under conditions that are suitable for the maintenance of life within these organisms. This is accomplished by the intervention of a special type of protein, the class of which is collectively referred to as *enzymes*. In very general terms, *enzymes* are biological *catalysts* that accelerate the vast number of reactions occurring in biological systems contained in a living organism. A *catalyst* is any substance that will accelerate the rate of a chemical reaction without undergoing any essential change in its own nature, thus, it is re-usable. Each group of enzymes will act upon a specific type of compound for which the enzyme will show either a complete or partial specificity. That is, some enzymes will be very limited in the type of compound with which they will interact while others will be less specific. An enzyme works in such a way as to lower the energy of activation for a given reaction to a point where it will be able to proceed under physiological conditions. In review therefore, an enzyme will act on a substance (*substrate*) in a reaction in such a way as to catalyze its conversion into some end product. Enzymes will catalyze reactions in which small precursor molecules are joined to form larger macromolecules (*anabolic reactions*) which may serve as stores of energy as in the case of carbohydrates and fats or as structural components as in the case of proteins, and protein compounds (lipoproteins, glycoproteins, etc.). Other enzyme catalyzed reactions involve the liberation of stored energy by either the *oxidative degradation* (*aerobic metabolism*—in the presence of oxygen, i.e., where oxygen is the terminal acceptor of electrons) or *anaerobic metabolism* (in the absence of oxygen, i.e., where some other suitable acceptor molecule is employed for the electrons released in the process). Such degradative reactions are referred to as *catabolic reactions.* Photosynthesis is an example of anabolic reactions in which carbon dioxide and water are synthesized into

Helium atom and a quanta of energy is released. The sun fuses some 657 million tons of Hydrogen into 653 million tons of Helium each and every second. The fusion processes transforms the missing 4 million tons of matter and releases it as energy. A major portion of this energy is radiated over the whole of the electromagnetic spectrum. The temperature of the sun at its core is estimated to be about 20 million degrees while the temperature at the surface is about 10,000 degrees, which is hot enough to melt nearly any material known upon earth. The gases of the sun are in the physical state known as plasma and continually fuse with the release of energy in the form of radiation.

Much of the solar radiation which travels across the 90 million miles to earth is filtered out by the atmosphere and gives earth its warm climate in a universe of absolute cold. A small portion of the spectral radiation actually reaches the ground surface. The centrality of the sun's energy to the function of life on earth cannot be over-emphasized. All of life on earth, from the simplest virus to man himself depends directly upon the sun for a habitable environment and indirectly for energy.

**The Absorption of Solar Energy by Plants**

Only the light radiation in the narrow range of 3900 to 7600 Angstroms ($10^{-9}$ m) is utilized by plants. Some organisms can utilize the additional ultra-violet and infrared portion of the spectrum and so expand the usable range from 3200 to 8000 Angstroms, Figure 3. The energy of the sun is thus transmitted as photons or quanta of light. Green plants utilize a pigment, chlorophyll to capture and stabilize the energy of these particles. They require mineral nutrients, compounds containing nitrogen, carbon dioxide, oxygen and water to convert this solar energy into the simple sugar, glucose, the first in a long series of organic, high-energy compounds.

The plant cells essentially reverse the process of respiration and give off oxygen. The energy trapped by chlorophyll is transmitted to the carrier compound Adenosine triphosphate (ATP) common to all living cells. The chloroplasts, the sites of chlorophyll, acts as self-sustaining factories to produce all the organic needs of the cell. The material cycle in the cell is repetitive while the energy is utilized to make up new chemical bounds and thus form new compounds. The maintenance of green plants is the only way to capture and sustain the energy of the sun in a form usable to animals and man.

Plants store a wide variety of organic compounds not the least of which are the proteins and lipids

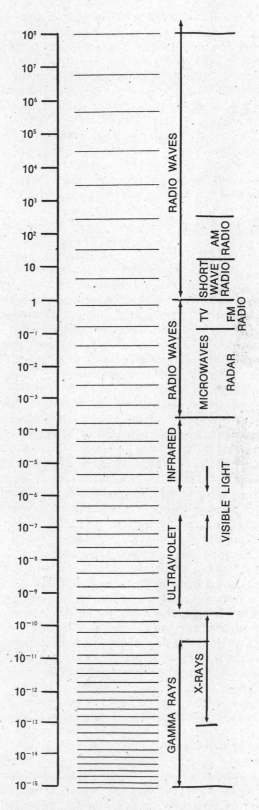

**Figure 3.** The Electro-Magnetic Spectrum.

which the animal and human cell cannot synthesize on its own. Experiments with radioactive tracers have shown that the plant cell works very rapidly so that in one experiment, tagged carbon dioxide molecules could be found attached to a variety of intermediate and end products of photosynthesis in only 30 seconds. The control and information storage ability of the cell is staggering to the imagination. Both slow and rapid chemical reactions are guided and catalyzed in the cell without the release of more than minimal heat. The capture, control and release of energy by organisms is performed in three specific biochemical cycles.

## The Carbon Cycle

In general, an ecosystem can be considered as the complex interplay between various plants and animals and their interaction with their non-living environment. The delicately balanced interactions of every structural and functional unit of an ecosystem make it difficult to consider the component parts of such a system separately. Attempts at fragmenting the unity of nature only lead to confusion. Conclusions that are drawn from artificially contrived model systems may seem logically sound within the isolated system being studied, but when such conclusions are re-examined in the context of the totality of nature, one finds many of them to be false. The total function of an ecosystem is dependent upon the activity of every subunit and the consequences of all peripheral interactions must be considered simultaneously to present a view which is as close to reality as is possible. Therefore, I would like to consider the carbon cycle not as a separate and autonomous element of an ecosystem, but in the light of its relationship to the overall energy-flow in the biosphere, the total living sphere, showing how life itself is dependent upon the functioning of this cycle. We will give consideration to the natural checks and balances in operation in the carbon cycle within the biosphere such that conditions suitable for the support of life are maintained. Interferences presented to this natural equilibrium as a consequence of man's activities in his rapidly advancing technological society will be discussed with respect to their possible effects upon the system. So then, it is possible to think of pollution in one sense, as interferences in the carbon cycle.

All living substance on earth is dependent upon the central structural element, carbon, which serves as the backbone for all organic material. What is it about carbon which makes it so unique,

which gives it this function within the biosphere where numerous complex carbon compounds are in a continuous state of flux from creation to decomposition?

A neutral carbon atom has four electrons in its outermost shell, which is capable of holding eight in all, thus it is half-way between 0 electrons and eight electrons and the tendencies for it to lose or gain electrons are somewhat balanced. Therefore, carbon forms covalent bonds in which a sharing of electrons can occur with other elements. X-ray analysis of pure carbon crystals reveals that a carbon atom shares four pair of electrons with adjacent carbon atoms. Referring to the molecular orbital theory, one can find that the actual nature of the bonding orbitals of the carbon atom is a symmetrical configuration in which the four pair of shared electrons can be thought to form the corners of a geometrical figure with the nucleus of the carbon atom at its center, Figure 4a. The highest form of symmetry involving such restrictions is that of a tetrahedron, a geometrical configuration with four triangular-shaped plane surfaces symmetrically intersecting each other so as to form four corners, a sort of rounded pyramid, Figure 4b. Lines drawn from the center of the tetrahedron to each corner intersect each other at the center at the angle of 109° 28'. It turns out that the carbon nucleus lies at the center and the four electron density clouds lie along the angles of a tetrahedron.[1]

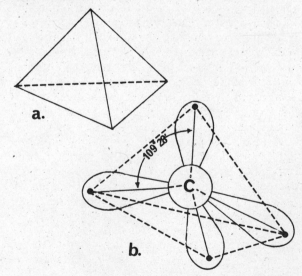

**Figure 4a.** Geometry of the Carbon Atom.

**Figure 4.** Tetrahedral configuration of bond-forming electrons of the carbon atom. The angle between the axis of each electron cloud distribution, bonding angle, is 109° 28'.

The covalent bonding capacity of up to four enables carbon to form long-chains from atoms with similar structure. It is this interesting property that one finds manifest in all living substances. In addition to long-chain carbon compounds, carbon is found to form a wide variety of different compounds exceeded in number only by those formed by hydrogen. Considering all the possible carbon compounds encountered in nature, one finds that living matter is based upon combinations of carbon, hydrogen, oxygen and nitrogen with trace amounts, infinitessimal quantities, of other elements mostly metals, serving special functional and structural purposes. Thus far as we know, these carbon-based combinations are the only group of compounds capable of supporting life as we understand it. This life process consists of a series of many complex, integrated and continuous chemical reactions between organic, carbon-based, molecules. Each step of this sequence involves slight structural changes in the intermediates and is carefully regulated with a state of near equilibrium being maintained throughout. Energy is transformed and stored in the intermediates of these reactions in a very efficient manner.

When burning a piece of wood, the basic structure of which is carbon, it can readily be seen that carbon compounds are rich sources of energy. In the case of the burning wood, the release of this energy in the form of heat and light, is rapid and un-regulated. On the other hand, grazing animals eat grass as well as other organic materials, the basic structure of which is carbon with the controlled release of the energy. Man in turn eats the flesh of these animals as well as vegetable matter and through the regulated release of the energy stored in these animals, man can carry out internal and external activities necessary for his life. Carbon compounds serve as one of the most important intermediates of energy transfer throughout the whole of the biosphere.

Carbon makes up only 0.03% of the earth's crust and the atmosphere contains a comparably small amount as carbon in the form of the gas carbon dioxide ($CO_2$), thus, one can readily see that the limited quantity of this element must be continually re-cycled within the biosphere. It's used and reused millions of times over without being changed or lost. This is the conservation of matter in living systems.

In order to live, organisms consume energy by means of oxidative processes in which degradative conversion of complex organic compounds leads to their most oxidized form, carbon dioxide. To maintain carbon stores and restore balance, photo-synthesis in the cells of green plants, reduces $CO_2$ to organic compounds with the production of molecular oxygen as a side product. Solar radiation serves as the source of energy to drive the photosynthetic process. Continued utilization of energy within the closed system of the biosphere would result in depletion of all energy in the system stored as carbon compounds, if it were not for the constant supply of radiant energy coming from the sun, replenishing these stores through the process called carbon fixation.[2,3,4]

The two major components of the carbon recycling unit are; (1) Free $CO_2$ gas found in the atmosphere and in water, and (2) bound carbon found in organic matter which is either living or stored in a non-living state. A constant interchange is occurring between these two compartments in the biosphere by means of various biological systems which effect the interconversion of the two.

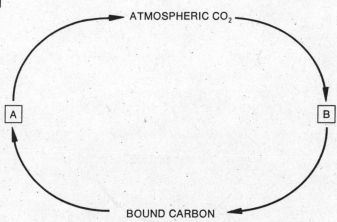

**Figure 5A.** $CO_2$ generating processes (oxidation). **B.** Carbon fixing processes (reduction).

The system is designed such that a dynamic equilibrium can be maintained under normal circumstances; The carbon dioxide generation is balanced by carbon fixation through photosynthesis.

$$\text{Bound Carbon (reduced)} \underset{\text{rate 2}}{\overset{\text{rate 1}}{\rightleftharpoons}} \text{Carbon dioxide (oxidized)}$$

where rate 1 $\simeq$ rate 2

The carbon fixing process occurs primarily through photosynthesis in which both land and aquatic plants, as well as photosynthetic organisms, use light energy to convert $CO_2$ and water into carbohydrates with the simultaneous release of life-sustaining oxygen into the atmosphere. This can be shown as follows;

$$6CO_2 + 12 H_2O \longrightarrow C_6H_{12}O_6 + 6O_2 + 6H_2O$$

where $H_2O$ is the H-donar (reductant) and all the liberated oxygen is derived from the water. $CO_2$ is reduced and incorporated into the carbohydrates formed. Certain photosynthetic organisms utilize other reductants;

$$6CO_2 + 12H_2 \longrightarrow C_6H_{12}O_6 + 6H_2O$$
$$6CO_2 + 12H_2S \longrightarrow C_6H_{12}O_6 + 6H_2O + 12S.$$

Nonetheless, carbon dioxide is fixed (reduced) in organic compounds in all of these reactions. In simpler terms, the process can be summarized as follows;

$$CO_2 + H_2O \longrightarrow C(H_2O) + O_2$$

(with "oxidized" bracketing $CO_2$ to $C(H_2O)$ above, and "reduced" bracketing below)

Only in plants or organisms that utilize $H_2O$ as the reductant is oxygen liberated.

Photosynthesis takes place in highly specialized organelles within the cells of plants called plastids which in green plants are known as chloroplasts. Electron microscopic studies reveal that

# ENERGY FLOW IN THE CARBON CYCLE

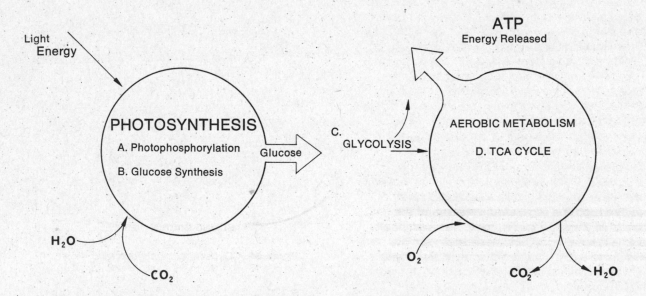

Figure 6. Energy Flow in the Carbon Cycle

**A.** During photosynthesis, light energy is captured by the chloroplasts and through a series of enzymatic reactions involving the photosynthetic pigments, light energy is converted into a usable chemical energy through the process of photophosphorylation. (Energy is transferred to high-energy phosphorylated intermediates.)

**B.** Using this chemical energy which has been stored in the form of ATP (adenosine triphosphate) the next phase of photosynthesis occurs in which $CO_2$ and $H_2O$ are reduced to glucose.

**C.** The energy required to form glucose is now stored in the glucose molecule. This energy is now available for use by either plant or animal metabolism where the first step of glucose degradation, glycolysis, involves the partial release of this energy and results in the degradation of glucose, a six carbon molecule, to two three carbon intermediates.

**D.** The two three-carbon intermediates resulting from glycolysis now enter into the TCA (tricarboxylic acid) cycle where they undergo aerobic metabolism, a series of oxidative degradations, where enzymatically catalyzed reactions oxidize the intermediates to $CO_2$ and $H_2O$, releasing the energy stored in them. This energy is again transferred to the high-energy phosphorylated intermediate, ATP, but the process is called oxidative phosphorylation in this case.

chloroplasts are composed of from 10 to 100 grana, cylindrically shaped membraneous bodies which have a lamellar structure, like plywood. It is around these lamellar bodies that one finds the photosynthetic agent in the form of pigments. Isolated broken-cell preparations containing these bodies have been shown to possess the ability to carry out photosynthesis on their own.[5] The overall size of a typical grana is in the order of 4000–6000 A in diameter and 5000–6000 A in height. It can be seen here that there is a close dependence between the macrocosmic or naked-eyesight level and the microscopic or sub eyesight levels where the photosynthetic apparatus brings carbon back into the mainstream of organic substances.

The process of photosynthesis is comprised of two main reactions occurring in the chloroplasts; a light reaction and a dark-phase reaction. Light absorbed during photosynthesis will supply the energy necessary to drive oxidation-reduction reactions in which water is oxidized to molecular oxygen and carbon dioxide is reduced to an organic form. Photosynthetic pigments capable of absorbing light, convert light to chemical energy by transfers of electrons, forming strong oxidizing and reducing agents. The strongest oxidant is used to oxidize water to oxygen and in doing so, becomes itself oxidized, yielding its electrons to an iron-containing, low molecular weight protein called ferrodoxin. The reduced ferrodoxin then can transfer electrons to carbon dioxide, reducing it also.

As photochemically produced oxidants and reductants release their stored energy during electron transfers, a portion of this energy is stored as chemical energy. One of the most important of all energy transfer mechanisms in the living cell involves the raising and lowering of the energy states of phosphorous in the molecules of ADP/ATP. Photophosphorylation occurs as electrons are transferred to lower energy levels, resulting in a conversion of adenosine diphosphate (ADP) and inorganic phosphate ($P_i$) to adenosine triphosphate (ATP). Simultaneously there is a reduction of pyridine nucleotides, another class of compounds associated with the transfer of electrons. During the dark phase of the photosynthetic process, carbon dioxide will be assimilated into carbohydrates using the energy stored in ATP that was formed during photophosphorylation. Both ATP and reduced nucleotides formed during the light phase of photosynthesis serve as a source of reducing power, transferring the necessary energy for the reduction of carbon dioxide. In addition to carbon fixing, plants do

respire when they are in need of energy other than that provided by sources of stored ATP. This is accomplished by oxidative degradation of stored carbohydrates, aerobic metabolism, with the release of carbon dioxide and water, (reverse of photosynthesis). The details of photosynthesis are covered in a number of recent reviews of the subject,[6,7,8] and working models of the photosynthetic process have been constructed.

The $CO_2$ producing element of the carbon cycle consists of all processes in which carbon is oxidized to form carbon dioxide and water. This will include not only oxidation of stored carbon sources such as fossil fuels, but also the biological oxidations of carbohydrates by animals which results in the release of $CO_2$ by respiration. When plants are consumed by an animal as a source of food, the animal oxidizes the carbohydrates, releasing the energy stored in these compounds during photosynthesis. This is done along with the consumption of atmospheric oxygen through breathing. Not all of the carbohydrates are oxidized directly and some will remain stored in animal tissue. These will ultimately return to the atmosphere as $CO_2$ and water either through the decomposition of animal wastes or through terminal decay of animal tissue after death. Plants respire as was mentioned previously and produce a certain amount of $CO_2$ while they are still alive. Plants too are not immortal and must die and with their death, the terminal oxidative degradation of their carbon substances during decay results in the release of most of the carbon that has been fixed by the plants during their life. Not all carbon fixed by photosynthesis is released through decomposition since some of this will be stored in the form of fossil fuel. Coal, oil and natural gas are such products. Although the fossil fuels we are using today were the products of photosynthesis in man's early prehistory, this process is in operation today and cannot be ignored when considering the carbon balance. When such fossil fuels are burned, $CO_2$ and water are liberated.

Carbon dioxide is also liberated during the weathering of mineral-bound carbon as in the case of limestone. Limestone as you may very well know, is carbon bound in the mineral form of calcium carbonate ($CaCO_3$). This is a somewhat different form of bound carbon than the usual organically bound carbon. It is nonetheless the result of living organisms which have the capacity to utilize carbon dioxide dissolved in water in conjunction with calcium to form their shells or exoskeletons, producing deposits of carbon as calcium carbonate as they grow and build bigger shells or larger skeletons. The extent of such activities can be seen in the vast deposits of calcium carbonate as limestone and in the extensive network of coral reefs found throughout parts of the ocean. Any decomposition of these carbonate deposits results in the release of $CO_2$.

Returning to the original scheme of Figure 5, it is now possible to diagramatically represent a more complete integration of the component parts of the carbon cycle, Figure 7.

Under normal circumstances, a natural equilib-

rium should be maintained and one would expect no significant changes in the overall levels of carbon dioxide and oxygen within the atmosphere over long periods of time. Up to this point, no mention has been made of man's contribution to this balance within the biosphere other than his natural respiratory and decompositional inputs. Man, through his intelligence, has been in the process of developing a very elaborate technology which has forced him to live in what may very well prove to be a most un-natural way. He no

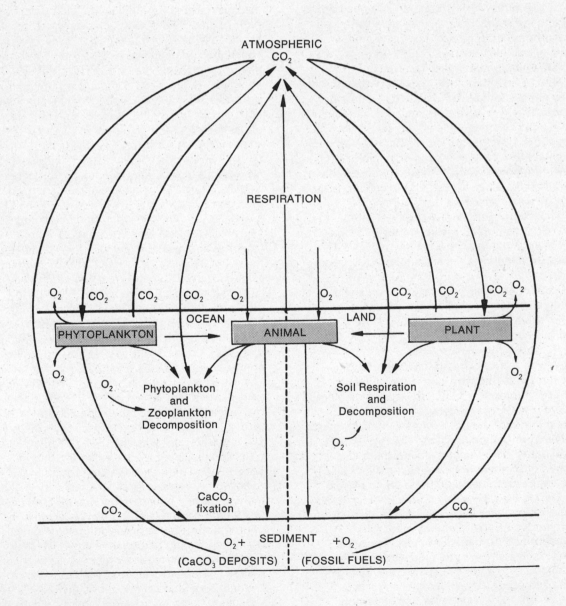

**Figure 7.** The movement of carbon within the biosphere as it is recycled in the carbon cycle.

longer "passes quietly" through nature without leaving a mark, but he stomps very loudly, leaving his destructive footprints on every aspect of his environment. Many of his technologically oriented activities are completely out of harmony with nature and through his efforts to alter his ecology to suit his own pleasures, modern man is creating a new ecology, but one which is steadily lowering the quality of life within it. His activities presented little problem in his early history since his population density was small and his ability to alter the environment was limited by the available techniques at that time. Life processes moved in relatively slow but steady and very carefully balanced cycles. With the advent of new techniques, man increased his ability to change his surroundings. The control of fire was probably one of the first steps in this process, giving man a means of supplying heat artificially to his environment, thus permitting him to extend his activities into areas which would have been otherwise too cold for his survival. His use of fire to clear vast areas of forest was his first eternal footprint, forever changing the balance of life. Carbon fuels, with their rich abundance of stored energy, were and are being exploited with increasing demand as more sophisticated technological innovations are developed. The greater energy requirements necessary for the maintenance of these systems has led to the indiscriminate destruction of vast stores of natural resources with a concommitant rise in pollution and a greater and greater imbalance in the carbon cycle.

If, man were to immediately curtail the "spewing forth" of short-term pollutants, that is, those which have a short duration in the atmosphere, the quantity of these substances would ultimately diminish to levels where their harmful effects would no longer persist. Although these substances are potentially dangerous, once the amount being released into the atmosphere is lowered, they will be reabsorbed or decompose. However, the consequences of burning petroleum and other fossil fuels (coal, wood, etc.,) at a rate that is so great that in a mere two centuries, significant quantities of the carbon dioxide that was fixed in fuel deposits over roughly six million years, are being released into the atmosphere.[9] There is compelling reason to believe that man's use of fossil fuel has caused atmospheric $CO_2$ levels to rise significantly enough to result in an alteration of the overall heat balance of the earth. Currently, the atmospheric content of oxygen is roughly 21% while that of $CO_2$ is 0.032%; a ratio of 650:1. Land clearing, the burning of fossil fuel and the production of cement from lime-

stone has resulted in a disturbance in the natural balance of these two odorless, colorless gases. The extensive use of fossil fuel is primarily responsible for this change, but this does not mean that the other factors should be ignored. It must be mentioned that changes in the atmospheric level of carbon dioxide have been extremely small and are almost at the limit of measurable sensitivity. Another difficulty added to this situation is that there is such a degree of over-specialization within the scientific community, with efforts directed towards isolated and specific goals, that the needed interdisciplinary effort for the precise measurement and recording of small, but significant long-term changes such as the raising of $CO_2$ levels in the atmosphere, has almost been impossible. Scientists use their speciality as an excuse to avoid any sense of responsibility, when their training in fact, should cause them to be crusaders, investigating the possible dangers that might result from any such changes in the environment in which we live and in which future generations must live.

Discernible differences in atmospheric carbon dioxide levels have been observed, with an increase being noted since 1860, when levels for $CO_2$ were estimated to be in the range of 282 ppm.[10,11,12] G. S. Callendar determined that the $CO_2$ level had risen to 325 ppm. by 1961, that is, an increase from 0.027% to 0.032%.[13] This seemingly small amount is actually a 15% increase over the estimated levels of $CO_2$ in the atmosphere in 1860. More accurate measurements of the amount of $CO_2$ in the atmosphere were begun in 1958 when the Scripps Institute of Oceanography began routine and precise measurements.[14] It was found that between 1958–1962, $CO_2$ content increased by 1.13% or 3.7 ppm. One might ask, so what? These are small changes and since the limit of human tolerance to $CO_2$ is in the order of 5000 ppm., it will take at least 5200 years before these limits are reached. This is of course assuming that all other factors remain constant. There is no guarantee to this effect and one must understand that any change in the balance of the carbon cycle will result in changes in many other cycles within the biosphere, among which are the nitrogen cycle, the oxygen cycle and the hydrologic cycle. One can only speculate as to what the consequences of such changes might be. One factor which would be most critical in the maintenance of a state of equilibrium is the rate of photosynthetic reduction of carbon dioxide.

Increasing atmospheric levels of $CO_2$ could very well cause a slow but steady increase in the mean global temperature. Carbon dioxide has an insulating property which causes it to create a "greenhouse effect."[15,16] Shortwave radiation impinging on the earth from the sun passes freely through the atmosphere which has an approximately even distribution of $CO_2$. This shortwave radiation is partially reflected by the earth's surface and again passes out of the atmosphere as it came into it. On the other hand, much of the incoming shortwaves are absorbed by the earth's surface and then radiated outward. In this process, the shortwave radiant energy is converted to longwave energy and can no longer pass through the atmosphere since $CO_2$ acts as an insulating shield. The result is that longwave energy is trapped within a zone near the earth's surface, serving to keep it warm. Not all of this energy remains, but the portion that does not pass through the $CO_2$ shield is sufficient to maintain temperatures suitable for the maintenance of life. Other factors such as ozone and water vapor have similar insulating effects. However, due to the somewhat variable and limited extent of water vapor distribution in the lower level of the atmosphere and the presence of ozone in the upper limits of the stratosphere, their influence is of little concern when compared to that of carbon dioxide.

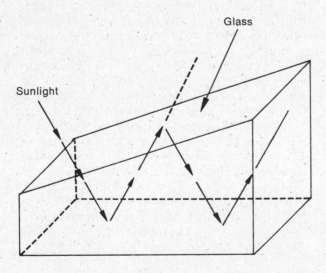

**Figure 8.** Short-wave radiant energy passes through the glass of the greenhouse and that portion which is not reflected as shortwave energy, is absorbed by the surfaces. In doing so, it is converted into long-wave energy and is radiated outward. No longer being able to pass through the glass, this energy remains trapped and causes the temperature to rise. Carbon dioxide in the atmosphere acts as does the glass of a greenhouse, creating a global hothouse.

The global hot-house phenomena created by carbon dioxide could become a problem when the amount of $CO_2$ in the atmosphere is significantly increased, since such increases will also increase the shielding effect. More energy will be trapped and temperatures will rise. As of 1961, evidence demonstrated that the world wide mean temperature had risen 0.9 degrees F., since 1865. Between 1940–1961, the mean winter temperature in the 40°–70° N. latitude increased 2.8° F. During that same period, 1940–1961, the mean temperature of the ocean waters of the North Atlantic had risen by 3.6°F. These data, although seemingly alarming at first sight, appear less so when one notes that for the same time interval, the mean global temperature had actually dropped 0.2°F. This drop could be explained in terms of dust and smoke emissions resulting from both natural sources (volcanic) and man's activities. Dust and smoke can enter the atmosphere where it will remain for a number of years. This can have an effect on temperature as can be seen from examples in which years immediately following the major eruptions of volcanoes have had lower mean global temperatures than years preceding the eruptions.[17] Therefore, careful studies in which global temperature, ocean temperature, atmospheric $CO_2$ levels, dust and smoke levels, and many other factors must be continued in the future so that any trends in progress might be accurately observed. One can falsely conclude that the rising temperature seen in the past 100 years may be in a reversing trend if this conclusion is drawn from data collected over a short period of time. Similar erroneous conclusions can be drawn about ocean water temperatures and numerous other factors if the data is not carefully analyzed. Data must be collected over long periods of time when considering trends that occur on the geologic time scale.

A word about the possible warming trend in ocean waters is in order at this time since any such warming will result in a decrease in the oceans' capacity to hold dissolved $CO_2$. Gas solubility in water is a phenomenon dependent on temperature; the higher the temperature, the less gas that can remain in any solution. This would be of particular significance in the colder waters of the oceans approaching the polar regions. Large amounts of $CO_2$ are found dissolved in these regions. It could be argued that increased ocean temperatures in polar regions would result in greater photosynthetic activities in these regions. This increase in photosynthesis could then serve to fix more $CO_2$ but the increase in temperature will result in a release of $CO_2$ dissolved in the water and the two factors may balance each other.

The slow and continued increase in global temperature and ocean water temperature, if the trend proves true, will have another effect. A continued recession of glaciers and the polar caps will occur with large amounts of water being added to the world's oceans. Although such a melting process would be slow, it has been estimated that if the warming trend were to continue at the rate it has been, by the year 2000, measurable changes in the polar caps will have occured. If such melting should occur and continue, there will be a massive redistribution of weight over the earth's surface. Such a redistribution could precipitate drastic equilibrium seeking geologic movements. Earthquakes and volcanoes would actively change the energy balance to a more static state and in doing so, could change much of the earth's surface as we know it.

Changes in the hydrologic cycle, the water balance of the earth, may also occur in response to increasing temperature and increasing free-water volume over the earth's surface. Global weather and circulation patterns may be altered. An increasing cloud cover, contributed to by dust and smoke released from volcanic activities could act as a reflector, diverting the needed solar energy from the earth's surface. Such an effect would cause decreases in photosynthetic activity and more importantly, would probably initiate a new period of glaciation.

The basic design and balance in the biosphere with its many integrated subsystems seems capable of resisting almost any outside effort to upset this balance. Man's activities will probably have little consequence so long as the natural machinery utilized in the maintenance of these systems is left intact. Extensive land clearing activities are destroying much of the remaining forests, with their photosynthetic capacity. Indiscriminate use of fossil fuel at ever increasing rates continues to issue $CO_2$ into the atmosphere. The use of rivers and the ocean as a garbage disposal with the assumption that there is no limit to the capacity of these waters to hold our wastes is threatening to destroy the photosynthetic machinery within them. Under normal circumstances, the oceans act as a somewhat closed unit within the carbon cycle, with $CO_2$ produced by aquatic sources being balanced by the carbon fixation occurring in the photosynthesis carried out in the ocean. The progressive destruction of land based carbon-fixing machinery will place increasing demands upon the ocean's ability to recycle $CO_2$.

As guardians of the earth that has been set before us, it is our obligation to protect it from the indiscriminate and destructive activities of those who are unaware of what they are doing. The almost indestructible nature of the earth and its naturally balanced systems permits us to be able to assume an attitude of optimism with regards to the future so long as we remain cognizant that there is the possibility of man's activities upsetting the balance that exists within the biosphere. All of the questions have not been answered and in most cases, many of the questions have not been asked. We must continue to speculate as to what our activities may do to the environment in which we live and in which our offspring will live. Although observations we make may not be sufficient to conclusively show major geologic trends, nonetheless, we must initiate such observations and set the guidelines so that future generations will be able to follow through with such projects. We must also curtail any activities which may be upsetting such systems as the carbon cycle even though we do not have definitive evidence implicating such activities. If the negative indication of much of the evidence is sound then we might not have time to develop acceptable alternatives if a crisis should develop. The time to plan and restrict is now.

## REFERENCES

1. Morrison, R. T. & Boyd, R. N., "Structure and Properties," *Organic Chemistry*, 2nd. ed. Boston (1966).
2. Yourgrau, W., & Vandermerwe, A., "Entropy balance in photosynthesis," *Proceedings of the National Academy of Science* (U.S.) 59:734 (1968).
3. Baldwin, E., "Carbohydrate production in the green plant," *Dynamic Aspects of Biochemistry*, 5th ed. Cambridge (1967).
4. Lehninger, A. L., "Energy Flow in the Biological World," *Bioenergetics*, Menlo Park, California (1965).
5. Bassham, J. A., Kirk, M., & Jensen, R. G., "Photosynthesis by isolated chloroplasts," *Biochim. Biophys. Acta.*, 153:211 (1965).
6. Bishop, N. J., "Photosynthesis," *Annual Review of Biochemistry*, 40:197 (1971).
7. Walker, D. A., & Crofts, A. B., "Photosynthesis," *Annual Review of Biochemistry*, 39:389 (1970).
8. Bassham, J. A., "The control of photosynthetic carbon metabolism," *Science*, 172:526 (1971).
9. Flawn, P. T., "Man as a geological agent," *Environmental Geology*, New York (1970).
10. Callendar, G. S., & Egarton, "The Amount of Carbon Dioxide in the Atmosphere," *Tellus.*, 10:243 (1958).
11. Bray, J. R., "An analysis of the possible recent changes in atmospheric carbon dioxide concentration," *Tellus.*, 11:220 (1959).

12. Koroleff, F., & Warme, K. E., "Carbon dioxide variations in the atmosphere, *Tellus*, 8:176 (1965).
13. Plass, G. N., "Carbon Dioxide and Climate," *Scientific American*, 201(28):41 (1959).
14. Keeling, C. D., & Bolin, B., "Large-scale atmospheric mixing as deduced from seasonal etc.," *Geophysical Research*, 68:13 (1963).
15. Plass ,G. N., "The carbon dioxide theory of climate changes," *American Journal of Physics*, 24:376 (1959).
16. Moller, F., "On the influence of changes in $CO_2$ conc. etc.," *Journal of Geophysical Research*, 68 (13), 3877 (1963).
17. Mitchell, J. M., "Recent secular change of global temperature," *Annual, New York Academy of Sciences*, 95:235 (1961).

## The Nitrogen Cycle

Among the numerous chemical compounds one finds of significance in biological systems are those containing nitrogen. Proteins and nucleic acids are two major classes of compounds upon which both animal and plant growth and reproduction are dependent. It is the bonded nitrogen in these compounds that determines their biochemical activity. Nitrogen is the most abundant element in the earth's atmosphere, but it is present in an unusable form. Diatomic nitrogen ($N_2$) is neither soluble or reactive, two properties which are necessary before either plants or animals can use it to build proteins.

In general, the nitrogen cycle consists of a series of integrated reactions in which molecular nitrogen (atmospheric) is converted into biologically active forms which are utilizable by living organisms and the subsequent conversion and return of this nitrogen to the atmosphere as the molecular form, Figure 9.

Although molecular nitrogen is biologically inert, it is the primary source of nitrogen in the formation of nitrogenous organic compounds. "Nitrogen fixing" organisms constitute a special category of

species in both the plant and microbial kingdoms that possess the capacity to convert nitrogen to ammonia. Among such organisms, one finds certain heterotrophic bacteria, both anaerobic (*Clostridium pasteurianum*) and aerobic (*Azotobacter vinelandii*); photosynthetic bacteria (*Rhodosperillum rubrum*); algae: and leguminous plants in symbiotic activity with a bacteria of the genus *Rhizobium*.[1,2] At this point, some explanations are in order. Heterotropic organisms are those which are not self-sustaining and require a reduced form of carbon for their metabolic activities. Anaerobic organisms are those which are capable of growth only in an environment that excludes free molecular oxygen, while aerobic organisms are those requiring molecular oxygen for their growth. Symbiotic organisms are two dissimilar organisms living in close association with each other. In the case of the legume *Rhizobium* symbionts, neither plant nor bacteria can fix nitrogen alone, but it is only through the combined activities of the two that nitrogen fixation is accomplished. Such a symbiotic association is said to be one of *mutualism* in that both of the symbionts derive benefits from their association with each other, and the end products of their efforts.

Before any further discussion of nitrogen and its role as a component part of an ecosystem, it will prove wise and helpful to define some of the basic terminology commonly used with reference to this topic. Nitrification refers to the reactions in which certain microorganisms are capable of oxidizing ammonium ion ($NH_4$) to nitrite ($NO_2^-$) and (or) the oxidation of ($NO_2^-$) to nitrate ($NO_3^-$). Denitrification is the reverse process, e.g. the reduction of ($NO_3^-$) and $NO_2^-$) to gaseous molecular nitrogen ($N_2$) or nitrous oxide ($N_2O$). Whenever dead material containing protein decomposes under the influence of microorganisms which convert the bound nitrogen to ammonia, the process is called ammonification. Although oxidation and reduction will be considered in a separate article within this book, brief mention of what the two processes have in common will be made here at this time. Oxidation is any process in which electrons are removed from a substance while the addition of electrons to any substance is termed reduction. Whenever one substance is being reduced, it implies that another substance is being oxidized as is the reverse procedure since electrons are neither created nor destroyed in a chemical reaction.

Nitrogen fixation by biological systems is an amazing biochemical capacity. Molecular nitrogen, $N_2$, is quite stable and resists reaction as can be

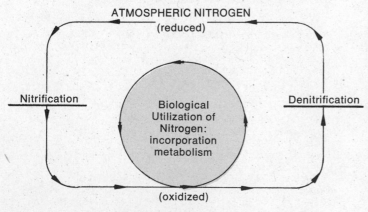

ATMOSPHERIC NITROGEN
(reduced)

Nitrification

Biological Utilization of Nitrogen: incorporation metabolism

Denitrification

(oxidized)

**Figure 9.** The Nitrogen Cycle.

seen from the bond strength of the triple bond in $N_2$ (226.8Kcal/mole). Drastic conditions are necessary to force its conversion into compounds in the laboratory while delicate nitrogen-fixing microorganisms can catalyze the same conversion under relatively mild conditions, with incorporation of atmospheric nitrogen in the order of several million tons per year.[3]

Recent developments employing cell-free nitrogen fixing systems from different organisms have enabled laboratory observations which are bringing about a better understanding of the sequence of events in nitrogen fixation.[4,5,6] Although the response of different systems to changing environmental parameters is varied, the end product of fixation in these cell-free preparations as in the intact systems is ammonia ($NH_3$).

Ammonia formation from molcular nitrogen is a two step process. Nitrogen activation, an energy-requiring reaction is the first step, in which molecular nitrogen is split into two atoms of free nitrogen. This is followed by its combination with three atoms of hydrogen, forming ammonia.[7]

$$(1)\ N_2 \dashrightarrow 2N \quad 160\text{Kcal/mole}$$
$$(2)\ 2N + 3H_2 \rightarrow 2(NH_3) - 13\text{Kcal/mole}$$
$$\overline{\quad N_2 + 3H_2 \rightarrow 2(NH_3) \quad 147\text{Kcal/mole}\quad}$$

The overall activation process requires a net input of 147Kcal/mole.

Microorganisms such as those represented by the chemotrophs of the genus *Nitrosomonas* utilize the ammonium ion as a source of energy, releasing some of the energy that is stored during the process of activation. In the presence of oxygen, ammonia is oxidized to the nitrite ion.

$$NH_3 + O_2 \rightarrow NO_2^- + H_2O - 67\text{Kcal/mole}$$

Nitrite is oxidized further to nitrate by another group of chemotrophic organisms found in the soil with the release of an additional 17kcal/mole.

$$2NO_2^- + O_2 \rightarrow 2NO_3^- - 17\text{Kcal/mole.}$$

At this point, nitrate can be used by plants and microorganisms to serve as; (1) a supply of nitrate for reduction and assimilation as ammonia with its eventual entry into cellular metabolism, or (2) a terminal acceptor of electrons, producing $N_2$, $N_2O$ or NO, none of which enter into any aspect of cellular metabolism.

The process of nitrogen assimilation from nitrate occurs in two steps in which nitrate is reduced to nitrite by the catalysis of a specific enzyme,

nitrate:NADH-oxidoreductase (E.C. 1.6.6.1) followed by nitrite reduction to ammonia by the enzyme, nitrite:NADPH-oxidoreductase (E.C. 1.6.6.3). Nitrate reductase activity is found in higher plants, microorganisms and fungi. It is an enzyme complex that is nicotinamide nucleotide linked with flavoprotein (FAD), molybdenum, cytochrome and other metal prosthetic groups necessary for its activity.[8,9,10]

Terminal reduction is carried out by microorganisms of the type represented by *Pseudomonas denitrificans*, an anaerobic bacteria found in the soil. In the absence of oxygen, these organisms are able to use either nitrate or nitrite as electron acceptors in the oxidation of organic compounds. When glucose is metabolized under anaerobic conditions in conjunction with denitrification, the energy yield is comparable to that obtained when glucose is reacted with pure oxygen.

$$(1)\quad \text{Glucose } (C_6H_{12}O_6) + (K)NO_3 \rightarrow 6CO_2 + 3H_2O$$
$$+ 6KOH + 3N_2O - 545\text{Kcal/mole}$$
$$(2)\quad C_6H_{12}O_6 + 6O_2 \rightarrow 6CO_2 + 6H_2O - 686\text{Kcal/mole}$$

The reaction of these organisms results in the reduction of nitrate and nitrite to nitrous oxide or to the elemental gaseous state. It is by denitrification that nitrogen is reduced and returned to the atmosphere. Denitrification occurs not only on land but also in the oceans, although the mechanisms have not been studied as extensively in the latter.[11]

Ammonia is the prime form of inorganic nitrogen utilizable by all living cells. By the operation of three major synthetic pathways, ammonia can be fixed by all organisms, at all phylogenetic levels, including mammals. Glutamic acid, glutamine and carbamyl phosphate are the results of these reactions. Once formed, these nitrogen-containing compounds can be incorporated into complex biological molecules such as the nucleic acids and proteins. Plants and many microorganisms can synthesize all the amino acids found in proteins by means of their ability to utilize ammonia in conjunction with carbon to fabricate the structures corresponding to each amino acid. Mammals, in contrast are only able to make approximately half of these carbon-nitrogen structures. This relative capacity for the synthesis of amino acids represents a major difference between the metabolism of plants and microorganisms when compared with mammals. It also demonstrates how mammals, although thought to be more highly developed in many respects, are metabolically deficient in their capacity to synthesize all the essential components in the formation of proteins which are the fundamental structural and functional blocks upon which mammalian existence is dependent. Higher forms are therefore dependent upon the dietary intake of these components from plant sources or from animal proteins which ultimately find their source of these amino acids in plants that are part of their diet. The food-chain cannot be broken without extensive loss of whole species which are mutually dependent.

Plants and microorganisms demonstrate another basic difference in that they manifest self-regulatory mechanisms enabling utilization of needed amino acids from their environment only to the degree required for the synthesis of nitrogenous cellular constituents necessary to maintain normal cellular function. Mammals differ in that similar regulatory mechanisms for amino acid intake do not exist and mammals are generally presented with an excess of these compounds arising from the digestion of proteins contained in their diet. Therefore, a mechanism for the removal of excess amino acids occurs in higher forms. This is accomplished by synthesis and excretion of urea (A) or by directing excess into oxidative or storage pathways of carbohydrate and lipid metabolism in the form of carbon-chain compounds (B), Figure 10.

A. $2NH_3 + CO_2 \longrightarrow CO(NH_2)_2 + H_2O$

B.

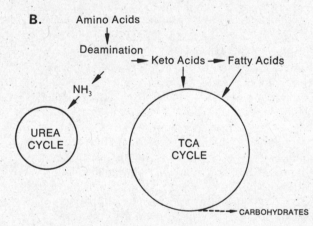

**Figure 10.** Carbohydrate Pathways.

Ammonia fixation by glutamic acid synthesis occurs almost universally throughout the plant and animal kingdom as well as in microorganisms. The basic reaction involves the action of an enzyme, glutamic dehydrogenase in conjunction with a pyridine nucleotide donor-acceptor system, Figure 11.

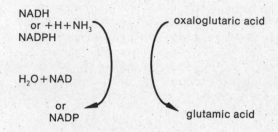

**Figure 11.** Glutamic acid Synthesis.

Glutamine synthesis is an important means of nitrogen fixation in plants, bacteria, and more significantly so in animals,[12] figure 12.

$$
\begin{array}{ccc}
\text{COOH} & & \text{CONH}_2 \\
| & & | \\
(\text{CH}_2)_2 & & (\text{CH}_2)_2 \\
| & \xrightarrow{\text{Mg}} & | \\
\text{CHNH}_2 & +\text{ATP}+\text{NH}_3 & \text{CHNH}_2 & +\text{ADP}+\text{P}_i \\
| & & | \\
\text{COOH} & & \text{COOH} \\
\text{glutamic acid} & & \text{glutamine}
\end{array}
$$

**Figure 12.** Glutamine Synthesis.

The enzyme involved in this reaction is glutamine synthetase. Glutamine synthesis exceeds all other forms of ammonia fixation in mammals since the amide nitrogen of glutamine can be transferred by various synthetic processes to numerous other compounds, including purines, histidine, NAD, and hexosamines. The physiological significance of this transfer in mammalian systems can be seen in that it enables nitrogen fixation to be accomplished with physiologically tolerable levels of free ammonia, which otherwise would not have been possible. Bacterial systems can utilize high concentrations of free $NH_3$ with no adverse effects while mammalian systems are extremely sensitive to free $NH_3$ which in too high a concentration becomes a deadly poison.

Carbamyl phosphate synthetase catalyzes $NH_3$ incorporation in the synthesis of carbamyl phosphate in both mammalian and bacterial systems.[13,14]

(A) Microorganisms—

$$CO_2 + NH_3 + ATP \rightarrow H_2NC\overset{\displaystyle O}{\overset{\displaystyle \|}{-}}O-P$$
Carbamyl phosphate

(B) Mammals—
$$H_2O + CO_2 + NH_3 + 2\,ATP$$

$$\xrightarrow[2ADP+P_i]{\text{N-acetylglutamate}} H_2NC\overset{\displaystyle O}{\overset{\displaystyle \|}{-}}O-P+$$

Once produced, carbamyl phosphate is used to carbamylate amino groups in the synthesis of additional nitrogen containing compounds.

$$H_2NC-O-P+R-NH_2 \rightarrow H_2NC-NH-R+P_i$$

This is the mode of synthesis used in the formation of citrulline, the pyrimidines and arginine.[15]

The ultimate fate of nitrogen which has been incorporated into plant and animal structure will be a return to ammonia as a consequence of bacterial decay in ammonification. The metabolic by products (wastes) of nitrogen-containing compounds will undergo a similar bacterial degradation to ammonia. This ammonia can be recycled either by incorporation into cellular metabolism or it can go into terminal reduction to molecular $N_2$ as was mentioned previously.

There are three additional components that must be mentioned to complete a discussion of the factors involved in the nitrogen cycle. A certain amount of nitrogen is fixed in the atmosphere by means of cosmic ionization. During thunderstorms electrical discharges serve to activate nitrogen oxidation and rain carries the oxidized nitrogen to the ground where it enters into various aspects of the cycle. There is also a small amount of nitrogen released from igneous rock during volcanic extrusion of molten rock. By far, the most significant factor not mentioned as yet is man's activity in his synthesis of nitrogen containing fertilizers. At present, the fixation of nitrogen by artificial synthetic means is almost equivalent to that occurring by means of soil organism fixation. Increased agricultural demands place increasing demands on soil nitrogen content and through supplimentation with fertilizers, man has been able to increase the amount of nitrogen incorporated into plant material. As increasing amounts of fertilizers are used, there will be larger amounts of nitrogenous compounds found in run-off and ground water.[16] The consequences of these increasing levels have not been fully explored. Future studies should be directed towards determining whether natural systems will be able to handle the increased nitrogen load, reducing it to its molecular form and returning it to its atmospheric source.

To summarize the activities of the nitrogen cycle, the following diagram shows the general aspects of the cycle with the integrated activities of each component as they relate to each other, Figure 13.

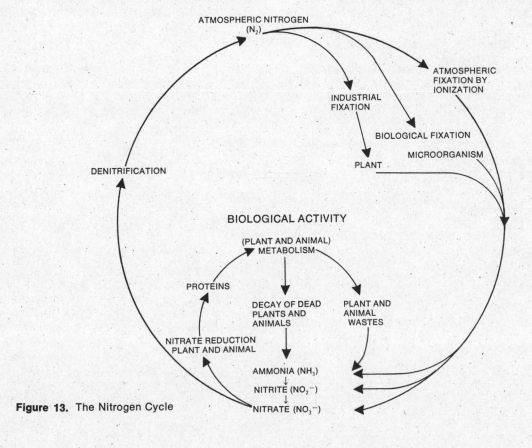

**Figure 13.** The Nitrogen Cycle

## REFERENCES

1. Carnahan, J. E., & Castle, J. E., "Nitrogen fixation," *Annual Review of Plant Physiology*, 14:125 (1963).
2. Bergersen, F. J., "Biochemistry of symbiotic nitrogen fixation in legumes," *Annual Review of Plant Physiology*, 22:121 (1971).
3. Hardy, R. W. F., & Burns, R. C., "Biological nitrogen fixation," *Annual Review of Biochemistry*, 37:331 (1968).
4. Carnahan, J. E., Mortenson, L. E., Mower, H. F., & Castle, J. E., "Nitrogen fixation in cell-free extracts of *Clostridium pasteurianum*," *Biochim. Biophys. Acta.*, 38:188 (1960).
5. Carnahan, J. E., & Castle, J. E., "Some requirements of biological nitrogen fixation," *Journal of Bacteriology*, 75:121 (1958).
6. Carnahan, J. E., Mortenson, L. E., Mower, H. W., & Castle, J. E., "Nitrogen fixation in cell-free extracts of *Clostridium pasteurianum*," *Biochim. Biophys. Acta.*, 44:520 (1960).
7. Delwiche, C. C., "The nitrogen cycle," *Scientific American*, 223(3), 136 (1970).
8. Garrett, R. H., & Nason, A., "Involvement of B-type cytochrome in the assimilatory nitrate reductase of *Neurospora crassa*," *Proceedings of the National Academy of Sciences*, *U.S.*, 58:1603 (1967).
9. Nason, A., & Evans, H. J., "Triphosphopyridine nucleotide-nitrate reductase in *Neurospora*," *Journal of Biological Chemistry*, 202:655 (1953).
10. Evans, H. J., & Nason, A., "The effect of reduced triphosphopyridine nucleotide on nitrate reduction by purified nitrate reductase," *Archives of Biochemistry and Biophysics*, 39(1), 234 (1952).
11. Northen, H. T., & R. T., *Indigeneous Kingdom*, Englewood Cliffs, N.J. (1970).
12. Baldwin, E., Transferases, *Dynamic Aspects of Biochemistry*, 5th ed. Cambridge (1967).
13. Pierard, A., & Winame, J. M., "Regulation and mutation of carbamyl phosphate in *Eschericia coli*," *Biochim. Biophys. Acta., Res. Commun.* 15:76 (1964).
14. Anderson, P. M., & Meister, A., "Evidence for an active form of carbon dioxide in a reaction catalyzed by *Eschericia coli* carbamyl phosphate synthetase," *Biochemistry*, 4:2803 (1965).
15. Hager, S. E., and Jones, M. E., "A glutamine-dependent enzyme for the synthesis of carbamyl phosphate for pyrimidine biosynthesis in fetal rat liver," *Journal of Biological Chemistry*, 242:5674 (1967).
16. Commoner, B., "Soil and fresh water: Damaged global fabric," *Environment*, 12(3); 4 (1970).

## The Oxygen Cycle

All matter within the biosphere is in a state of flux, changing from one form to another. Many of these changes involve mere alterations in the physical state of a substance such as the transition from ice to water vapor or steam. On the other hand, many changes involve actual differences in chemical structure. It is this latter case in which one finds oxidation-reduction reactions occuring. The term oxidation refers to any reaction in which there is a transfer of electrons. When a loss of electrons occurs, it is oxidation, while the acquisition of electrons is reduction. Every time an oxidation occurs, it must be accompanied by a simultaneous reduction since there can be no loss of matter in conventional chemical reactions. In an oxidation-reduction reaction, the oxidant is the electron acceptor while the reductant is the electron donor. The oxidant becomes reduced while the reductant becomes oxidized. Although the terminology is somewhat confusing, a simple device I have found helpful in keeping the two processes straight is the following; "*Leo* goes *Ger*!!" Leo,—*L* oss of *E* lectrons is *O* xidation. *G* ain of *E* lectrons is *R* eduction.

This can be shown in the laboratory using reactions in which metals are oxidized at an electrode.

$$Cu^{O} \rightleftharpoons Cu^{+} + e$$
$$Fe^{++} \rightleftharpoons Fe^{+++} + e$$
$$Mn^{++} \rightleftharpoons Mn^{+++} + e$$

or in general terms;

$$Me^{n} \rightleftharpoons Me^{n+1} + e$$

Reductions would be the reverse reactions.

$$Me^{n+1} + e \rightleftharpoons Me^{n}$$

Oxidation of inorganic substances is much easier to visualize than a less obvious situation such as the case of $H_2 + \frac{1}{2}O_2 \rightleftharpoons H_2O$. In the case of the water molecule, hydrogen is covalently bonded to oxygen and electrons are shared between the two atoms. Although shared, a pair of electrons resides closer to the oxygen nucleus than to the hydrogen nucleus, hence, hydrogen is oxidized since it has partially lost electrons to oxygen, which is in turn reduced. The actual situation is made clearer if we write HOH or even $_H O_H$ instead of the familiar $H_2O$.

When dealing with organic compounds, the situation becomes more difficult. Many organic substances undergo an oxidation process which is known as dehydrogenation. Such reactions involve changes in hydrogen content and in many instances, oxygen removes hydrogen without increasing the oxygen content of the compound involved. The oxidation of an aliphatic alcohol such as ethanol to its respective aldehyde, acetaldehyde, is an example of an oxidation involving a dehydrogenation.

$$CH_3 - \overset{\overset{\displaystyle H}{|}}{\underset{\underset{\displaystyle H}{|}}{C}} - OH \rightarrow CH_3 - \overset{\overset{\displaystyle H}{|}}{C} = O$$

Many biological oxidations involve the transfer of the hydride ion, $H^+$, one proton and two electrons, to an acceptor molecule. The oxidation of ethanol in terms of an acceptor molecule can be shown as follows;

$$CH_3 - \overset{\overset{\displaystyle H}{|}}{\underset{\underset{\displaystyle H}{|}}{C}} - OH + A^+ \rightleftharpoons CH_3 - \overset{\overset{\displaystyle H}{|}}{C} = O + AH_2$$

where $A^+$ is the oxidized acceptor and $AH_2$ is the acceptor in the reduced form.

The significance of the dehydrogenation reaction can be seen in that most biological oxidations significant in physiological systems occur in this manner. Oxygen is usually the terminal electron acceptor with one or more intermediate hydrogen carriers transferring electrons to oxygen. As a carrier accepts hydrogen, it is reduced. It is then reoxidized by transfer of its hydrogen to another carrier, thus returning to its original oxidized state in readiness to accept another hydrogen. The second carrier is reduced as it accepts hydrogen, and so down the line. Another possible variation in which a carrier can be reoxidized after it accepts electrons in the case where it transfers these electrons to an oxidized substrate which acts as the acceptor.[2]

A general representation of the transfer process occurring in dehydration is depicted as follows:

| Substrate−$H_2$ (red.) | Carrier #1 (oxid.) | Carrier #2 (red.) | ½ $O_2$ (oxid.) |
| --- | --- | --- | --- |
| Substrate (oxid.) | Carrier #1 (red.) | Carrier #2 (oxid.) | $H_2O$ |

In the maintenance of life, all organisms must derive an energy supply from the free energy of their surroundings. This will be in the form of light in the case of photosynthesizing autotrophic organisms or in the form of stored chemical energy, which found its ultimate source in light. Carbohydrates, lipids and proteins consumed as substrates by heterotrophic organisms serve as a source of energy and during respiration, the free energy stored in these substrates is released through a controlled oxidation. A portion of this energy is utilized to phosphorylate adenosine-5'-triphosphate (ATP), an energy rich phosphorylated intermediate which is used as an almost universal source of energy for the maintenance of many biological reactions. Some of the energy is stored in the intermediates of electron transfer when a carrier is reduced as it accepts electrons (hydrogen) or as it is oxidized when it releases electrons (hydrogen).[3]

Dehydrogenation is carried out by a class of enzymes called oxidoreductases which are characterized by carriers utilized with the particular enzyme. Nicotinamide coenzymes, (NAD, NADP), flavin nucleotides (FMN, FAD), cuproproteins and cytochromes are some examples of these carriers. Such carriers play a specific energy related role in the transfer of electrons (hydrogen) to oxygen during the oxidation of a substrate. The redox potential, that is, the oxidation-reduction potential ($E_O$), which is the potential difference recorded at an inert electrode when an oxidation-reduction is presented in solution at the electrode, determines the appropriate carrier which must be associated with a substrate. Redox potentials are expressions of the ability of various reducing agents to lose electrons. Table 1 lists a number of redox potentials of biological substances.

Conditions favoring the transfer of electrons (hydrogen) from a substrate to a carrier require that the redox potentials of each are in the same range. Many intermediates of carbohydrate metabolism have relatively low redox potentials, while oxygen has a high potential. To bridge the gap between these two extremes, a series of electron transfers occurs. The first is from the substrate to the oxidized nicotinamide coenzyme, which has a comparably low $E_O$. The substrate is oxidized while this carrier is reduced. The reduced nicotinamide coenzyme then transfers its electrons to an oxidized flavin nucleotide coenzyme which has an intermediate $E_O$.

| | |
|---|---|
| $H_2O/\frac{1}{2}O_2$ | 0.82 volts |
| $NO_2^-/NO_3^-$ | 0.42 |
| Cu protein/Cu protein | 0.35-0.60 |
| $Cu^+$ (aq)/ $Cu^{++}$ (aq) | 0.35 |
| $H_2O_2/\frac{1}{2}O_2$ | 0.30 |
| Cytochrome a ($Fe^{++}\rightarrow Fe^{+++}$) | 0.29 |
| Cytochrome c ($Fe^{++}\rightarrow Fe^{+++}$) | 0.23 |
| Cytochrome b ($Fe^{++}\rightarrow Fe^{+++}$) | 0.12 |
| Ascorbate/dehydroascorbate | 0.080 |
| succinate/fumarate | −0.031 |
| alanine/pyruvate+$NH_4$ | −0.130 |
| Flavoprotein−$H_2$/Flavoprotein | −0.185 average |
| Lactate/pyruvate | −0.190 |
| NAD(P)H/NAD(P)$^+$ | −0.320 |
| Malate/pyruvate+$CO_2$ | −0.333 |
| $\frac{1}{2}H_2/H^+$ | −0.420 |

The flavoprotein is thus reduced and the nicotinamide coenzyme is reoxidized; ready to accept another pair of electrons (hydrogen). Subsequently, the reduced flavoprotein transfers electrons to either cytochromes or cuproteins which have redox potentials that are high and comparable to that of oxygen. The final transfer occurs when oxygen is reduced to water and the carriers (cytochromes or cuproteins) are reoxidized, figure 14.

Nicotinamide adenine dinucleotide(NAD$^+$) and nicotinamide dinucleotide phosphate(NADP$^+$) can accept electrons from many substrates during dehydrations due to the nicotinamide portion of this molecule which can undergo repetitive oxidation-reduction.

*Nicotinamide Adenine Nucleotides and NAD(P)-linked Dehydrogenases*—For a long time, these carriers have been referred to as diphosphopyridine nucleotide (DPN) and triphosphopyridine nucleotide (TPN) and in the older literature, you will see NAD and NADP designated as DPN and TPN respectively. At present, either designation is acceptable. Nicotinamide adenine nucleotide (NAD)(P) occurs in all cells and its actual role can be compared to that of ATP. ATP is a universal phosphate carrier and NAD(P) is a universal electron carrier.

The pyridine-linked dehydrogenases will transfer two reducing equivalents of the substrate (which is being oxidized) to the oxidized form of NAD(P); one as a hydrogen atom which reduces NAD(P) to NAD(P)H and the other appears as a free hydrogen ion H$^+$ which is released into the medium.

$$\text{Reduced substrate} + \text{NAD(P)}^+ \rightleftharpoons \text{Oxidized substrate} + \text{NAD(P)H} + H^+$$

It should be noted that NAD is found in greater relative concentrations within the mitochondria of the cell where it is primarily concerned with dehydrogenase reactions in which the transfer of electrons from a substrate to molecular oxygen occurs via the respiratory sequence. On the other hand, NADP is found in greater relative amounts in the cytoplasm, the "soluble" portion of

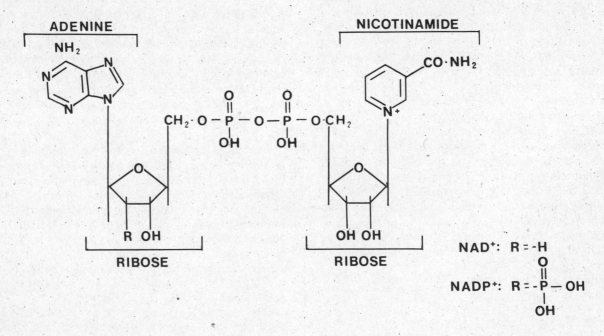

**Figure 14.** Oxygen Reduction, Carrier Reoxidation.

the cell. Electrons derived from NADP-linked dehydrogenases acting on substrates in catabolic degradations are transferred to reductive reactions involved in biosynthetic pathways (anabolic reactions).

As a pair of electrons is transferred to the pyridine nucleotide and the oxidized form of this molecule is converted to the reduced form, there is a characteristic shift of the absorption spectra. A brief description of an absorption spectra will prove most helpful in understanding what this means. When a light is passed through a solution of a substance, a certain amount of this light will pass through the solution, but another portion of this light will be absorbed, that is, prevented from passing through the solution. That portion which is absorbed will be dependent upon the characteristic structural arrangement of the electrons within the molecules of the substances contained within the solution. If one uses a device which can carefully define the wavelength of the light which is passed through the solution in such a way that the wavelength can be varied in a continuous fashion from one point to another, with a detecting device to measure the light passing through the test solution (transmitted light), it is possible to deter-

mine at what wavelengths a substance will absorb. Such an instrument is known as a spectrophotometer, and the resulting recording wavelength versus transmitted light is called an absorption spectra. By using a spectrophotometer, it is possible to follow the oxidation-reduction changes in many intermediates as a function of their own characteristic spectral changes which are in turn related to changes in their molecular arrangement, Figure 15.

The flavin coenzymes received electrons from reduced substrates or from reduced nicotinamide coenzymes through the intervention of specific flavoprotein enzymes. FMN and FAD can accept electrons or be electron donors due to the oxidation-reduction of the isoalloxazine ring.

*Flavin-linked Dehydrogenases*—FAD(FMN) serve as the coenzyme for the flavin-linked dehydrogenases. The flavin nucleotide is tightly bound to the dehydrogenase and in most cases does not leave the enzyme during the reaction process. The oxidation of the flavin coenzyme is usually referred to as a one step transfer of two electrons, but it has been shown that the process is one in which two one-electron transfer steps occur. The transfer of one electron to FAD(FMN) results in the formation

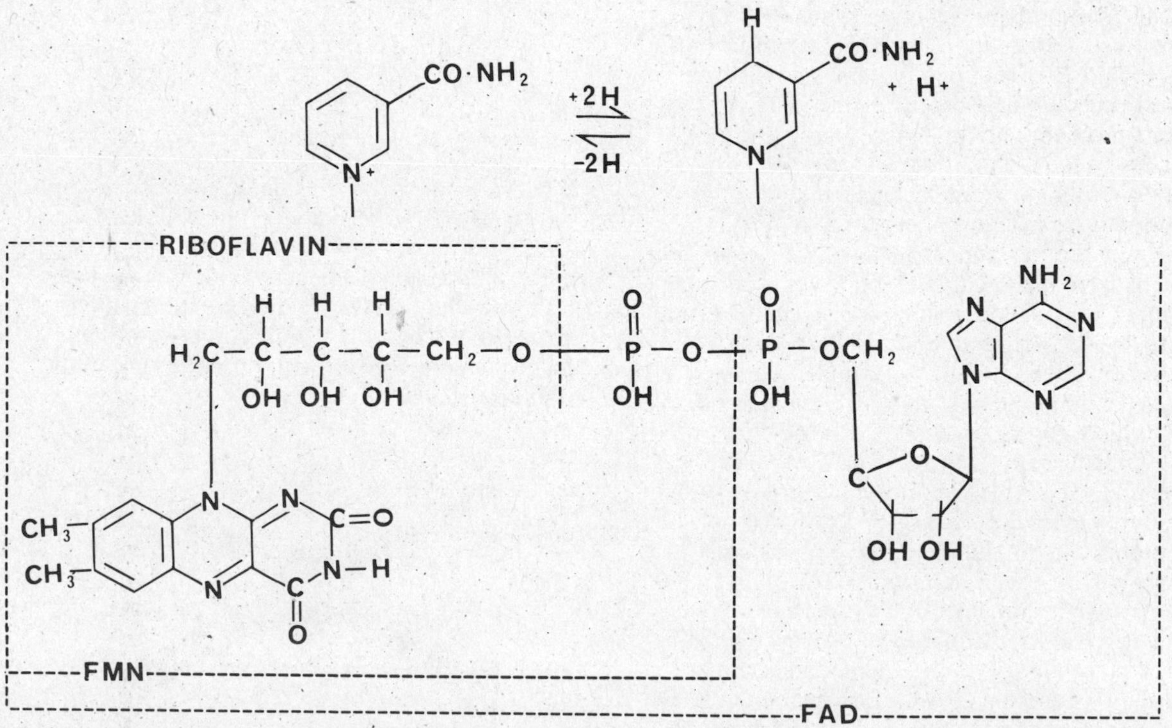

**Figure 15.** Flavin mononucleotide (FMN)—riboflavin phosphate
Flavin adenine dinucleotide (FAD)—riboflavin, phosphate attached by a pyrophosphate linkage to adenosine monophosphate.

**Figure 16.** Isoalloxazine Ring of FMN and FAD

of a chemical species known as a *free radical*, that is, a form in which there is one unpaired electron. This semiquinine is unstable and its life span is very short. In fact, only evidence from absorption spectra indicate that this intermediate exists. The semiquinone then goes on to accept the second electron, thus reducing the flavin nucleotide completely. Flavin dehydrogenases have characteristically intense colors in their oxidized forms but when reduced, a bleaching of this color occurs. This property enables one to readily follow oxidation-reduction reactions involving the flavin nucleotides, figure 16.

Cuproproteins and cytochromes owe their reactivity in oxidation-reduction systems to the copper (Cu) atoms and the iron (Fe) atoms within the molecule. By accepting or releasing electrons, the metal ions are reduced or oxidized. Cytochromes are heme-containing proteins and iron (Fe) is found in most of them, but there is a group of non-heme cytochromes which depend on other metals for the transfer of electrons.

The cytochromes serve to sequentially transfer electrons from the flavin nucleotides to molecular oxygen. In studies with mitochondria isolated from plant cells and higher animals, at least five cytochromes have been found. Four of the five are tightly bound to the membrane while one is less tightly bound. Electrons are transferred sequentially from cytochrome b to cytochrome $c_1$; from cytochrome $c_1$ to cytochrome c; from cytochrome c to cytochrome a-$a_3$; and finally from cytochrome a-$a_3$ to molecular oxygen. As each cytochrome accepts a pair of electrons it becomes reduced, then in transferring the electrons to the next member of the sequence, it becomes reoxidized. The terminal cytochrome, cytochrome oxidase, is a complex of two of the members of the group, cytochrome a and $a_3$. It has the unique property of being the only cytochrome capable of being reoxidized by molecular oxygen.

The active prosthetic group involved in the reversible changes of valance that occur during oxidation-reduction is the iron-porphyrin group. This moiety is similar to the porphyrin base structure found in hemoglobin and myoglobin. Porphyrins are ring structures, the base of which is a tetrapyrole.

Porphin

Each class of porphyrins differ in the substituent groups which may be added to each side-chain. The four nitrogens of the porphyrin structure form metal complexes with such metal ions as that of iron, with the iron taking up a position within the tetrapyrole ring. This can be represented as follows where only the four nitrogens (N) of the ring are shown.

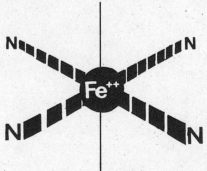

Protoporphyrin IX is the form most abundant in many of the cytochromes.

CH₃ CH=CH₂

Protoporphyrin IX

released, is derived through the oxidative phosphorylation process. Thus glucose and other sugars are the bearers of most of the energy utilized by organisms and one of the most essential classes of chemical substances for life.

Thus, iron within the structure of the cytochromes can undergo reversible oxidation-reduction, as an electron acceptor-donor, ($Fe^{++}$ $+e^- \rightleftharpoons Fe^{+++}$) serving to transfer the electrons sequentially to molecular oxygen.

It is the ability to transfer hydrogen (electrons) that makes the carrier complexes so important in substrate oxidations during the dehydration process. Through cyclic reduction and oxidation, continued reuse of a relatively small amount of these carrier substances (relative to the amount of these carrier substances (relative to the of material that is turned over through the system) is possible with the efficient transfer of energy to a readily usable form. By this mechanism a cell or fine organ can turn over very large quantities of energy for its size and all under precisely controlled conditions.

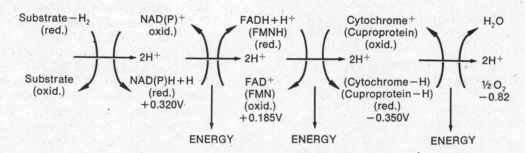

**Figure 17.** Dehydrogenation of a Substrate.

In general, the hydrogen from dehydrogenation (oxidation) of a substrate can be transferred to oxygen in graded oxidation-reduction steps such that the stored free-energy in these substrates is released in an efficient and controlled manner. This allows sufficient energy for all the complex functions of life without burning up, boiling or electrocuting the cell or organism.

The above representation is a limited and incomplete summary of what is known as the respiratory chain in which oxidative phosphorylation occurs. There are a number of other carriers involved in the sequence between each of those shown and many of the details of this process are still uncertain. The interested reader should refer to recent reviews of the subject for further details.[4,8,10] It should be mentioned that the energy released in the process of oxidation is transferred through phosphorylation of adenosine diphosphate (ADP) with inorganic phosphate ($P_i$) to form ATP. In the oxidation of glucose, a carbohydrate that serves as the primary substrate in the metabolism of most organisms, approximately 90% of the energy

## REFERENCES

1. Clark, W. M., *Oxidation-Reduction Potentials of Organic Systems*, Williams and Wilkins, Baltimore (1960).
2. Baldwin, E., Oxioreductases: Dehydrogenase systems, *Dynamic Aspects of Biochemistry*, 5th ed. Cambridge (1967).
3. Wang, J.H., "Oxidative and photosynthetic phosphorylation mechanisms," *Science*, 167:25 (1970).
4. Lehninger, A. L., *The Mitochondrion*, Benjamin, N.Y. (1965).
5. Lehninger, A. L., "Electron transport and oxidative phosphorylation, *Biochemistry*, Worth Publ. N.Y. (1970).
6. Harrow, B., & Mazur, A., *Textbook of Biochemistry* chapt. 5, 9th ed. Saunders, Philadelphia (1966).
7. Neims, A. H., & Hellerman, L., "Flavoenzyme catalysis," *Annual Review of Biochemistry*, 39:867 (1970).
8. Wainio, Walter, *The Mammalian Mitochondrial Respiratory Change*, Academic Press, N.Y. (1970).
9. Lardy, H. A., & Ferguson, S. M., "Oxidative phosphorylation in mitochondria," *Annual Review of Biochemistry*, 38:991 (1969).
10. van Dam, K., & Meyer, A. J., "Oxidation and energy conservation by mitochondria," *Annual Review of Biochemistry*, 40:115 (1971).

Ronald F. Albano, M.A.sc.
Department of Pharmacology,
Research Division of Hoffmann-La Roche

# Chapter 5

# The Utilization of Substances by Plants

The existence of all living organisms, including man, depends upon the utilization of energy. In another sense, life is a form of the expenditure of energy. To stay "alive," the living system must continually consume, store and release energy; the quality and quantity of life being controlled by the amount of energy available for consumption. When the energy source is removed, life rapidly ceases.

## Energy Requirements

Only one form of energy, chemical energy, is utilized in driving the metabolic processes which result in growth and activity in living organisms. This energy is stored in the bonds of chemical compounds, the formation of which requires a potential energy. The breaking of these bonds releases this energy. In compounds held together by stable bonds, chemical energy may be stored for long periods of time and is measured in calories in the equivalent amount of heat energy. Living organisms have the ability to break the bonds in certain types of stable compounds, and release this energy in an even, controlled manner for their own use.

The amount of energy required by a living organism to grow and carry out its normal activities is large. Take, for example, the average man, who in order to carry out his normal activities of work, recreation etc., and maintain his body in a healthy condition, at a static weight, requires 3200 kilocalories per day (1 kilocalorie, Kcal or Cal. =1000 calories). A man with less than average weight or activity requires fewer kilocalories and a man who either performs heavy labor or who is above average in size requires correspondingly. A man's daily food intake must furnish him with all the energy he requires. If all the food that man ate was converted to usable energy, he would need to consume about 500 to 800 grams of food or, in other words, about 1 to 1½ pounds per day (dry weight), with the exact amount depending on the food source. Although the energy requirement is the greatest, man also needs substantial quantities of protein and other essential substances such as vitamins and minerals. The protein requirement for the average man comes to about 50 grams per day.

Man's food intake is far greater than that directly required for his energy and protein needs. Food substances, such as sugars, starches, fats, oils and protein are readily digestible, but, there is also much fibrous material, made up of cellulose and lignins, which man does not digest. This indigestible material can make up about 50% of his diet. This passes through the digestive tract and is voided. Also, from food material such as vegetables, a great deal of material is disposed of, either during harvesting because only certain plant parts, such as seeds, may be utilized, or during packaging and preparation when unsightly, spoiled or diseased material is discarded.

In underdeveloped countries, a pound of grain per day may be all that is consumed to meet a person's minimal energy requirement. About 4 to 10 times this amount of plant material is needed to grow that one pound of grain, the remainder being composed of straw and roots. In developed countries the food demands are much greater. While less than half a pound of grain may be consumed directly, a very high proportion is consumed indirectly in the form of meat, milk or eggs. Therefore, the average American requires close to five pounds of grain per day to meet his food demands. The American may consume more tasty and nutritious food but he has sacrificed a great deal of energy in the process. This is because he chose to convert his energy through highly inefficient animal conversion to something more palatable such as oats to bacon or grass to beefsteak. Thus, a person can greatly increase his demand for energy by consuming a high proportion of animal products. His actual energy intake may not have increased but he has greatly increased the amount of energy needed to produce his food. Whereas a person in a poor country consumes an average of 360 pounds of grain per year, the average American in the same period directly and indirectly consumes 1600 pounds of grain, Figure 18.

**Figure 18.** A person living on a diet composed almost entirely of plant products requires far less plant material to sustain him than a person living on a diet containing a large proportion of animal protein. A person living mainly on wheat products, as is the case in many underdeveloped countries, can maintain himself on 360 pounds of wheat per year, compared to 1600 pounds of grain that are needed by a person on a high protein diet as is the case in North America. Most of this grain is fed to cattle which are then consumed.

This same pattern of direct and indirect energy consumption exemplified by man is repeated for all animals. Whether an animal eats plant or animal material, it is constantly deriving its energy directly or indirectly from green plants. This idea is demonstrated in Figure 19. Whether grain is fed to cattle, for man's consumption, or whether an insect feeds on grass leaves and is consumed by a sparrow, which is consumed by a snake, which in turn could be consumed by a hawk, the original energy source is the green plant. There is no way at the moment for either man or animal to escape this dependence on green plants.

The most basic and fundamental relationship man has to other living organisms is through his food requirements. There are, however, many other human dependencies on these organisms. Man utilizes plants and animals in hundreds of ways. For example, the assimilated carbon that woody plants deposit in their trunks and branches in the form of cellulose serves many useful purposes. The strength of the wood makes it useful structurally for houses, furniture, boxes, etc. The wood is also used in the manufacture of paper and cardboard. Antibiotics are extracted from organisms. Animal fibres or skin are used for clothing. Plant products are used for beverages, cosmetics, fuel, and manufacture of rubber, glue, etc. The list is endless. However, man is making himself less dependent on these products by switching to synthetic substances. As well, the aesthetic value of plants in the home, outside the home, in parks and wild areas must not be overlooked. No doubt, plants also play a very important role in preserving man's mental well-being. Man is more closely linked to his environment than he realizes or wishes to acknowledge.

By providing energy, green plants play a unique and crucial role in maintaining life on this planet. The green pigments, called the chlorophylls, present in the leaves of the plants enable the plant to capture and retain some of the light energy that strikes the leaf. Within the chlorophyll molecules, the light energy is absorbed and concentrated in an electron which becomes "excited" and moves into a higher orbit. In the structure of the chloroplast, the excited electron is led away by a series of electron-carrier enzymes, and in the process, some of the energy from the electron is transferred and conserved in a chemical form.

Some of this energy is utilized in the chain of events to form high energy bonds in the molecule ATP (adenosine triphosphate). ATP plays the role of an energy carrier and is able to transfer the stored energy to chemical reactions taking place in living systems. It is the most important compound involved in energy transfers in all living cells. Some of the "excited" electrons are passed on to the electron carrier NADP (nicotinamide adenosine dinucleotide phosphate). The electrons that are lost from the chlorophyll are replaced by electrons derived from the splitting of water, the reaction that also releases oxygen into the atmosphere. This excitation and transfer of electrons, driven by light and absorbed by chlorophyll, is the basic process supplying all the energy for the plant. The two energy carriers, ATP and reduced NADP, are highly reactive compounds and pass on their stored energy very readily. The next important step in the plant is to transfer this energy to more

**Figure 19.** The many pathways of energy flow from plants to animals and man. Vegetation is eaten by herbivores (plant-eaters), such as the meadow mouse (2), deer (4), insects (7), small fish (8), sheep (6) and cow (5). The herbivores in turn are consumed by carnivores (meat-eaters), such as the weasel (10), bird (3), frog (11) and large fish (9). Other carnivores in turn may consume the first level carnivores, such as the snake (12) eating the frog or the hawk 13) eating the bird or snake or mouse.

Man (1) is in a special position in that he eats both plants and animals (carnivore). He also uses both plant and animal products for clothing, shelter, fuels, etc. Another group of organisms lives below the soil surface living off dead and decaying plant and animal matter. These organisms are called decomposers and are important in returning nutrients to the soil for plant use. The sun supplies all the energy to run the whole intricate system.

stable compounds. In the chloroplasts, the capture, transfer and storage of this energy are closely connected. The ATP and reduced NADP provide the energy to drive the reaction in which carbon dioxide is incorporated via a series of steps into glucose. The fixation of carbon dioxide is a process requiring a great deal of energy and this is provided by the conversion of light energy to chemical energy. The overall reaction of converting light energy to chemical energy and storing this energy in glucose has been summarized as follows:

$$\text{light} \quad H_2O \quad \longrightarrow e-$$
$$\boxed{\text{chlorophyll}} \longrightarrow \begin{array}{l} \text{ATP} \\ \text{NADP}_{red} \end{array} \longrightarrow \begin{array}{l} \text{Enzymatic} \\ \text{Reactions} \\ CO_2 \end{array} \longrightarrow \text{glucose}$$
$$+O_2$$

$$6CO_2 + 6H_2O + light \longrightarrow C_6H_{12}O_6 + 6CO_2$$
$$\text{glucose}$$
$$\text{(673 Kcal stored)}$$

Now the energy is stored in a stable high energy compound from which the energy can be released again at any time in a controlled manner by all living organisms in the process of respiration. All life revolves around this process of energy absorption, storage and utilization involving the energy stored in the C-C bond of glucose. The glucose also provides the building blocks for all the other compounds that the plant needs to synthesize, grow and carry on its life functions. The glucose may be incorporated into starch for energy or it may be converted to oils or fats. Glucose may also be used in the production of amino acids, which form the building blocks of proteins. The proteins, in turn, may form the enzymes so vital for the metabolism in the cells or they may be incorporated into the structures of the many membranes in the cells. In the plant, a large proportion of the glucose is used in the synthesis of cellulose and lignins which make up the basic structures of plant cell walls. In woody plants especially, large amounts are stored more or less permanently in this form in the trunks, roots and limbs of trees. Many other compounds are synthesized by plants: nucleic acids, hormones, pigments, vitamins, alkaloids, etc. Since this process of photosynthesis supplies all the energy for life, we are interested in just how much light energy the plant can convert to chemical energy. It has been found that in natural ecosystems, the efficiency of energy conversion is very low: only in the order of 1 or 2%. In many systems it is much lower than this. In food production, it has been the

aim to try and improve on this efficiency and in crops such as corn, sugar beet and sugar cane, we can obtain efficiencies of 5 or 6% but rarely above 8%. The conversion of light energy places the first limit on how much life the biosphere can support, a limit which we will probably never approach because of other factors which become limiting before the light conversion process does.

### Factors Limiting Photosynthesis

In most areas, light energy is not generally considered to be the primary limiting factor in plant growth and production. Under cloud cover, light could become limited and evidence also suggests that the reduced light intensity at the earth's surface in many areas, caused by a polluted atmosphere, may be suppressing photosynthesis.

Other possible limiting factors are the two substances directly utilized in photosynthesis, namely water and carbon dioxide. Although a lack of sufficient water can indirectly limit photosynthesis, it probably is never directly a limiting factor, because of the high percentage of water in living cells. Carbon dioxide, however, appears universally to be a retarding factor. The atmosphere contains only slightly more than 0.03% carbon dioxide and experiments have shown that in many cases the rate of photosynthesis can be increased by an increase in the carbon dioxide concentration. Since carbon dioxide is universally low, in relation to what the plant can use, little can be done to increase it except locally in controlled enclosures such as green houses. The concentration of carbon dioxide has been increasing steadily in the atmosphere since the beginning of the industrial revolution. This increase has caused a great deal of concern amongst scientists because of the possible impact these increased levels might have on world climate. At present, there is no evidence to indicate that the increased amounts have resulted in increased crop production. It is postulated that the beneficial effect of increased carbon dioxide on plant growth is being offset by the decrease in light intensity from air pollution.

In plant metabolism and the synthesis of various compounds in the plant, many other elements, besides carbon, hydrogen and oxygen, are required. In excess of twenty mineral elements are essential in varying amounts for normal efficient plant growth. A deficiency in any one of these could reduce plant growth and photosynthesis. The last one hundred years has seen great strides in the elimination of nutrient deficiencies in crop plants. But, economics does not allow for the use of fertilizers in large natural systems, such as forests where certain nutrients may also be limiting plant growth.

### Cycling of Nutrients as a Limiting Factor

Animals eat plants for food and thus satisfy their energy requirements and need for proteins and other essentials. In consuming plant material they also take in all the nutrients that were incorporated in the plants. As the animals break down the carbon compounds to carbon dioxide and release this gas back into the atmosphere, the nutrients become more concentrated in the animal. The excess nutrients or those the animal does not require are discharged as wastes back into the soil. The herbivore feeding on plants may fall prey to a carnivore and more nutrients are discharged or some may become more concentrated.

When plants or animals die, a whole host of organisms await to consume the dead bodies. Many bacteria, fungi, protozoa and invertebrates utilize the carbon compounds from the dead remains as their energy source and concentrate and release the excess nutrients back to the soil. Thus, we have a continual chain of "eat and be eaten," with energy assimilated by the plant being utilized by many animals in succession. Energy is lost as heat at each step along the chain. During this process, the carbon dioxide is released back to the atmosphere and the water and nutrients are released back to the soil. In natural systems, a balance is established and maintained between the nutrients taken in by the plant and those released again by animals and microorganisms. Thus, nutrients are continually cycled among the soil, plants, animals and decomposers and back to the soil. The energy, however, does not cycle but comes in continually from the sun and is lost as heat in metabolism.

For an ecosystem to continue to function at a high and efficient level, the nutrients must be continually cycled within the system. Otherwise, plant growth will decrease, less energy is assimilated as less life can exist in the system. Strangely enough, the growth of plants is dependent on the plants being eaten and decomposed. Man all too often has paid little attention to these natural processes or has been ignorant of them. It seems that the more civilized man becomes, the more he breaks or ignores the laws of nature. With the population becoming more and more concentrated in huge metropolitan areas, more food has to be brought in from greater distances. How much of the nutrients in the food

is returned to the land? Virtually none. Most are disposed of in the quickest way possible and flushed into rivers and lakes all to end up in what is being treated as big cess pools—the oceans. Meanwhile the land is deteriorating. These nutrients required in large quantities, such as nitrogen, potassium and phosphorous have been added to agricultural soil for many years. More recently, other nutrients are becoming deficient and more and more has to be added from supplies that are not inexhaustable. When trees are removed from forests either by man or by physical events such as forest fires, nutrients are removed with the trees and the soil is often laid bare to erosion and more nutrients are removed. But, these nutrients are not replaced by man. Short term economic returns do not permit these areas to be fertilized, and their production decreases as a result. Nature's laws can not be broken without penalty. With an increasing world population, we can not afford to let plant productivity decrease, that is, let the world energy converting mechanism run down.

### Climate as the Limiting Factor

Of the sun's energy striking the earth's surface, only a small fraction is used to run the life systems. The remaining light energy is not without useful function. Although some of the light is reflected and returned to outer space, more than 50% is absorbed by the earth's surface or by the atmosphere and is converted to heat. It is this heat energy that is the driving force behind the world air circulation pattern, which determines the rainfall distribution and modifications in the earth's temperature. The light intensity is highest in the equatorial regions where the highest temperatures occur. The polar region receives the least light and is correspondingly colder. The air-flow pattern, the proximity to bodies of water and the location in relation to mountains, influences both the temperature and the rainfall. The result is a great variety of temperature and moisture combinations, ranging from very hot and dry or very cold and dry, to very hot and wet or cold and wet with all shades of intermediate conditions, Figure 20. The combination of temperature and rainfall has a controlling influence over the type and amount of vegetation that exists in a given area, Figure 21. Under hot and humid conditions, very luxuriant growth, such as in tropical rain forests, occurs with very high productivity. Under hot and dry conditions, vegetation is very sparse and stunted with very low productivity. These areas of desert occupy some 14% of the earth's land surface. Over most of the land's surface, either the

**Figure 20.** Lush green vegetation of subtropical rainforest in Australia. Productivity in these forests is high as both the temperature and rainfall provide ideal conditions for plant growth.

**Figure 21.** A dry interior region between mountain ranges where the temperature is relatively high but rainfall is sparse. The trees are widely scattered, sage brush has filled in where the grass was over-grazed. Productivity is low in this region.

**Figure 22.** An area almost totally denuded for miles due to the poisonous sulphur dioxide fumes emitted by nickel smelters. Much of the soil has been eroded away and productivity here is low and will be low for many years even after the sulfur dioxide is removed from the air.

temperature, the rainfall, or the seasonal distribution of these elements curtails the productivity of the land.

Man has not been able to alter the earth's temperature purposely, with the exception of limited frost control in cold areas. His success in attempting to increase rainfall in dry areas has been very meagre, but, he has been able to increase the moisture in limited areas by damming rivers and diverting the water to be used for irrigation. Although man for years has been dreaming of and even planning to alter the world's climate, any major change is apt to result in catastrophe over wide areas. His greatest impact appears to be in altering climate inadvertantly through pollution of the atmosphere from industry, aircraft, or other activities. Climate will continue to be the major limiting factor in plant growth and productivity.

### Pollution: a new limiting factor

The detrimental effects of atmospheric pollutants on the health and vigor of plants have been observed for many years, Figure 22. In localized areas, such as around smelting operations, the devastating effects of sulphur dioxide on vegetation are well known. The large denuded areas are a silent testimony to man's abuses of his plant resources. With the increase in the number of industries, the increased number of internal combustion engines and the concomitant increase in the use of fossil fuels, a whole new dimension has been added to the air pollution problem. Much of

industrialized North America is covered by a haze or smog. The smog is composed of many hundreds of chemicals and substances spewed out by man's machines. But some of the most toxic to plants and animals are those chemicals that are produced from the interaction of these chemicals in the air in the presence of sunlight. Some of the most active and detrimental that have been identified are ozone ($O_3$) and peroxyacetyl nitrate (PAN). The importance of these chemicals as phytotoxic agents (substances toxic to plants) was established during the mid 1950's. The impact of these substances reaches far beyond the large metropolitan areas over which they are mostly produced. They may be carried a hundred miles or more downwind and these toxic effects can be seen on agricultural crops and natural vegetation, such as forests, many miles from any major city. In many areas of North America, it has become difficult to grow some of the more sensitive crops and damage to agricultural crops has been estimated to run into the billions of dollars annually. Effects on forests are obviously very difficult to assess in terms of economic losses. Here we have a major influence on productivity and thus, on the amount and quality of life that can exist on earth. It also is an influence that man can control and must bring under control. The consequences of prolonged or even more intense air pollution are unpredictable.

Many aspects of life and how it functions within the environment are still not understood. The great concern among many scientists is the unknown impact that environmental changes brought about by man's activities will have on the life of earth. This concern has spurred greater activity in gathering more information to acquire a better understanding of how ecological systems function. Only when we understand fully how they function will we be able to predict what changes in natural systems will occur under certain man-made stresses. To define natural systems more accurately, information is being gathered by many specialists in the fields of biology, hydrology, meteorology and others, for the purpose of inte-grating all the information and using it to build model systems. The use of the computer has allowed scientists to integrate data on a scale never before possible. Once a workable model has been constructed, it can be used to help predict changes that will occur should conditions in the environment change by altering various variables in the model and noting the results. To gather sufficient and correct information and to put this together in a workable model presents a formidable challenge. Models must be continually improved and new data incorporated to make them more useful. To get governments and peoples to respond to impending imbalances may be an even greater challenge.

Other studies are being conducted on major ecosystems where environmental pollutants are having a major impact. The whole pattern of energy flow through the system can be altered if certain major plant species are selectively eliminated from the forest, as is happening in the San Bernadino mountains and other areas. The long range, either in terms of the functioning of the system as a natural system or in terms of changes In economic value of the system, are largely unknown. How far can man change and manipulate ecosystems without destroying them? How much abuse can nature take before it has been harmed irreparably? How much of our remaining natural resources can man use before their depletion becomes a reality? These and many other questions must be answered and answered soon, if the flow of energy through living systems is to continue at a level to sustain life as we know it.

The amount of energy available for life on earth is limited. For life to continue, man must ensure that his activities do not reduce the efficiency of light energy conversion. Our demands on plants for food and other uses are high and our efficiency of usage can and must be improved.

Prof, Gerald Hofstra, Ph.D.
Ontario Agricultural College,
Department of Environmental Biology,
University of Guelph,
Guelph, Ontario, Canada

# Chapter 6

# The Metabolism of Animals and the Effects of Pollutants

Pollutants in today's environment affect the utilization of many substances by organisms. However, all organisms do not react similarly to various chemical and physical conditions of their environment. Through evolutionary processes, various organisms have adapted and are adapting to various combinations of environmental factors. Shelford's Law of Tolerance states that there is both a minimum and a maximum value of any environmental parameter which an organism can tolerate, Figure 23. It should be remembered that different organisms have different tolerance ranges. Using temperature as an example, some organisms withstand only a small range of temperatures. These are termed *stenothermal.* Those which can withstand a wide range of temperatures are termed *eurythermal.*

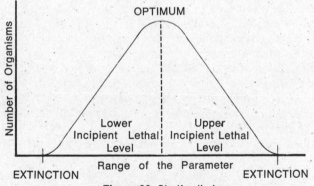

**Figure 23.** Shelford's law.

Many forms of pollutants enter our rivers, lakes, oceans and air every day. Each may act independently or in combination with others. If the result is greater than the sum of the two acting independently, this is termed *synergism.* In some cases, the effect of the one pollutant lessens the toxicity of a second. This is known as *antagonism.* One of the substances may in itself not be harmful to an organism, but when combined with another pollutant it greatly enhances the second, for example temperature and chlorine. The toxicity of the chlorine in DDT to grass shrimp is greater at high temperatures than at low.[1]

Toxicants may act in a variety of ways on organisms. Some will block biochemical reactions. Others will effect nerves, reproduction, respiration, energy production or enzyme synthesis. Depending on the substance and its effect on the organism, different manifestations may result. For example, radioactivity may not kill an organism or even change its appearance, yet it may change the genetic make up so the organism's offspring will be different or born with birth defects. Other toxicants may kill outright or may just effect reproductive potential, for example, DDT or other chemical pesticides.

Since the publication ten years ago of Rachael Carson's famous book *Silent Spring,* much attention has been devoted to study of herbicides (chemicals which are used to control weeds) and pesticides (chemicals which are used to control insects). Because of its wide-spread use, DDT (dichloro-diphenyltrichloroethene) has been studied more extensively than any other single pesticide. Abbott et al.[2] reported that the persistant hydrocarbons are readily transported by air; in fact, detectable concentrations of DDT have been reported from Antarctic snow.[3] While many states have put severe restrictions on the use of chloronated hydrocarbons (the group of pesticides that includes DDT, Dieldrin, Aldrin), they will continue to be a worldwide problem because of their use in most underdeveloped countries.

DDT's effect on organisms is quite varied. One of the best known effects is on the reproductive cycle of animals. Butler[4] reported that the gonads of oysters concentrated DDT to a greater extent than any other organ. Hunt[5] found a greater concentration of DDT and DDE (a breakdown product of DDT) in the ovaries than in testes. One possible action is that DDT causes the breakdown of estrogen which in turn effects the calcium metabolism in birds.[6] This would be one reason why many birds effected by DDT produce weak egg shells and a greater mortality rate of chicks results. The egg shells of Osprey fed DDT were 25% thinner than the untreated control group.[7] These workers also determined that there was a correlation between the amount of DDT residue and the thickness of sea gull eggs. Similar results have been reported for the eggs of prairie falcons when the adults were fed chlorinated hydrocarbons.[8]

Holden[9] believes the primary effect of pesticides upon fish is on the central nervous system which results in sluggishness, instability and respiratory difficulties. Both Walker[10] and Lemke and Mount[11] report damage to fish gills from poisons. Walker reported the herbicide Endothal produced gill injury which increased with temperature. Alkyl benzene sulfonate causes a thickening of the lamelle of bluegill sunfish. This subsequently led to decreasing the fish's tolerance in water of low dissolved oxygen concentrations. In conclusion, herbicides and pesticides can effect many metabolic functions, the interrelations of which we have not yet begun to understand. Recent results obtained from experiments indicate that sub-lethal concentrations of a wide variety of pesticides depress the filtering rate of the common water flea, *Daphnia magna.*

The action of pesticides on the metabolism of organisms is not just limited to effects on reproduction. Lowe reported that oyster beds treated with only 10 ppb (part per billion) DDT had significantly increased the occurrence of mycelial fungus infections. This would indicate that the oyster's defense mechanism to this parasite was broken down. Chlorinated hydrocarbons may disrupt the osmoregulatory mechanisms of some marine bony fishes. DDT has been shown to inhibit ATP in the intestinal mucosa of eels. This led to an impairment of their fluid absorption.[12] Yet another important body function that is interfered with by pesticides is the drug metabolism enzyme produced by the liver. This enzyme helps to protect the body from chemical poisons. Stimulation and inhibition of this enzyme are caused by a wide variety of contaminants. Foutes[13] found that DDT actually enhanced the rate that some poisonous substances were removed from the body. However, this may make the liver more susceptible to damage from other chemicals such as carbon tetrachloride.

Oil pollution has recently come to the attention of the public. While several studies have shown various oils to be toxic to a wide variety of organisms. Unfortunately only a few workers have studied the mechanism by which oil kills organisms. Galtsoff[14] noticed that when oysters were exposed to carbonized sand containing oil, the volume of water pumped per day decreased from 207 to 310 liters during the first six days of the experiment to only one liter per day on the 14th day. This would mean that the oyster's feeding and respiratory rates would also be reduced. Some oils have been shown to depress respiratory activity and have an anaesthetizing effect on a variety of organisms from single celled amoeba to the brine shrimp *Artemia.*

Heavy metals have recently become a major environmental problem. There have been several major occurrences of mercury poisoning in recent years. Eighty-three people who ate fish from the Minamata Bay in Japan were effected either fatally or with permanent disabilities. The source of contamination was traced to the effluent of a plastics factory. A family from Alamagordo, New Mexico, experienced severe sickness from eating mercury contaminated pork. Sweden had a major problem of mercury pollution both mercuric fungicides and the phenylmercury from pulp and paper industry. The latter caused a change in the processing of paper.[15] While the United States Environmental Protection Agency estimates that industry has decreased their discharges of mercury by 94%, they are still potentially dangerous. Mercury exists as an elemental metal and occurs both in an organic and an inorganic source. Mercury vaporizes easily and may subsequently be absorbed through the skin, by ingestion or by inhalation. In the body, mercury becomes dissolved in blood lipids, oxidized and transported throughout the body.

The most toxic forms of mercury occur as organic compounds. Bacteria may cause methyl mercury to form from inorganic mercury. This process simply adds a carbon molecule to the inorganic mercury. Absorption of inorganic mercury is relatively low (2%) compared with 90 to 95% absorption of methyl mercury. Organic forms are used for seed treatment and as a wood preservative.[16]

When pregnant mice were administered methyl mercury dicyandiamide, there was an increased number of dead feti and resorption sites compared to normal mice and mice who received mercury before mating.[17] Mothers in the Minamata disaster who showed no symptoms of mercury poisoning gave birth to infants with methyl mercury poisoning.

Mercury may also readily accumulate in the brain. Inorganic mercury does not generally get into the brain, but methyl mercury is often found there in considerable quantities. Organic mercury causes damage to the cerebellum, as well as to the visual cortex and hearing centers.[18] In monkeys and rats, mercury iodide produced intense and wide-spread morphological brain damage. Friberg[19] and Berlin *et al.*[20] found high concentrations of mercury in the brain of rats. The state of the mercury appears to be important. Berlin *et al.* reported that uptake of mercury by rabbit and rats when exposed to vapor was ten times higher than in animals injected intravenously.

Recently there has been some evidence that mercury in very small quantities may cause chromosome damage. There was a significant increase in chromosome breakage in lymphocytes white blood cells according to Lofroth.[21] He further showed that mercury may cause disjuncture (an unequal number of chromosomes going to each daughter cell during meiosis) by interferring with spindle formation in the nucleus during cell division.

Excretion of mercury once it is absorbed is again a function of the form in which it was ingested. Inorganic mercury is more readily excreted than is organic. Gage[22] showed in rats that generally excretion of mercury was rapid when exposed to vapor and was complete except for a small amount of residual mercury in the brain. Northseth[23] observed that 40% of the mercury excreted upon exposure to methyl mercury salts was inorganic. He then traced half this biotransformation to occur in the intestine, with a limited amount transformed in the liver which accounted for a large portion that appeared in the feces.

Thermal additions to our lakes, rivers and streams from industrial sources are rapidly increasing. In fact by the year 2000, 6% of all the heat entering the Great Lakes will be from man-made sources.[24] The largest single source of waste heat will be nuclear and fossil fueled electric generating stations. The thermal properties of the animal's environment play a major role in its ability to metabolize food sources. Since the greatest effect of thermal pollution is on aquatic systems, it is first necessary to explore some basic properties of water.

The maximum density of water occurs at 4°C, which means that any increase or decrease in the temperature makes the water less dense. Consequently, water is the only substance which in its solid state is less dense than its liquid state and will float. Thus, when ice forms on the surface, this tends to insulate the remainder of the water body from further loss of heat. This prevents our lakes and streams from freezing solid and permits aquatic life to be carried over from year to year.

The second important property of the thermal structure of lakes is that of summer thermal stratification. Seasonal increase in solar radiation annually warms surface waters in the summer. This heating of the water results in the establishment of three well known layers: (1) *epilimnion* or surface layer, (2) *hypolimnion* or lower layer, and (3) *thermocline* (metalimnion) or the sheer plane between

the above two layers. During the period of stratification, each layer behaves relatively independently, and considerable water quality differences may develop, Figure 24.

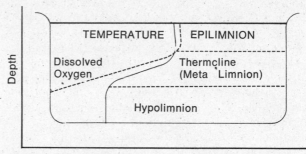

**Figure 24.** Diagram showing the three layers of a typical lake in mid summer.

Temperature fluctuations may affect aquatic organisms in the following ways: (1) change in behavior, (2) control of metabolism, (3) control of rate of assimilation of food, (4) control of rate of maturation of eggs, (5) act as depressant, (6) lethal. The lethal temperature is that temperature which will kill 50% of the fish, in an indefinite period of time. Within the zone of tolerance, Fry *et al.*[25] has described three ranges which we need to be concerned with. They describe an upper and lower zone of thermal resistance and a central zone of thermal tolerance. When fish are exposed to temperatures in the zone of tolerance, they will only survive for some finite period of time.

The lethal temperature has been found to vary with many factors including acclimation time, diet, activity, environmental stress and general health of the organisms. Consequently it is very difficult to determine the exact lethal temperature in the field since the lethal temperature varies somewhat with each of the above factors. Of particular importance is the acclimation temperature, Figure 25.

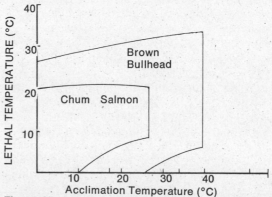

**Figure 25.** Influence of Acclimation Temperature on the zone of tolerance of Chum Salmon and Brown Bullhead. (After Bretl 1956).

the relationship between acclimation temperature and the incipient lethal level. It will be easily noted that as the acclimation temperature for chum salmon increases from 10 to 24°C, the lower incipient lethal limit increases from essentially 0°C to 5°C, while the upper incipient limit increases from 21 to 22°C. While the values are somewhat higher and the ranges wider, a similar pattern may be observed for bullhead catfish. The time necessary for thermal acclimation varies among species. Fry[26] acclimated speckled trout at the rate of 1°C per day. The small fish called roach may withstand acclimation at the rate of 1.0°C per day, Alabaster and Downing,[27] whereas Sprague[28] working with several different crustaceans found they could be acclimated at the rate of 2.3°C to 4.6°C per day.

Lethal temperatures are again very variable as has been shown by a number of different authors. Some Pacific salmon will not tolerate temperatures above 24°C while the top-minnow, *Fundulus* will tolerate 42°C. Just as important as upper lethal levels are lower lethal levels. During the winter a sudden cooling of thermal plumes during the shut down of power generating stations has caused massive mortalities of fish in the Great Lakes.

Another significant effect that temperature plays in the life of aquatic organisms is on their reproductive behavior and duration of incubation of eggs. The temperature requirements for successful reproduction of many species may be narrower than for other metabolic functions. For example, Farley[29] has reported that for successful spawning of striped bass, the temperature must be between 16.1 and 20.6°C. Further, the stimulus to begin the reproductive period may require a temperature change of only 1 or 2°C Brett.[30] Cold temperatures are known to delay spawning seasons, whereas high temperatures will bring the onset earlier in the year Brett,[30] Further, these organisms appear to adapt to different environments. For example, largemouth bass in Minnesota begin spawning when water temperatures reach 15°C[31] whereas the same species in Alabama does not spawn until temperatures reach 20–24°C.

Decreases in temperature or fluctuations in temperature are just as important in the survival and successful spawning as are maximum or minimum temperatures. Such fluctuations are likely to occur when thermal discharges are to estuaries or rivers. Smallmouth bass have been observed to abandon their nests when water temperature drops from 58 to 48°F. Further Lydell[32] showed that decreasing the water temperature from 65 to 45°F caused mortality of smallmouth bass eggs. Eggs

from American smelt were killed with daily fluctuations in temperature of only 12°C.[33] However once again this suceptibility to water temperature appears to vary widely with species. Albrecht[34] found striped bass could withstand daily temperature fluctuations of 11.5°C.

Warmer temperatures up to some limit generally cause fish to grow at a faster rate, however above this optimum, continuing to heat water actually retards the growth of fish. This is because of the increase in metabolic activity that is using up the fish's energy supply at a greater rate than it is consuming energy. Faster growing fish will mature at an earlier age and produce more eggs.

Probably most research to date has examined effects of thermal additions to fish. However fish are not the only important organisms in water. Just as important as the fish are the organisms that serve as food for fish. It is rather obvious that if we kill the food organisms, eventually the fish themselves will die. Since many benthic organisms are relatively sessile, they are unable to move out of the influence of a thermal plume. It is very likely that most aquatic insects can stand temperatures in excess of those that will kill fish. However, one major area of concern is that heated discharges cause insects to emerge prematurely. Stoneflys in thermal plumes have been observed to hatch in January instead of May. Obviously the adults freeze to death upon emerging into sub-freezing air. Still another problem is that different temperatures favor some species over others. Consequently, there is a shift in the composition of the community with shifts in water temperatures. It should be remembered that while high temperatures may directly or synergistically (for example, heat and changes in salinity) kill bottom organisms such as oysters. Probably the most pronounced effects are on the organism's physiology, metabolism, development and reproductive potential. For example, studies have shown that above 15°C, the shells of European oysters remain open nearly 24 hours per day.

The effect of temperature on the behavior and metabolism of zooplankton is not well studied. Burns and Rigler[35] found that *Daphnia* fed most rapidly at 20°C. Any increase or decrease in temperatures above or below this temperature resulted in a lower feeding rate. Further work by Patalas showed an increase in the production of a Polish pond which was heated by thermal discharges. Heinle[36] found that *Acartia tonsa* grew fastest at 27°C, and the upper incipient lethal limit is approximately 31°C.

In summary, it is clear that both chemical and thermal pollutants effect many organisms in many ways. Besides being lethal, pollutants may control metabolic reactions, enzymatic pathways, reproductive success, species composition and a whole host of other biological parameters. Scientists have not yet begun to understand the ecological effects resulting from physiological processes in individual organisms.

## REFERENCES

1. Eisler, N., "Acute Toxicities of Insecticides to Marine Decapod Crustaceans," *Crustaceana* 16:302-310 (1969).
2. Abbott, D. C., Harrison, R. B., Tatlon, J. O., and Thompson, J., *Nature* 211:259 ff. (1966).
3. Tatlon, J., & Ruzicka, J. H., "Organo-chlorine Pesticides in Antarctica," *Nature* 215:346-348 (1967).
4. Butler, J., "The Significance of DDT Residues in Estuarine Fauna," Miller & Berg (eds.) *Chemical Fallout*, Charles C. Thomas, Springfield, Illinois (1969).
5. Hunt, E. G., Azevedo, J. A., Woods, L. A., Castle, W. T., "The Significance of Residues in Pheasant tissues etc." Miller & Berg (eds.) *Chemical Fallout*, Charles C. Thomas, Springfield, Illinois (1969).
6. Wurster, C. F., "Chlorinated Hydrocarbon Insecticides and Avian Reproduction," Miller & Berg #5 above.
7. Hickey, I. J., & Anderson, D. W., "Chlorinated Hydrocarbons and Eggshell Changes in Raptorial and Fish Eating Birds," *Science* 162:271-273 (1968).
8. Enderson, J. H., & Berger, D. D., "Pesticides: Eggshell Thinning and Lowered Production of Young Prairie Falcons," *Bioscience* 20:355-356 (1970).
9. Holden, A., "Contamination of Freshwater by Persistent Insecticides and Their Effects on Fish," *Annual of Applied Biology* 55:332-335 (1965).
10. Walker, C. R., "Endothal Derivatives as Aquatic Herbicides in Fishery Habitats," *Weeds* 11:226-232 (1963).
11. Lemke, A. E., & D. I. Mount, "Some Effects of Alkyl Benzene Sulfonate on the Bluegill, *Lepomis macrochirus*, *Transactions of the American Fish Society* 92:372-378 (1963).
12. Janicki, R., & Kuster, M., "DDT Inhibition of Intestinal Salt and Water Absorption in Teleosts," *American Zoologist* 10:540 ff. (1970).
13. Foutes, J. R., "The Stimulation and Inhibition of Hepatic Microsomal Drug Metabolizing Enzymes with Special Reference to Effects of Environmental Contaminants," *Toxicol. Appl. Pharmacol.* 17: 804-809 (1970).
14. Galtsoff, P. S., Personal Communication (1969).
15. Johnels, H. G., & Westermark, T., "Mercury Contamination of the Environment in Sweden," Miller & Berg #5 above.
16. Hunter, D., Bomford, R. R., Russell, D. S., "Poisoning by Methyl-Mercury Compounds," *Quarterly Journal of Medicine* 9:193-207 (1940).
17. Loforth, G., "A Review of Health Hazards and Side Effects Associated with Emission of Mercury Compounds into Natural Systems," *Swedish National Science Research Council* 2:49-50 (1970).
18. Takalata, N., Hayashi, H., Watanabe, S., & Anso, T., "Accumulation of Mercury in the Brains of Two Autopsy Cases with Chronic Inorganic Mercury Poisoning," *Folia Psych. New. Japan.* 24:59-69 (1970).
19. Friberg, L., "Studies on the Metabolism of Mercury Chloride and Methyl-Mercury dicyandiamide," *Archives of Industrial Health* 20:42 (1959).
20. Berlin, M., Fazackerly, J., & Nordberg, J., "The Uptake of Mercury in the Brains of Mammals Exposed to Mercury Vapor and Mercuric Salts," *Archives of Environmental Health* 18:719-729 (1969).
21. Loforth, G., "Methyl-Mercury," *Ecological Research Council Bulletin 4*, Swedish National Research Council (1969).
22. Gage, J. C., "The Distribution and Excretion of Inhaled Mercury Vapors," *British Journal of Ind. Medicine* 18:287-296 (1961).
23. Northseth, T., "Studies of Intracellular Distribution of Mercury," Miller & Berg #5 above.
24. Denison, P. J., & Elder, F. C., "Thermal Inputs to the Great Lakes, 1968-2000," *Proceedings of the Thirteenth Conference of Great Lakes Research*, 811-828 (1970).
25. Fry, F. J., Hart, J. S., & Walker, K. F., "Lethal Temperature Relations for a Sample of Young Speckled Trout, *Salvelinus fontinalis*," *University of Toronto Stud. Bulletin Ser. #54* (1946).
26. Fry, F. J., "Effects of Environment on Annual Activity," *Ontario Fish Research Laboratory* 68:1-62 (1947).
27. Alabaster, B., & Downing, A. L., "A Field and Laboratory Investigation of the Effects of Heated Effluents on Fish," *Fish Investigations*, Ser. 1. Ministery of Agriculture, Fish, and Food, U.K. (1966).
28. Sprague, J. B., "Resistance of Four Freshwater Crustaceans to Lethal High Temperature and Low Oxygen," *Journal of the Fisheries Research Board of Canada* 20:387-415 (1963).
29. Farley, T. C., "Striped Bass, *Roccus savatilis* Spawning in Sacramento-San Joaquin Delta," *Bulletin of the Department of Fish and Game of California*, 136:28-42 (1966).
30. Brett, J. R., "Some Principles in the Thermal Requirements of Fishes," *Quarterly Review of Biology*, 31:75-87 (1956).
31. Krammer, R. H., & Smith, L. L., "Formation of Year Classes in Largemouth Bass," *American Fisheries Society Transactions*, 91:29-41 (1962).
32. Lydell, D., "Increasing and Insuring the Output of Natural Food Supply of Smallmouth Black Bass Fry in the South," (Georgia), *American Fisheries Society Transactions*, 29:129-153 (n.d.)
33. Rothschild, B. J., "Production and Survival of the Eggs of the American Smelt, *Osmerus morader* in Maine," *American Fisheries Society Transactions*, 90:43-48 (1961).
34. Albrecht, A. B., "Some Observations on Factors Associated with the Survival of Striped Bass Eggs and Larvae," *California Fish and Game Bulletin*, 50:110-113 (1964).
35. Burns, C., & Rigler, F. H., "Comparison Rates of Filtering Rates of *Daphnia rosea* in Lake Water and Suspensions of Yeast," *Limnology-Oceanography*, 12:492-502 (1967).
36. Heinle, D. R., "Temperature and Zooplankton," *Chesapeake Science*, 19:3-4 (1969).

Prof. Harold V. Kibby, Ph.D.
United States Environmental Protection Agency
Washington, D.C.

# Chapter 7

# Ecosystems

The concept of the ecosystem is the center of today's professional ecology and the most relevant concept for understanding man's environmental problems. The author's recent text carries the following formal definition: "Any unit including all of the organisms (i.e., the "community") in a given area interacting with the physical environment so that a flow of energy leads to a clearly defined trophic structure, biotic diversity, and material cycles (i.e., exchange of materials between living and non-living parts) within the system is an *ecological system* or *ecosystem*."[1]

As recently as ten years ago the theory of the ecosystem was rather well understood but not in any way applied. The applied ecology of the 1960's consisted primarily of efforts directed to managing the components of an ecosystem as more or less independent units. Thus we had forest management, wildlife management, soil conservation, pest control, etc., but no ecosystem management and no applied human ecology. Practice has now caught up with theory. In the past two years the public has seized on the root meaning of ecology, namely "oikos" or "house," to broaden the subject beyond its previously narrow academic confines to include the whole environmental house, as it were. There is now an ongoing and historic "attitude revolution"[2] in the way people look at their environment. This is simply because man for the first time is faced with ultimate rather than merely local limitations. It will be well for us to remember this as we face the controversies, false starts, and backlashes that accompany man's attempts to stop the vicious spiral of uncontrolled growth and recover from the resource exploitation of past decades.

No one has expressed the relevance of the ecosystem concept to man better than Aldo Leopold in his essays on the land ethic. He emphasized the need for an ethical overview,[3] stating that "Christianity tries to integrate the individual to society, democracy to integrate social organization to the individual. There is yet no ethic dealing with man's relation to the land" which is "still strictly economic, entailing privileges but not obligations." Thus man is continually striving, with but partial success so far, to establish ethical relationships between man and man, man and government, and, now, man and environment. Without the latter what little progress has been made with the other two ethics will surely be lost. Garrett Hardin says it another way when he points out that technology alone will not solve the population and pollution dilemmas; ethical and legal restraints are also necessary.[4]

## A Short Historical Review

The British ecologist A. G. Tansley first proposed the term ecosystem about 1935,[5] but the idea of the unity of organisms and environment (as well as the oneness of man and nature) is quite ancient. This thought has been a basic part of many religions; however, as pointed out by the historian Lynn White[6] it was not quite as marked in Christian thinking. George Perkins Marsh wrote a classic work[7] on the independent development of human culture and the natural environment, forecasting doom for modern civilizations unless man adopted a comprehensive environmental planning approach. His work was largely based on an analysis of the causes of decline of ancient civilizations. Biologists began to write about the unified functioning of nature in the late 1800's; interestingly enough, this occurred about the same time and in a similar manner in German, English, and Russian language writings. The works of Karl Mobius[8] on oyster reef organisms, of the American S. A. Forbes on lakes, and the two Russian scientists V. V. Dukachaev and G. F. Morozov[10] are outstanding in this regard. Biologists were already putting together primitive conceptual schemes of the organism-environment relationship.

## Two Approaches to Ecosystem Study

G. Evelyn Hutchinson (1964) contrasted the two longstanding ways in which ecologists have studied lakes and other large ecosystems. He discussed the *merological* approach, which is a part-oriented approach in which the different parts of the system are studied separately and the overview of the whole system is then constructed from this information. Hutchinson cites the work of Forbes[9] as exemplary of this type of study. The second approach is whole-oriented; the study concentrates on the functioning of the entire system as a whole, with emphasis on inputs and outputs and without special reference to individual parts of the system. This second procedure he called the *holological* approach. According to Hutchinson, the work of E. A. Birge[11] on the heat budget of lakes was a pioneer study in the use of this approach. Both approaches are still very much in use today. The part-oriented approach tends to dominate the thinking of ecologists who are concerned with the study of specific species. The whole-oriented approach is more popular with ecologists and engineers concerned with larger questions.

The environmental crisis has speeded up the application of systems analysis to ecology. This formalized, mathematical modeling approach is now known as systems ecology. It is rapidly becoming a science in its own right, for two reasons: (1) extremely powerful new tools are available in terms of mathematical theory, computers, and the like, and (2) formal simplification of complex ecosystems provides the best hope for solutions of man's environmental problems, which can no longer be trusted to trial-and-error or one-problem-one-solution procedures. This modern application still retains the old whole-orientation vs. part-orientation contrast. Some systems ecologists start at the level of an individual component in an ecosystem and "model up." Others "model down," stating from the system as a whole. Some very interesting studies have been completed using laboratory-experimental ecosystems, that is, self-contained ecosystems in the laboratory.

### The Ecosystem's Components

It is convenient for analysis and modeling to recognize two functional layers, six structural components, and six processes as comprising an ecosystem; these are given in Table 2. These divisions are somewhat arbitrary, but hopefully they are a useful beginning for an understanding of these extremely complex phenomena. Although all of the components are inseparable, in practice different methods are required to study structure on the one hand, and to measure rates of function on the other. The ultimate goal is to understand the relationships between structure and function. In this brief article, the major emphasis will be on function.

### ECOSYSTEM COMPONENTS

#### Layers
- Green belt or autotrophic layer, in which the organisms, plants fix light energy and utilize simple inorganic substances, the buildup of complex substances from simpler material predominates.
- Brown belt or heterotrophic layer, in which the organisms utilize, rearrange, and decompose complex materials. The breakdown of complex substances predominates.

#### Structural Components
- Inorganic substances, carbon (C), nitrogen (N), carbon dioxide ($CO_2$), water ($H_2O$) etc. involved in the materials cycles.
- Organic substances and compounds, proteins, carbohydrates, lipids, humic substances and the like that link the biotic and abiotic.
- Climatic system, including temperature, rainfall, etc.
- Producer organisms, autotrophic, primarily green plants which are able to synthesize foodstuffs from simple substances and light energy.
- Consumer organisms, phagotrophic, animals that ingest particulate organic matter or other organisms.
- Decomposer organisms, saprotrophic, primarily bacteria, protozoa and fungi, and the like, which break down complex substances with the release of both organic and inorganic products which are then recycled by plants or provide energy sources and have a regulatory effect on other biotic components.

#### Processes
- Energy Flow.
- The food chains or trophic relationships.
- Diversity patterns, spatial and temporal.
- Mineral, cycling of nutrients aside from food.
- Development and evolution.
- Control or cybernetic aspects.

### Energy Flow and Nutrient Cycling

The flow of energy through the ecosystem and the recycling of its nutrients are the two overriding processes determining the state of the system. The ultimate source of energy is the sun. Most of the incoming solar energy is dissipated in driving the water cycle and the other natural cycles, in driving the weather systems, and in maintaining a temperature gradient. Until recently man has taken all this "free" work in nature for granted. In general between one and five percent of the solar energy is fixed by the green plants, which make new plant material or *primary production*. This plant material is the ultimate food source of the animals in the ecosystem. The series of transfers of food energy from one group of organisms to another is called the *food chain*. At each transfer there is much energy lost to the system as heat in the process of metabolizing the food or respiration. Between ten and twenty percent of the energy of an animal group is transferred to the group feeding upon it. It is convenient to consider two classes of food chains. The *grazing food chain* is the type in which living plants are consumed directly by animals. The *organic detritus food chain* on the other hand has the plant material consumed only after it dies and falls. A give ecosystem is likely to have both types of food chains.

## General Conclusions from Systems Analysis

The mathematical models used to describe the environment to date are crude. Although they have limited predictive value at present, the models have given the following results from studies which cover the whole of nature.

1. Available energy declines with each step in the food chain, so that a system can support more plant eaters than meat eaters; thus, if man wishes to remain a meat eater, there will have to be fewer people supported by the food base.

2. Specific materials on the other hand, may actually become more concentrated higher in the food chain. Some of these materials, such as DDT, may be harmful pollutants and toxic to man.

3. A high biological productivity, that is, the energy taken in and absorbed and used to produce more living material, has natural limits. In agriculture and in some natural systems, the really high levels of productivity are maintained by introducing large amounts of energy into the system from outside. This increases the overall energy available for producing new material. These increases are very significant, because in most ecosystems a great deal of the system's energy must be used to maintain the system itself. Such introductions of outside energy may be in the form of fertilizer or labor in a field of crops, rain and wind in a rain forest, tidal energy in an estuary, etc.

4. Harvesting in a system, removing materials from it tends to reduce the overall energy available to an ecosystem for its own maintenance. Pollution has the same general effect. Both activities greatly increase the stress on an ecosystem.

For this reason it is dangerous to try to force too much production from an ecosystem. It may result in an ecological catastrophe. Some of the results of such over-exploitation are:

• pollution from overuse of fertilizers, insecti- cides, and consumption of fossil fuels.

• unstable or varying conditions created by a one-crop systems of agriculture.

• disease to new lines of high-yield crop plants.

• social disorder as people leave rural areas to become city dwellers.

5. The balance between production of organic material and its decay in nature has never been exact. In the history of the earth, production has perhaps exceeded decay. However, man is now tending to reverse this trend by the burning of fossil fuels.

6. Such burning has raised the carbon dioxide content of the atmosphere. Small changes in this content can have significant effects on the overall heat budget of the earth.

7. The diversity of organisms dwelling within an ecosystem seems to vary directly with the stability of the system. Thus, the more diverse the system is, the more stable also. On the other hand, the more productive the system, the more unstable it may be.

8. There will always be a long time lag in the effects of crowding, pollution, and resource exploitation as they act to dampen the growth of human populations. This means that human populations will overshoot their safe levels unless intelligent planning is instituted before the effects are actually felt.

9. The cycling of water and chemicals in nature is called biogeochemical cycling. The pathways of different materials vary tremendously. There seem to be at least four major ones, which vary in their importance in different ecosystems.

a. recycling via microbial decomposition.

b. direct recycling from plant back to plant via helping microorganisms.

c. recycling via animal excretion.

d. chemical recycling, with no organism involved.

The recycling of a particular chemical substance may involve more than one of these pathways.

10. It may be tentatively theorized that in an industrialized society it is not energy such as power, food, etc., that is limiting, but the pollution and other consequences of the use of the energy and other resources which determine the bounds of the system.

## Ecosystem Development

Ecosystems change in time. These changes are termed ecosystem development or ecological succession.[13] Generally speaking, ecosystems go through a rapid series of changes leading to a relatively stable mature state called the *climax*. The early stages are often strikingly different from the climax. Typical early stage characteristics are: high productivity, high ratio of absorbed energy (production) to lost energy (respiration), short food chains, a low diversity of organisms, small size of organisms, instability, and open nutrient cycles. The climax is characterized by: low productivity, high diversity of organisms, stability, complex system of food chains. The major flow of energy seems to be from productivity to maintenance (respiration). The principles relating to ecosystem development are highly important to the planning of land use. This can be seen in the application of the following model contrasting young and mature ecosystems.

*Young systems emphasize*

|  |  |
|---|---|
|  | *Mature systems emphasize* |
| Production | Protection |
| Growth | Stability |
| Quantity | Quality |

Both extremes cannot be included at exactly the same time and place, and yet all six properties are desirable.

There are two possible solutions to the dilemma that this poses for land use. First, there can be a compromise providing moderate quality and moderate yield on all of the landscape. The second approach is to employ a mixed strategy by compartmentalizing the land into predominantly productive and predominantly protective units. Most of the desired multiple land uses are conflicting; it is thus possible to enact such a multiple use strategy under only one of the above schemes. The fact is, the uses of land are often mutually exclusive.

## REFERENCES

1. Odum, E. P., *Fundamentals of Ecology*, 3rd ed. Phila. W. B. Saunders (1971).
2. Odum, E. P., "The attitude lag," *Bioscience*, 19:403 1969); and "The attitude revolution," *Crisis of Survival*, pp. 9-15, Glenview, Ill. (1970).
3. Leopold, A., "The conservation ethic," *Journal of Forestry*, 31:634 ff. (1933).
4. Hardin, G., "The tragedy of the commons," *Science*, 162:1243 ff. (1968).
5. Tansley, A. G., "The use and abuse of vegetational concepts and terms," *Ecology*, 16:284 ff. (1935).
6. White, L., "The historical roots of our ecological crisis," *Science*, 155:1203 ff. (1967).
7. Marsh, G. P., *Man and Nature: Or Physical Geography as Modified by Human Action*, (1864) reprinted, Harvard University Press, Cambridge, Mass. (1965).
8. Mobius, K., "Die Auster und die Austernwirtschaft," Berlin (1870) translated and reprinted, *Report of the U.S. Fish Commission*, 1880:683-751 (1880).
9. Forbes, S. A., "The Lake as a microcosm," *Bull. Sc. H. Peoria, Ill.* reprinted in *Illinois Natural History Survey Bull.*, 15:537 ff. (1925).
10. see especially: Sukachev, V. N., "On the principle of genetic classification in biocoenology," translated and abstracted in Raney, F., and R. Daubenmire, *Ecology*, 39:364 ff. (1944).
11. Birge, E. A., "The heat budgets of American and European Lakes," *Transactions of the Wisconsin Academy of Arts and Letters*, 18:166 ff. (1915).
12. Odum, H. T., "Biological circuits and the marine systems of Texas," Olson, T. A., & Burgess, F. J., eds. *Pollution and Marine Ecology*, N.Y., (1967) and #2 above.
13. Odum, E. P., "The strategy of ecosystem development," *Science*, 164:262-270 (1970).

Prof. Eugene P. Odum, Ph.D.
Director: Institute of Ecology
University of Georgia
Athens, Georgia

# Chapter 8

# Population Ecology

In the study of living things it is never sufficient to examine the organism in isolation. The studying of the individual organism as enmeshed in a complex set of organism populations is part of the new science of systems ecology. A major portion of most organisms' environment consists of other organisms. They affect its behavior, its health, its chances of living, reproducing and dying. As you will see, population ecology is in its infancy. But its first screams are already a warning that it cannot be ignored if people want to understand their world and how to conserve it.

### Single species populations

All sexual individuals and probably all others too are at least influenced by members of their own species. If one assumes they are not, one is soon led to an absurdity called permanent exponential population growth: Suppose each individual of a species reproduces at a net rate $r$ young per year. This net reproductive rate, $r$ is composed of individual birth rate minus death rate. Then the whole species grows at a rate $rN$ where $N$ is the number of individuals. Soon the Earth is chest deep in that species. It doesn't matter how small $r$ is as long as it is greater than zero. Even elephants, notoriously slow reproducers (elephantine gestation time alone is almost two years) would soon carpet the earth, figure 26. In fact, in about two centuries the elephant population in order to maintain that very slow individual rate of reproduction; would be exploding out into space at a speed faster than light. This is clearly absurd.

The most important cause of populations ceasing to grow is competition within the species or intraspecific competition. As more and more individuals of a species are packed into an environment, each individual finds less of the resources it needs to live and reproduce, so its birth rate falls and its death rate rises; thus $r$ declines. Soon, so many individuals have been added that $r$ is zero and the population stops growing. The size of the population which results when $r$ reaches zero because of intraspecific competition is properly thought of as a balance of nature. Such balances are robust, not delicate, and

**Figure 26.** In time even a slowgrowing elephant population, if it maintains a constant individual growth rate, would carpet the earth.

**Time in Elephant Generations**

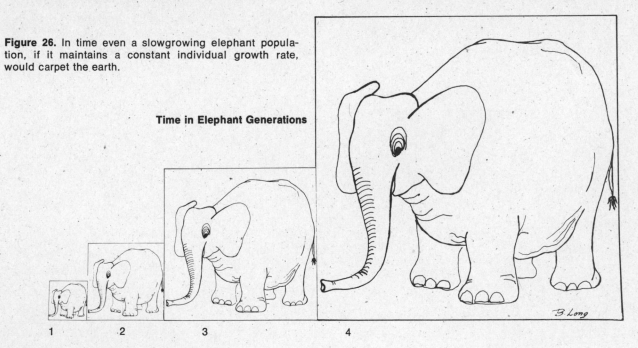

1    2    3    4

can restore themselves provided the resources upon which the species depends are still available. The strength of the balance lies in the fact that a population declines if larger than steady-state since it has so little resources per individual that death rates exceed birth rates; whereas a population smaller than steady-state increases because its resources are in excess.

A small population growing in an abundant environment at first shows little or no sign of resource limitation. Its size increases at steeper and steeper rates. Soon however the rate begins to fall and the population levels off. This pattern of population dynamics is called "s"-shaped or sigmoidal growth. The ring-necked pheasant population of Destruction Idaho, gave a classic exhibition of sigmoidal growth after being introduced there in 1937.

Several mathematical equations have been devised to describe sigmoidal growth. The most famous is over 100 years old and is called the logistic equation. A German scientist, Verhulst, first invented it, but early in the 20th century it was independently reinvented by the American population biologist, Raymond Pearl. In this equation $r$ is the individual net reproductive rate with no competition, $K$ is the steady-state size of the population and dN/dt is the symbol which means change in population size per time unit (e.g. day, year). The logistic equation is

$$\frac{dt}{dN} = rN\,(K-N)/K.$$

This equation has many limitations including the fact that it is rarely accurate when a population is near its steady-state. But it has been quite useful as a shorthand generalization for sigmoidal growth.

In a large proportion of the world's species, a new individual is different from one of any other age. The same might be said of one any number of time units old. Youngsters (including seedlings) might have very little chance of living. Oldsters' resistance to death might be lessened. Reproduction may be confined to a small part of the life span. And even fertile individuals of different ages are likely to have different birth rates. In such a population the simple analysis of the preceding paragraphs might seem wrong and indeed in some details it is, but the general principles are valid for most populations.

One new property of such a complex population is its age distribution: what fraction of the population is any given age? As an example, we might

have 40% new borns, 30% one-year olds, 20% two-year olds and 10% three-year olds. Knowing the age distribution of a human population is extremely important since humans at different ages provide different amounts of work and require different amounts and qualities of service. Fast growing populations have a high proportion of young; stable or declining populations have more middle and older individuals. If a population is growing at any constant rate (including zero) its age structure will eventually stabilize. Furthermore that stable structure can be predicted from knowledge of the birth and death rates of each age.

Growing populations with an age-structure have a kind of momentum. Reasonable methods take a long time to stop their growth. In a growing human population for example, there are more children present than there were at any previous time. So, barring wholesale catastrophe, there will be more parents present 20 years later and more grandparents in 40 years. In fact it would take just about 70 years, one average human life-span, to stop a human population from growing if *every* woman quit at two children. Of course population growth can be stopped faster, but that would be much like stopping a car by running it into a stone wall; brakes are somewhat smoother, less dangerous to the occupants and leave the machine in appreciably better condition. Bounded by natural means any population will tend towards a steady-state, figure 27.

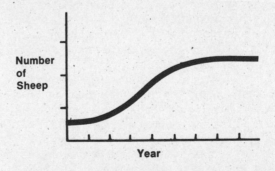

**Figure 27.** The sheep population of Tasmania, Australia grew sigmoidally until it reached a steady-state in about 1850. After that the population fluctuated gently about its steady-state.

Populations with age structures may grow beyond their steady-state at first. For example, a human new-born doesn't require as much resources as he will at age 14. Thus although he may represent the realization of the steady-state, his population will continue to grow until more of his resource demands are felt. This is an example of time-lag. His population overshoots its steady-state, then later returns toward it. Because of time-lag the population may also overshoot on the way down. Gradually the population oscillates about its steady-state in the sigmoidal pattern getting closer and closer to the point of no over-shoot in either direction with each pass. In this case the steady-state is said to be a stable focus. In the simpler case where the population smoothly approaches its steady-state, the state is called a stable node, figure 28.

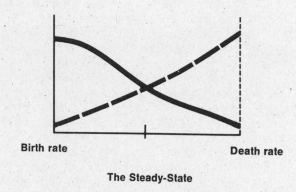

**Birth rate**

**Death rate**

**The Steady-State**

**POPULATION SIZE**

**Figure 28.** As population size increases, per capita birth rate declines and per capita death rate increases, both owing to shortages of resources.

Sometimes intraspecific competition is expressed quite simply through shortage. Red grouse in Britain eat heather shoots. But sometimes there are so many birds that the shoots become deficient in nitrogen. At that point the ecosystem's nitrogen is tied up in grouse. Grouse cease reproduction and death overtakes enough birds to reenrich the soil with bird-bound nitrogen. The heather returns to good condition and the grouse return to repro-ductive activity. Experimental addition of nitrogen to the soil prevents the grouse from ceasing reproduction. A very similar story of nutrient quality can now be told about the relationship of Alaskan lemmings, phosphorus and grass.

The amount of food rather than the quality can affect birth and death rates. The number of eggs laid by tawny owls declines with the decline in the mouse supply. The number of eggs laid by various song birds declines with the lowering of the insect supply. Increases in deer populations are often accompanied by overbrowsing of deer food and massive starvation of deer. One experiment on the Kaibab plateau (the north rim of the Grand Canyon) so increased the deer beyond their steady-

state that 2/3 of the population died within a few years, figure 29. Occasionally a species is limited by the supply of a non-food item like den space. Hole-nesting species of birds cannot reproduce without a suitable cavity. Erecting artificial nest-sites (just nest boxes with an appropriate sized hole) produces increases in the steady-states of such birds (e.g. wood ducks).

Frequently the resources of a population are quite predictably variable throughout the year. Summers may be rich with food, winters lean. Populations may adjust to this by having high birth

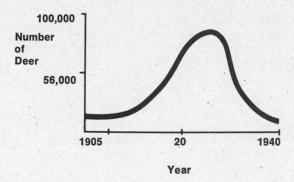

Actual numbers for plotted values:

| Year | Deer |
|---|---|
| 1906 | 4,000 |
| 1918 | 50,000 |
| 1920 | 62,000 |
| 1922 | 78,000 |
| 1923 | 91,000 |
| 1923¼ | 94,000 |
| 1923½ | 97,000 |
| 1923 | 99,000 |
| 1924 | 100,000 |
| 1926 | 40,000 |
| 1929 | 30,000 |
| 1930 | 25,000 |
| 1931 | 20,000 |
| 1939 | 10,000 |

**Figure 29.** Mule deer increased rapidly on the North Rim of the Grand Canyon after their preda-tors were "controlled." The population overshot its carrying capacity and destroyed its food plants. Starvation among the fawns was first recorded in 1920. The population irrupted during the winters of 1924 and 1925 and leveled off during the 1930's.

rates so that the few remaining survivors of winter can rapidly take advantage of increasing resources. Hibernation and seasonal migration are other ways of compensating for fluctuating resources. Some migrating birds travel thousands of miles twice a year to avoid rigorous seasons. Populations may be limited directly by something other than their resources. Cannibalism, territoriality and stress are such limits and are appropriate to discuss in this section because they all emanate from within the population. Exploitation and interspecific competition are others; but they involve outside species and will be discussed in the next section.

The only well studied example of cannibalism occurs in flour beetles. These insects live in stored grain products but never eat themselves out of house and home because their population is always low. The reason is that adults eat all the eggs and pupae they find. Flour beetle larvae are ravenous egg eaters too and will also eat a few pupae (the pupa is the developmental stage of life during which the insect is almost immobile and defenseless). A mutant strain of flour beetle that lacks cannibalism attains high population sizes and does deplete its flour home and larder.

The individuals of many populations defend part or all of their accustomed homesites from intruders of their own kind. When they do, the defended portion is called a "territory." Territories are rarely fought over. Usually an individual communicates his right to the place by some ritualized signal such as a song or feather display, an odor, a posture or a drumming of feet. If territories include more resources than are needed by their owners, then the species' population will be limited by the territoriality: more individuals could exist since some resources are being wasted. No one has ever adequately demonstrated this wastage however. But some studies have shown that territoriality influences the breeding rate of populations. For example a kind of crow called a rook has many "landless" males. These stay in a bachelor flock together. When a breeding male dies, a recruit from the bachelors takes over his territory and commences breeding. An experiment in Maine spruce forests designed to measure the impact of small songbirds on insect pests showed that the songbirds could be repeatedly shot out of large stretches of forest without much lowering the bird populations. Bachelor males and females quickly replaced those removed from their territories.

In the past several decades investigators have become aware of a striking relationship between population size, stress and breeding behavior. A group may have plenty to eat and drink, but its membership stops growing because it either ceases sexual activity to a large extent or because maternal behavior and physiology goes awry. In the latter case, mothers may abort spontaneously or not lactate or even destroy and eat their own young. In laboratory experiments with mice and rats, it has been determined that these abnormal phenomena are produced by the stress of too many contacts with other members of one's species. Other manifestations are aggressive behavior, endocrine abnormalities, high blood pressure and other cardiovascular diseases. All these problems together are called "stress syndrome," or "general adaptational syndrome (GAS)." Although its discoverer, Dr. Hans Selye, has much evidence that humans suffer from GAS and that many fatal heart attacks are caused by it, it is not known what the relationship of GAS to population size in humans is. It is not clear whether riotous and aggressive human behavior or human baby-battering (by parents) has anything to do with GAS or population. These relationships will undoubtedly be the subject of much important research.

People have wondered if these non-resource types of population control could have evolved to protect the species. Perhaps individuals declare wastefully large territories, cannibalize their fellows and turn off their reproductive systems in order to preserve their food supplies? But as far as science knows this idea is false. Natural selection does not protect the species. It maximizes the ability of individuals to survive and reproduce. Whereas this is usually beneficial for the species, if it produces too many individuals, it is dangerous for the species. The reason species aren't constantly exterminated by natural selection is that the living resources of the species are being selected simultaneously for their ability to avoid being used. The deer is a good escape artist, the mammal has antibodies to avoid infection, the insect lays small eggs, the plant produces antibiotic poisons. Thus each species grows as fast as it can because of natural selection. If flour beetles are cannibalistic, it is because the moisture they derive from their fellows is scarce in a flour bin and because the egg they eat as a larva might have grown up to eat them when they pupate. If rats under stress don't breed it is either because they haven't yet adapted to their laboratory colonies or because mild stress in most environments demands a mild form of GAS in order to preserve life. Certainly no one knows all the answers yet, but whenever a species has been closely examined, scientists have found it growing just as fast as its environment would allow. So far man has been no exception to this generalization.

## Population interactions

When two populations co-occur they may influence each other's growth rates. Adding more of a species might depress (hurt) the average net reproductive rate of the other's members or it might enhance (help) that average. This fact allows us to specify the three fundamental population interactions.

Let there be two species. If each helps the other the interaction is said to be *mutualism*. If each hurts the other, they are involved in *competition*. If one helps or supports its fellow, but in return is hurt, the interaction is called *exploitation* (or predation).

Population interactions influence the density and the stability of populations. Because of them, a population may be smaller or larger than otherwise. It may fluctuate violently or achieve an almost perfect balance. Most of the surprises and paradoxes of ecology emanate from this study.

Experiments have been done that prove the existence of population interactions. One that defies easy classification was done on animals that live between the high and low tides of the Pacific Ocean. In this environment on the northwest U.S. coast, live 15 species of barnacles, limpets and sea snails. Seven of them depend for their very existence on a starfish that eats them. If the starfish is removed the seven disappear and two of the rarest of the 8 that are left become superabundant. Undoubtedly a mixture of interactions occurs here but surely there is at least one competition and one exploitation. Another interesting study has been performed on the giant cowbirds and their associates in Panama. Cowbirds do not raise their own eggs. Instead they lay eggs in the nests of host birds. In some places these hosts when nestlings are attacked by a botfly. The larva of the fly burrows under the baby bird's skin and eats away. Seven such larvae are sufficient to destroy the nestling. Cowbird nestlings however are active and alert. They pick off the larvae before the larvae can burrow under. Even though the foster-mother works hard to feed the gluttonous cowbirds, she can raise more of her own young with cowbird protection. Therefore, she does not attempt to evict them. In such places mutualism reigns and the cowbirds usually make no attempt to conceal their visit to dump eggs in host nests.

Mutualistic populations are stable and result in high populations for each species. Where there are no botflies however, giant cowbirds are exploiters. A fostermother never raises as many of her own young as she would without the burden of the cowbird. Consequently she attempts to evict any cowbird eggs from her nest. Through the mechanism of natural selection, the cowbirds have responded by attempting to conceal their activities and also by producing eggs which mimic in color, shape and size, the eggs of their host.

In exploitative interactions the density of the victim is always reduced below what it would be if only its resources were present. For over 100 years humans have taken advantage of this in attempting to reduce the pests of their fields and orchards. A recent survey showed that 160 (over ⅓) of the insect pests in the USA are foreign creatures unwillingly brought here and released without any of their natural exploiters. Investigators have returned to the countries where these pests originated, found their natural enemies and imported them. This method, called classical biological control has succeeded in controlling about half of these pests at very little cost and with no ecological damage. A remarkable example of this concerns a prickly-pear cactus which became a pest when introduced to Australia. It covered vast acres of sheep range ruining them for forage. Introduction of its serious exploiter, *Cactoblastis cactorum*, a moth whose catepillar eats the cactus, has reduced the prickly-pear population to a few fugitive patches. Each patch sends out colonists to new places, but eventually *Cactoblastis* catches up with it and wipes it out.

Successful biological control agents are quite specific in attacking the pest species. When a general exploiter is introduced, great ecological disasters can occur. In the Antilles, introduced rats were ruining sugar cane. To control them, people brought in a mammalian carnivore, the mongoose. But the mongoose preferred to eat birds and has greatly reduced populations of many desirable species without much harming the rats.

Very rarely, a successful control agent is discovered in a place different from the original home of the pest. Australia has been particularly unfortunate in receiving many foreign pests, but has used this method to control one of the most serious, the European rabbit. The rabbit used to take much of the range forage that ranchers needed for their sheep, but it was controlled by a virus disease of the Brazilian cottontail. The cottontail is hardly bothered by the virus, but it kills a high proportion of European rabbits.

Sometimes men try to improve on the performance of natural exploiters by attacking pests with biocides (e.g. DDT). Besides being more expensive and dangerous than biological control, this method has produced important failures owing to the agriculturalists' lack of understanding of the exploitative interaction. If a pesticide is applied to a system of predator and victim, and the pesticide kills a fraction of both populations, then often, the farmer gets a surprise: the control agents decline in population size and the pests increase! This either thwarts the farmer in his attempt to control the pest or it produces a new pest whose population was previously negligible because it was under excellent biological control. A particularly well studied example of this principle occurred after the introduction of biocides into the vineyards of California. Vine-eating mites which had caused no problem before, began to destroy a significant share of grape production. Experiments have shown this occurred because the biocide also killed the mite's exploiters, carnivorous mites of various species. It takes about 5 years for a vineyard which has been repeatedly treated with biocides to recover a population of control mites sufficient to restore its natural health. Although "common sense" dictates that something which kills both the pest and its exploiter, should also reduce both populations, the interaction is capable of preventing this. The circumstances which allow the surprise, arise when the pest (or victim) population is controlled exclusively by the exploiter's ability to reproduce by eating the pest. The biocide, by increasing the exploiter's death rate, reduces this ability and allows the pest to grow. The total depressive effect of the biocide is thus transferred to the exploiter. Despite the fact that the basic mathematics showing how this could occur was first developed 50 years ago by two well known ecologists, the principle has been ignored.

These same ecologists introduced the world to the stability problems associated with exploitation. Because the victim is harmed, its population should decline when its exploiter is growing. The victims become too rare and the exploiter population declines. Then the victim can increase. This increase in turn supports an increase in the exploiter. Populations of exploiter and victim therefore tend to oscillate. At first it was thought these oscillations should persist, but now we know that the conditions which allow their persistence are restricted and special. In general, either the populations zero in on some equilibrium state or they deviate further and further from their balance. The first is a stable equilibrium and it helps maintain nature's balance. The second is unstable and unless one of the special conditions arises to maintain the oscillation, the populations of both exploiter and victim are in danger of becoming extinct.

A population's equilibrium in a simple two-species exploitation is stable if the victim-species involved exhibits intraspecific competition. It is unstable when the victims are benefited (as individuals) by the addition of more victims. The latter case is called intraspecific mutualism or the *Allee effect*, after its discoverer. Allee effects can be overt: there might be social cooperation among the individuals such as in herds of animals and in schools of fish. However, the exploitation itself provides a subtle opportunity for an indirect Allee effect. Predators can be satisfied: there can be so many victims that the predator's exploitative machinery is working as fast as it can. Added victims don't make the predator's life easier. In this case, adding victims divides the same exploitive pressure among more potential victims. We might think of this as sharing the burden of the exploitation; it is a simple kind of Allee effect. If the predator keeps the victim so rare that the victim has abundant resources, it is likely that the Allee effect will be stronger than any added competition from the new victims. In such a case victims and exploiters are involved in an unstable equilibrium and may be in danger. Laboratory experiments have shown that such subtle Allee effects and the instability they cause are both real.

The Allee effect may be at the root of paradoxical artificial eutrophication problems. Man, by dumping beneficial and important resources (e.g. phosphates, nitrates) into bodies of water, sometimes destroys most forms of life in them. It is quite possible that some or all of these problems are caused by loss of exploitative stability: adding resources weakens the competition but leaves the Allee effect intact. The observations that have been made on enriched ecosystems support the possibility that this is so. Mathematical and computer studies also have led us to worry. This concern is especially important since men hope to increase food production in the sea by pumping nutrients up into the lighted or photic zone where plants need them. That will have to be done carefully (if at all) in order to avoid exploitative destabilization and the resultant eutrophication.

For a long time ecologists raged over the question: do species ever compete with each other? Some ingeniously simple experiments have shown they definitely do. They are called perturbation studies, because they involve forcing populations away from their steady-states. If two species compete, then depressing either should improve the growth rate of the other. Permanently eliminating one should yield a permanent increase in the other. One of the path-finding perturbation studies was performed on intertidal zone barnacles of two species. One species lives high in the zone where it is usually dry. The other lives in the lower, more often submerged part of the zone. By excluding these latter, Connell found that dry-zone barnacles could also succeed in the wetter zone. They are normally excluded from it by competition from the faster-growing lower-zone barnacles which literally grow so fast as to either smother or pry up the losers from the rocks. Exclusion however is only one possible outcome of the competitive interaction. There are two others. It is possible that the two competitors will coexist, each at a density lower than it would attain in the absence of competition. And it is possible that one or the other would succeed but on an unpredictable basis: whichever by chance gets to be relatively common first will succeed. One should note in both this case and the case of competitive exclusion that a species which is physiologically perfectly competent to survive in its environment, cannot because of a population interaction. This is another example of a system's ability to surprise us.

The unpredictable outcome has never been well demonstrated. However, there is much evidence which helps us to understand how it might occur and makes us believe that it too is real. This evidence concerns antibiosis or allelopathy. Allelopathy refers to the production of an antibiotic substance which inhibits the growth, germination or survival of one's competitors. In a recent survey of the flora of coastal Washington, most of the 40 common plant species studied were found to produce antibiotics. Most of these appeared to be gaseous which leads to the belief that such toxins are waste products or modified waste products. Antibiotics are also known to be common among small algal species. If two species each produce a substance toxic to the other then whichever is common first will poison the environment more effectively against its competitor: this of course would produce the undiscovered competitive outcome. Often an organism is its own downfall in competition. By having a high population, it produces just the right environment for its competitor which then takes over. The best example

of this is forest succession based on light requirements. In this process, plants needing much light in which to germinate invade a bare patch of soil. These plants are usually grasses and weeds. Soon they are so dense that their seeds cannot germinate. Their weedier, brushier competitors do germinate and grow to shade out the grass. The process continues until after a century or two there remain only the tree species which can thrive as seedlings in dense shade and the similarly tolerant set of forest-floor herbs.

What causes competitive exclusion? How do competitors avoid it to coexist? These questions have been among the most bothersome to ecologists of the past decade and they will no doubt continue to inspire more important study. We now know that exclusion is the result of species being too similar in the resources they require. It is also known that these resources must be depletable like nutrients or food or else they must be occupiable on an exclusive basis like space and time. But it is still not known how similar is "too similar."

These criteria of exclusion arise from the criterion for coexistence. In order to coexist, each species must interfere with itself more than it does with it competitor. That is, adding a new individual of one species depresses the net average individual reproductive rate of that species more than it depresses the similar statistic of its competitor. Clearly, if each species has some special resource(s), it interferes mostly with itself by being common since no other species requires that resource(s). Therefore when it is very common it declines whereas its rarer competitor can increase. There are but few major categories of specialization which allow for coexistence. Two species may utilize different habitats either in time or space, or they may use different resources or they may be resources which are used differentially. The first two specializations are collectively known as habitat selection, the second two as resource allocation.

An example of habitat selection has been studied by Bovbjerg who worked on two species of freshwater crayfish. The two live in the same region of Iowa but one prefers the fast moving water of shallow riffles, the other lives in more stagnant waters. Experiments showed that the riffle-dweller actively drives out the other form from its territories. Removing the riffle-dweller allows its competitor to move in to the riffles where it does well. The riffle-dweller cannot live in stagnant water because of the lower concentrations of oxygen there.

Equally good experiments have never been done for temporal habitat selection, but many, many instances are known which can be interpreted as such. Many species of plants which flower in spring have neighbors which flower in the fall or summer. Cactus mice estivate; they do not stay active in the hottest, driest part of summer. At this time their probable competitors the pocket mice, are actually breeding. The pocket mice in turn, hibernate. Ponds and lakes often show a regular, predictable progression of plankton throughout the growing season. Sun-loving invaders of bare patches of soil must wait for a fire to clear an area and each

species of the succession must wait until time produces the proper habitat for it. Perturbation experiments are more difficult to do for resource allocation. To work they require much more time because natural selection must be allowed to interfere.

An organism's selection of resources depends upon itself. A large carnivore takes large prey; a small one smaller prey, Figure 30. Seed eaters often show a similar set of preferences. The currently accepted explanation for this is that each individual must maximize the rate at which it collects energy. If finding food is fairly easy, then the individual who attacks too small a package

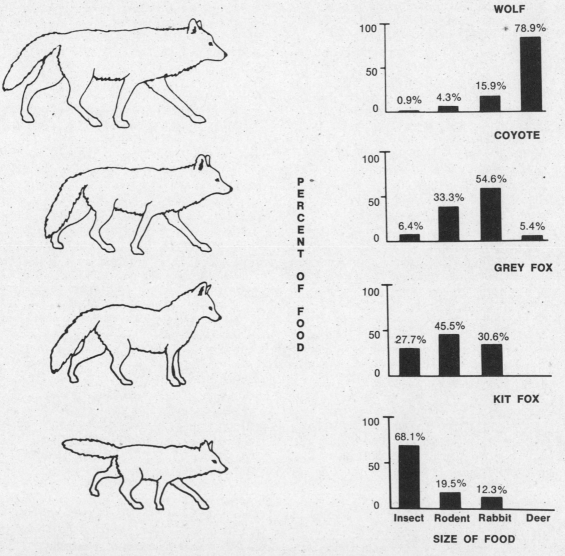

**Figure 30.** Members of the Dog Family take foods roughly in proportion to their body size.

is wasting time; so is the individual who tackles an impossibly large one. Foxes don't eat moose except as carrion; wolves don't eat mice unless a near continuous stream of mice practically runs into the wolf's mouth (In mouse plague years, this does indeed happen). Similar examples are known for birds (e.g. falcons), snakes, lizards, fish and similar examples probably exist in toothed whales, spiders and other meat eaters. If an investigator removes a competitor, he hasn't changed the ability to capture food of those species remaining. Nor has he changed their metabolic requirements. Thus they should more or less continue their old choices. In order for us to observe the change in resource selections, we must wait for natural selection to produce (e.g.) smaller or larger individuals to exploit the open opportunity. Even though perturbation work is difficult, some very fine observations of resource allocation have been made. A particularly fine set was made on the mites that suck the nutritive liquids from the feathers of birds. To accomplish this the quill mite must be large enough to penetrate the surface of the feather shaft. Each female quill mite tries to find an unoccupied suitable feather and fill it with her own young. Because all the population of the feather are usually her descendants, she can stop reproducing when the space available has just been filled. Overfilling the feather would cause a higher death rate and lower her net reproductive rate. When the quill mites' resource is molted, the females must find a new one. Smaller females can do this better. Therefore there are several species of quill mites differing in size. The large ones can use all feathers, but are not as effective at finding them as small mites. The small species is the best disperser, but it cannot use anything but the smallest feathers. Of course, since quill mites actually live for a time on their resource one might argue effectively that this is also a case of habitat selection. But the semantics are not very important as long as the ecology is understood.

If species coexist because they are resources which are allocated, then perturbation work to prove it can and has been done. One of the earliest perturbation studies involved excluding meadow voles from patches of fields in England. These mice eat prodigous amounts of grass and their removal spelled doom for the mosses which were soon crowded and shaded out. Once, a fence rusted out, the mice returned and within a year so did the mosses. A similar experiment has recently been completed on the shores of New Zealand. Here starfish are particularly fond of a certain mussel. When the starfish are removed, 7 of the 20 species of algae, sea snails, limpets, and barnacles become extinct because the mussel population increases rapidly covering about 80% of the available space.

Recently it has become apparent that a predator may need to switch its preference for resources and that this alternation may produce coexistence. If the predator needs to hunt each of two species in a special manner and it is more efficient to stick to one strategy at a time, then the predator will need to specialize on the more common species. In this way the common species hurts itself more by attracting all the predators; the two competitors therefore satisfy the criterion for coexistence. This phenomenon, called simply "switching" is a very special case of resource allocation that has not yet been discovered in nature. But it is known that some opportunistic species do prefer to utilize only common resources whatever they may happen to be at the time.

**The richness of life**
What determines the number of species at any given time and place? Why not just one species? Why not 10 billion? This is obviously a complicated question. It involves knowledge of population dynamics and interactions. It also involves some appreciation of the processes by which one species evolves to produce more than one, and the processes leading to extinction. Instead of discouraging them, this complexity has proved magnetic to population ecologists. Evolutionists know at least a few processes by which species split and multiply. But these are beyond the scope of this chapter because ecologists do not yet know how the number of species influences the rate at which these processes proceed. Ecologists have been more concerned with the causes of extinction.

We have already seen that extinction can proceed from either competition or exploitation. Surely it can also proceed if a species' environment disappears. Finally, it can occur if statistical sampling accidents should befall a very rare species. Perhaps this last seems difficult to understand, but it is not. Each individual has a chance in any given time period of dying by accident. That is why nature's equilibria must be steady-states: they must have the power to return from accidental (but very common) perturbations. Environmental accidents cause steady-states to show some fluctuation; the perfectly unchangeable population size occurs only in a computer, a world of unrealistic stability and determinism. Thus if a population is exceedingly small, it has a fairly large chance of disappearing by accident. Many ecologists who believe that, just as the size of most populations is at steady-state, so the number of species is at steady-state. In this view, extinction rates match speciation rates. Although it is far from being a settled issue, much fragmentary evidence both empirical and mathematical indicates this may well be true for many groups of species.

Careful enumeration of the number of kinds of mammalian fossils in the Americas has shown that the rate of loss of forms just about matches the rate of gain. This data extends over a period longer than ten million years and it includes many more or less catastrophic events, e.g. the ice ages, the appearance of the isthmus of Panama and the appearance of man himself owing to the exposure of the Bering land bridge from Siberia to Alaska. (This bridge was exposed when sea level dropped during the ice ages owing to the fact that glaciers had tied up a significant share of the Earth's water.)

Enumerating the known families of fossils in the whole world also supports the steady-state hypothesis. After the first appearance of life on land, it took only about 50 million years for the land forms to diversify. For hundreds of millions of years after that, the number of families inhabiting land didn't rise or fall appreciably. The same is true for a much longer time with respect to oceanic families; they were steady for about 600 million years. It must be emphasized that during this time there was much extinction and replacement. Evolution was far from static on land or at sea. In only three periods do we see much change: During the colonization of the land, during the beginning of the Paleozoic era when backbones and other fossilizable material first became widespread and during the past hundred million years, the number of known families grew with time. In each of the latter two cases, the growth in number is easily explained by the reasonable assumption that we have more knowledge of the life forms at the end of the growth period than at the beginning. The first, the increase brought about by the occupation of terrestrial environment, is thus the only known case of a growth in diversity on a large scale.

How does this steady-state arise if it exists? Mostly it is understood in terms of the competition interaction. (The role of overexploitation, if it plays a role at all, is not understood.) As a new species enters an environment, two fundamental things might happen; it might coexist or it might produce an exclusion. If it coexists, it is because it pioneers at least a slightly new specialty. When that happens, some resources are taken from some of the established species. Furthermore, by a process called character displacement, natural selection may strengthen the tendency to specialize. Character displacement also forces older species to specialize more narrowly. This makes species rarer and it makes more probable the disappearance of their special environment. Both changes enhance extinction rates. Furthermore, in a biota where a lot of species already exist, it is difficult to find a pioneering role; it is even more difficult to evolve by accident and expect to be suited to a pioneering role, but that is actually what must happen. It should be no surprise then, that ecologists feel competitive exclusion becomes more and more probable as an invader faces a more and more diverse set of established competitors. Evidence that this is so comes from the fact that most attempts to introduce an exotic (foreign) species into a biota result in failure.

Sometimes character displacement can be seen by carefully examing living forms in different places. A group of birds (that helped inspire Charles Darwin) display character displacement.

These birds are called Darwin's finches, and all but one species is found on the Galapagos archipelago. Quite a few species are seedeaters and have the conical bill of a finch or a sparrow. Three species seem to be especially close competitors, but have different bill sizes. There is evidence that this produces resource allocation by forcing the birds to specialize on appropriate-sized seeds. On some islands, however, not all three species are present. In these cases, the birds that are present have bills which are inappropriate to their species. Some of the "small" species have intermediate bills and some of the "large" do too. This pattern has been understood as a character displacement: only on the islands with all three species is each forced into a specialized morphology and role.

An important caveat exists here. Character displacement can also be caused by the need to be able to recognize one's own kind. When members of two species interbreed, their progeny may be inviable or sterile. To avoid this, species have special recognition devices. David Lack showed that for Darwin's finches, the bill's size is the only recognition mark, figure 31. Thus far not one case can be cited in which the mechanism of character displacement can be surely shown to involve competition and not species recognition.

**Figure 31. Darwin's Finches,** Drawing courtesy of the **Scientific American**, p. 2, April (1953)

Despite that caveat, it does seem to be true that species are less specialized in less diverse associations. Their individuals also appear more flexible and opportunistic. There is no question they tend to be more common. It is probable that the general understanding of the steady-state in number of species is correct for most cases.

Although ecologists have far to go in understanding the richness of life, they have been able to describe several interesting patterns in that richness. One of the most controversial patterns is called the *latitudinal gradient*: diversity is inversely proportional to latitude. This is not true for all forms (e.g. land mammals don't display it), but most forms exhibit their peak diversities in tropical areas and their minimal diversities in polar regions. One reasonable explanation of this gradient is that the tropics are more predictable and stable environments and cause fewer accidental extinctions.

But other equally credible explanations exist. One traces high diversity to lack of seasonality. Indeed, in the tropics it is possible to find evergreen forests and forests whose canopy is deciduous because of a dry season. The evergreen understories of each forest have about the same number of bird species, a number well in excess of a temperate zone habitat. But the deciduous canopy holds only about the same number of species as a temperate zone canopy, whereas the evergreen canopy is far richer. Probably the correct explanation of the latitudinal gradient involves more than one phenomenon.

One pattern of diversity which is well understood is found on islands. The farther an island is from the mainland, the smaller the number of species it harbors. Furthermore, large islands have more species than smaller ones. This pattern appears to be the result of a steady-state produced when the immigration rate of new species to an island counterbalances the rate at which species become extinct on the island. The steady-state exists because colonization should decline as diversity increases, whereas extinction rates should go up. Colonization declines because each added species depletes the pool of possible new colonists. Extinction increases for much the same reasons as previously discussed except that on islands, rarity (and the accidental extinctions it brings) are thought to be largely responsible. That is why smaller islands are felt to harbor smaller diversities: the small island supports fewer individuals of each species when their populations are at the same average density as those on a larger island. The reason for the distance effect is simply that potential colonists have more probable success if they need only reach a nearby island. Several experiments and observations tend to confirm all these theories.

Ruth Patrick created artificial islands of microscope slides in streams and observed the diatom communities that settled on them. (Diatoms are algae whose cell walls are silica.) At first, colonization exceeded extinction, but soon a steady-state was reached. New species continued to move in, old ones continued to become extinct. But both processes occurred at the same rate.

Simerloff and Wilson have actually depopulated small islands in the Florida keys and observed their recolonization. The islands return to their steady-states in a few years or less. A famous case of natural depopulation and recolonization occurred around the turn of the century when a volcano destroyed all life on Krakatoa. Except for the snakes, the island was back at its steady-state within a few decades. (Because of their low colonizing ability, snakes should take longer.)

It is necessary to emphasize that these are steady-states. Species *are* being replaced. Grinnell censused the Channel Islands near Los Angeles in the first decade of this century. Diamond, returning some fifty years later, found many species gone and many new ones present. But the net diversity change in each case was close to zero.

The island patterns teach us an important lesson about conservation. City parks and small state parks may harbor many species. But they are insufficient by themselves to maintain the diversities one would like them to hold. They are really only small islands in a concrete sea. We shall always need vast stretches of wildernesses to act as "mainland" reservoirs for our more accessible preserves. In turn, the integrity and diversity of each little piece of wilderness depends upon its membership in a large contiguous population of such pieces. To cut one out of a herd destined for slaughter only prolongs its execution. Wildernesses cannot be preserved if only small samples are protected.

Prof. Michael L. Rosenzweig, Ph.D.
Department of Biology,
University of New Mexico
Albuquerque, New Mexico

# Chapter 9

# Natural Selection and Artificial Control of Human Population

Garret Hardin has written: "Nobody ever dies of overpopulation." His tongue was in his cheek, however, since he proceeded to show many cases in the recent past in which people had died by the hundreds of thousands, with overpopulation directly implicated as the cause. For example, the terrible destruction wrought by the hurricanes in Bangla Desh's deltaic farmlands was predictable. The farmers know the area is prone to such calamities on a regular basis, but they live there anyhow because that is the only place left to find a farm. Taking one's chances with typhoons is better than certain starvation in the overpopulated interior.

This is not the place to review the limited resources of the Earth and guess how many people civilization should ask it to support. Suffice it to emphasize that those resources are limited and that science and technology can only convert resources from forms useless to man into forms which are useful. In doing so, technology often (but not always) takes resources from other species and so reduces their populations in order to increase that of mankind. Physicists, such as Albert Einstein have proven there is a limit to such conversions. Excellent treatises by the Ehrlichs and by H. Borgstrom have plumbed the depths of this limit.

In the preceding article on "Population Ecology," it was pointed out that it takes a full human lifetime to stop human population growth. Also from that article, it should be clear that populations in nature reach their steady-states by a combination of increasing death rates and decreasing birth rates. Since we need food, clothing and shelter which must be derived from resources, we too are a population in nature. Anytime we have reason to believe that our limit may be reached within seventy years, we had better start soon to halt our population's growth. The alternative is to allow our death rates to rise and our birth rates to fall by deprivation, as all irrational beings must. Since there is a chance that we are rational, perhaps we can avoid higher death rates. We can do this imposing upon ourselves low birth rates when our population is small

enough so that its citizens still enjoy a relatively bountiful existence. This is the argument for artificial population control. This argument does not include a threat of human extinction for that is unlikely to happen. Why should we become extinct just because we reach our own limit? Nothing has been able to outcompete us. We have devised good means for protecting ourselves from the diseases and pests that attempt to exploit us. We are too widespread and numerous to have to worry about accidental extinction. And we have proved to be the most adaptable, flexible species the world has yet known. If there is life on this planet, we are likely to be leading it. Other species may disappear, but we are likely to go on.

Why introduce the concept of natural selection to a discussion of the methods of population control? Because both things involve reproductive rate. Natural selection is the reason that most species are forced to live at high density in a deprived state. Because of natural selection, most proposed methods of population are foredoomed to fail.

The theory of natural selection must not be thought of as a wild guess. Its assumptions and deductions have been as well tested as any scientific proposition can be, and they have been found accurate. Scientists are as sure of little else. The theory is deceptively simple and may be summarized from Darwin. If individuals in a population tend to look, to act, to be constituted differently, and if these different constitutions have a tendency to reproduce at different rates and if these different constitutions are at least partly heritable, then the rate at which individuals reproduce any given environment will increase. This occurs, of course, because the rapidly reproducing constitutions tend to become a larger and larger fraction of the population. Such a change in the average heritable appearance of the members of a population is organic evolution and natural selection appears, from all accounts, to be its chief moving force.

There is much evidence to indicate that human beings fit this theory well; there is none to indicate the contrary. We are certainly differently constituted and we pass many of the differences on to our children. Furthermore, we reproduce at different rates as evidenced by our differing family sizes.

Evidence exists that family size does indeed have a heritable component independent of culture. The family sizes of four generations of British peeresses during Victorian times whose cultural background was fairly similar during any particular generation, exhibit this tendency well. These data are for thousands of individuals. For each child in excess of the average that a mother had, her daughter had one-fifth of an extra child and her granddaughter one-tenth. Since a grandchild has four grandparents and only two parents, the dilution (by half) of the effect is just what one expects. If fathers also have some influence on the constitution of their daughters, then it is possible that the tendency to have more or less children than average is heritable to a 40% extent. Such percentage-heritabilities are quite similar to those known for the reproductive rates of other vertebrates, for example, songbirds. Thus it appears if we attempt to educate people in one way, we will still observe heritable variation in family size.

Knowledge of natural selection enables us to dismiss immediately the most frequently suggested means of population control, pure voluntarism. In this method, society merely attempts to educate all people as to the need for population control and the consequences of doing without it. It backs this up only by ensuring that various effective methods of birth control are readily available. Although no one can fault these two items (they are obviously necessary for any method of population control), they are not sufficient. Natural selection will increase the reproductive rate and those people who have a tendency to follow the dictates of conscience will also tend to disappear. The people who will remain will be reproductive renegades, either unable to understand the issues or unwilling to plan for the future of their children and grandchildren.

In fact, in order to achieve even temporary success with this method, society would have to demand extraordinary sacrifices from those willing to volunteer. It would have to say something like: "We have a population problem and in order to solve it each woman should limit herself to two children. But we know in advance that we cannot expect full cooperation and in fact we should expect diminishing cooperation after a while, so please, if you care, stop at zero or one to make up for the excesses of others." Such sacrifices might

be noble if they could be expected to produce long-lasting effects, but since natural selection will soon eliminate whatever temporary good they might bring, it might be considered wrong even to suggest them.

Recognizing this problem, some have suggested that economics be harnessed for the good of population control. Either legislation would be passed to reward couples who postpone and limit births or it would be written to penalize (with extra taxes) those who have too many births. Unfortunately, until it actually deprives babies enough to kill them, this method has the same fault as pure voluntarism. Few families grow because the parents have assessed their resources and decided another heir could be provided for. Few stop growing because they are impoverished. If an economic program of population control were to be tried, those that did respond would be selected against. Money would become an even more irrelevant aspect of the decision-making process than it is today. It is a well-documented fact that the poorer people of any given culture and of the whole world have larger families than the more affluent. "Money is simply not as important as babies," rules natural selection.

We are the first species to understand selection, but we must study it carefully in order to defeat it. To control population at a low density we must break one of the links of natural selection. We must either prevent family size from being an inherited tendency, or eliminate variations in family size, or produce a population in which almost no one who has more than two children rears them all. The last is clearly heinous. It involves raising the death rate, and that is what we are trying to avoid. But the first may be possible and so may the second. To achieve lack of inheritance, everyone must agree to have whatever number of children a computer tells him to have, the numbers to be assigned to women strictly at random. On the other hand, to eliminate variation effectively each woman merely bears the same number of young.

With either of these plans, natural selection is defeated, but neither plan is simple. Each requires universal cooperation, at least as universal as paying income taxes. Thus a massive educational program must precede it; enough people must want population control for legislation to be enacted. Effective legislation would almost certainly have to guarantee citizens the right to the bountiful lifestyle they were helping to preserve. It would surely provide people with free birth control of whatever sort they found physically and morally acceptable.

Probably most people would prefer everyone had equal birth rights rather than accept the dictates of a computer. At first, when we would be trying to halt population growth, that would work. But once we got to the stage of needing to maintain a steady population, the number 2 would no longer do; equal rights would mean perhaps 2-1/5 children per mother. The whole number 2 would lead to decreasing population owing to infant mortality and adult fertility problems. The extra births needed to bring the average to 2-1/5 (or whatever turns out to be needed) could then be distributed at random to those women who wanted more than two children.

It is important to note that the means for effecting such a plan are already available. Birth control techniques are sufficient as long as vasectomy, tubal ligation and abortion are kept as ready reserves to be used at the discretion of couples. (They will need to decide on one in advance of a failure in conception control.) It is also worth noting that any political entity which has the power to control immigration across its borders can embark on population control. If they have the technology and natural resources to be self-sufficient, and adequate population to maintain their industry and defenses, population control will probably increase their political and military power relative to their overabundant neighbors. The day is long past when a nation's achievements depended on the total gross weight of its citizenry.

Prof. Michael L. Rosenzweig, Ph.D.
Department of Biology,
University of New Mexico
Albuquerque, New Mexico

# Chapter 10

# Methods in Ecology

Environmental issues of late have become subjects of concern and the term ecology has become a household word. Such widely discussed topics as the productivity of the sea and of agricultural lands, the concentration of pesticides and heavy metals in food and animal populations, the growth and regulation of animal populations including the harvesting of fish and the hunting of game animals, all involve basic ecological concepts. When reading about such topics, one must wonder how the ecologist determines the productivity of ecosystems, estimates the density of populations of plants and animals, and learns something about the interactions of organisms and the environment. To study the organisms and the physical aspects of their environment, ecologists can modify the techniques of physics, chemistry, and physiology. But to investigate ecosystems and populations, ecologists have had to develop specialized methods. It is these methods, rather than the ones used to measure the physical and chemical aspects of the environment that are stressed in this article.

## Methods of Estimating Production

The study of the functioning of ecosystems involves the investigation of the productivity of ecosystems. Of primary interest are the rate at which the energy of the sun is fixed by plants, the accumulation of biomass, and the transfer of energy and materials from one feeding group to another. Knowledge of such information is basic to understanding the productivity of agricultural and forest ecosystems and their management and to the maintenance of natural ecosystems. Estimating the rates of energy fixation in an ecosystem is difficult and the task of measuring all aspects of energy flow is enormous. Several techniques for determining primary production estimate energy fixation indirectly by relating the amount of materials, oxygen, or carbon dioxide released or used.

## Light and dark bottle

The light and dark bottle method is commonly used in aquatic environments. It is based on the assumption that the amount of oxygen produced is proportional to gross production, the total amount of energy fixed by the green plant, since one molecule of oxygen is produced for each atom of carbon fixed. Two bottles containing a given concentration of phytoplankton are suspended at the level from which the samples were obtained. One bottle is black to exclude light; the other is clear. In the light bottle a quantity of oxygen proportional to the total organic matter fixed or gross production is produced by photosynthesis. But at the same time the phytoplankton, the microscopic plant life, is using some of the oxygen for respiration. Thus the amount of oxygen left is proportional to the amount of fixed organic matter remaining after respiration or net production. In the dark bottle oxygen is being utilized but is not being produced. Thus the quantity of oxygen utilized, obtained by subtracting the amount of oxygen left at the end of the run (usually 24 hours) from the quantity at the start gives a measure of respiration. The amount of oxygen in the light bottle added to the amount used in the dark provides an estimate of total photosynthesis or gross production. This method has several shortcomings. Estimations of production are confined to that portion of the plankton community contained within the sample bottles. The method fails to take into account the metabolism of the bottom community. And the procedure is based on the assumption that respiration in the dark is the same as that in the light, which is not necessarily true. A modification of this method employs the whole aquatic ecosystem as a light and dark bottle. Daytime represents the light bottle, nighttime the dark. The oxygen content of the water is sampled every two or three hours during a 24-hour period. From this the rise and fall of oxygen during the day and night can be plotted as a diurnal curve. This method, which is most useful in the study of running water communities, takes into account the photosynthesis of bottom plants, as well as the oxygen exchanged between air and water and between water and bottom mud.

## Carbon Dioxide

While the light and dark bottle method is the most useful in aquatic ecosystems, the measurement of the uptake of carbon dioxide and its release in respiration is better adapted to the study of terrestrial ecosystems. In this method a sample of the community, which may be a twig and its leaves, a segment of the tree stem, the ground cover and soil surface or even a portion of the total community, is enclosed in a plastic tent. Air is drawn through the enclosure and the $CO_2$ concentration of the incoming and outgoing air is measured with an infrared gas analyzer. The assumption is that any $CO_2$ removed from the incoming air has been incorporated into organic matter. A similar sample may be enclosed within a dark bag. The amount of carbon dioxide produced in the dark bag is a measure of respiration. In the light bag the quantity of carbon dioxide would be equivalent to photosynthesis minus respiration. The two results added together indicate gross production. A metabolism chamber may also be employed, figure 32.

Another approach to the estimation of productivity by the measurement of $CO_2$ is the so-called aerodynamic method. It involves the periodic measurement of the vertical concentration of $CO_2$ by means of carbon dioxide sensors arranged in a series from the ground to a point above the canopy of the vegetation. The reduction of the carbon dioxide in the profile represents the amount of carbon dioxide consumed by the foliage of each layer of the profile. In many ways this technique is still in the developmental phase particularly in the design of instruments. One shortcoming of this and other techniques involving the measurement of $CO_2$ is the inability to distinguish between the soil and the plant contributions to the $CO_2$ flux and to determine the uptake of carbon dioxide by the roots and its transport to the chloroplasts.

### Chlorophyll

An estimate of primary production may be obtained by using the measurements of the concentration of chlorophyll and the intensity of light. This technique is based on the discovery by plant physiologists that a close relationship exists between chlorophyll and photosynthesis at any given light intensity. This relationship remains constant for different species of plants and thus communities, even though the chlorophyll content of plants varies widely, because of nutritional status and the duration and intensity of the light to which the plant is exposed. This method, useful in aquatic ecosystems, involves several steps. The amount of chlorophyll in the plants, expressed as per gram or per square meter, must be determined. The total daily solar radiation reaching the surface must be measured, as well as the extinction

coefficient of visible light in the water, and the photosynthesis of the plant population at light saturation. From such information total daily production within a body of water can be estimated. The chlorophyll method is less useful in terrestrial ecosystems. Productivity is limited more by the lack of nutrients than by the quantity of chlorophyll. The concentration of chlorophyll is greater in plants of shady habitats than in plants growing in the sunlight, yet the productivity of such plants is low.

### Harvest Method

Widely used in terrestrial ecosystems is the harvest method. It is most useful for estimating the production of cultivated land and range where production starts at zero at seeding or planting time, becomes maximum at harvest, and is subject to minimal utilization by consumers. A modification of the method is used in estimating forest productivity. Briefly the technique involves the clipping or removal of vegetation at periodic intervals, drying to a constant weight, and expressing that weight as biomass in grams per square meter per year. Caloric values of the material can also be determined. This biomass then is converted to calories and the harvested material expressed as kilocalories per square meter per year. To be accurate the plant material must be collected throughout the growing season and the contributions of each individual species determined. In some determinations only the above ground portions are harvested, and the root matter goes unsampled. Other studies attempt to estimate the root biomass. Sampling the root biomass is difficult at best. While the roots of some annual and crop plants may be removed from the soil, the problem becomes more difficult with grass and herbaceous species and even more difficult with forest trees. The task is laborious; except for annual plants, the investigator faces the almost impossible task of separating new roots from older ones. He has the added problem of estimating the turnover of short-lived small roots and root materials and the variability of the sample. Because plants of different age, size and species make up the forest and shrub ecosystem, a modified harvest technique known as dimension analysis is used. This involves the measurement of the diameter at breast height (DBH) and the height of trees, and the diameter growth rate of the trees in the sample plot. A set of sample trees is cut, preferably at the end of the growing season; the height to the top of the tree, DBH, depth and diameter of the crown,

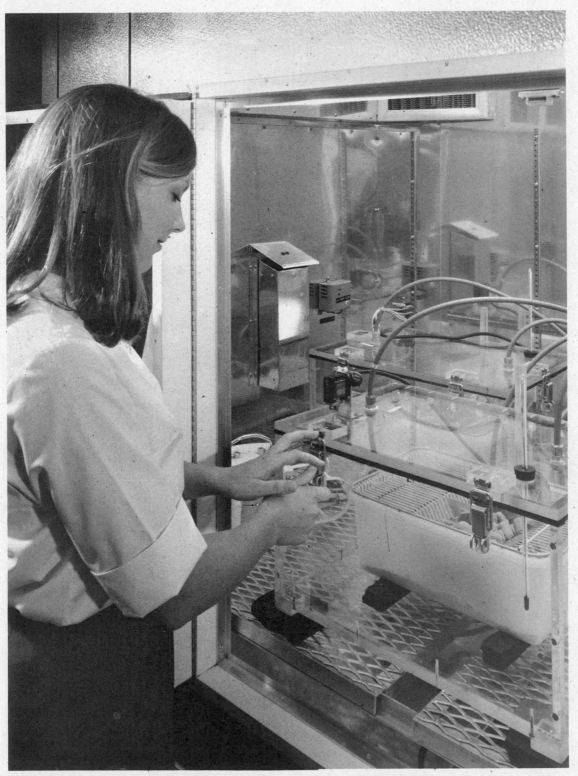

**Figure 32.** Metabolism chamber in which animals are placed for $CO_2$ measurement. Photo courtesy of Oak Ridge National Laboratory.

and other measurements are taken. Total weight, both fresh and dry, of the leaves and branches is determined, as is the weight of the trunk and limbs. Roots are excavated and weighed. By various calculations the net annual production of wood, bark, leaves, twigs, roots, flowers and fruit is obtained. From this information the biomass and production of the trees in the sample unit are estimated and then summed for the whole forest. The biomass of the ground vegetation is also determined, as well as litter fall, estimated from material collected in litter traps. By the use of such techniques the biomass and nutrient content of various trees and forest stands can be obtained.

## Radioactive Tracers

Productivity can be measured by determining the rate of uptake of radioactive carbon ($^{14}C$) by plants. This is the most sensitive technique available to measure photosynthesis of aquatic ecosystems under field conditions, figure 33. A quantity of carbon-14, usually in the form of a bicarbonate such as $NaH^{14}CO_3$ is added to a sample of water containing its natural phytoplankton population. After a short period of time to allow photosynthesis to take place, the plankton material is strained from the water, washed and dried. Counts of radioactivity are taken, and from them calculations are made to estimate the amount of $CO_2$ fixed in photosynthesis. The estimate is based on the assumption that the ratio of activity of $^{14}C$ added to the activity of phytoplankton is proportionate to the ratio of the total carbon available to that assimilated. In common with other techniques the $^{14}C$ method has certain inherent weaknesses. The method does not adequately measure changes in the oxidative states of the carbon fixed. All of the carbon fixed is not retained by the producers. Some tend to seep out of algal cells as water soluable organic compounds used by bacteria. The various primary producers have different abilities to utilize available light. The amount of carbon fixed is influenced by the species composition of the plankton community.

**Figure 33.** Sampling for stream benthic organisms. Photo courtesy of Oak Ridge National Laboratory.

### Estimating Consumer Production

To follow the fate of energy fixed by plants, the ecologist has to estimate its utilization, direct or indirect, by the various consumer groups, the herbivores, carnivores, and decomposer groups. Methods involve the determination of food consumption, energy assimilation, heat production, maintenance requirements, and growth.

Food consumption can be determined in the laboratory or estimated in the field. Laboratory determinations involve feeding the animal a known quantity of its natural foods, allowing it to eat over a period of time usually 24 hours, then removing the food and weighing the remains. The amount of food consumed equals the amount fed minus the amount removed. The caloric value of the food consumed is determined by burning a sample in a calorimeter. If the activity periods of the animal and the weight of the food its stomach will hold are known, then consumption can also be accurately determined by multiplying the activity periods by the mean weight of observed contents of the stomachs from a sample of animals taken from the population. Activity periods are used, since most animal activity is concerned with feeding.

Assimilation of food consumed can be determined by subtracting the energy voided in feces from energy consumed. The energy assimilated is used for maintenance and growth. Energy used for maintenance is lost. The cost of maintenance can be determined by confining the animal to a calorimeter and measuring the heat production directly; or the energy used in maintenance can be determined indirectly by placing the animal in a respirometer and measuring the oxygen consumed or carbon dioxide produced. These results are then converted to calories of heat. To do this one must know the respiratory quotient, the ratio of the volume of carbon dioxide produced to oxygen consumed. The respiratory quotient varies with the type of food utilized in the body. To estimate accurately the heat production of a population from laboratory determinations, one must also know the daily activity periods, the weight distribution of the population, and the environmental temperature.

Production or storage of energy is estimated by weighing individuals fed on a natural diet in the laboratory or by weighing animals each successive time they are caught in the field. An indirect and usually more useful method is based on the age distribution of a population, the growth curve for the species, and the caloric value of the animal's tissue. Growth curves must be obtained for each population under investigation and for each season under study. Once a growth curve has been constructed and the age distribution of the population is known, the weight of tissue produced in a given period can be estimated for each age class. The weight gain is then converted to caloric equivalents.

Radioisotopes are also used to determine secondary production, particularly of insects. The rate at which a radioisotope is ingested can be converted to food consumption as long as the concentration of the isotope in the food is known. The method is based on the fact that insects feeding on plant material tagged with a radioactive tracer such as cesium-137 accumulate the isotope in their tissues until there is a steady state equilibrium. The intake through the consumption of plants is balanced by the loss through biological elimination.

### Investigating Biochemical Cycles

The fixation and transfer of energy is one function of the ecosystem. Another is the cycling of nutrients. To gain some insight into biogeochemical cycling and the budgets of nutrients in ecosystems, ecologists employ another set of techniques. One of the most useful is radioactive tracers such as cesium-137 or phosphorus-32. By tagging plants, animals, soil or water with a radioactive tracer, the investigator can follow a specific nutrient within an ecosystem. Aquatic ecosystems can be tagged by spiking the system or some enclosed portion of it with a radioactive isotope. The isotope will be taken up by the phytoplankton, transferred from terrestrial ecosystems, the tag may be sprayed on the leaves of plants or injected into the trunks of them to other organisms in the food chain. In trees, figure 34. The tracer may be added to fertilizer and applied to the soil and litter. Once in the ecosystem the tracer may be followed by drawing samples of plant and animal tissues and determining their radioactivity with a Geiger, liquid scintillation, or gamma radiation counter. By this means the ecologist can determine the movement and ultimate distribution of the element—what portion becomes fixed in sediments and soil and removed from further circulation; what portion is taken up by plants and transferred through the food chain; what amount is returned to the substrate and recycled. He can obtain information on the pathways of food chains. He can estimate turnover rates, the amount of an element taken up by an organism that is released into the ecosystem for further uptake in a given period of time. The same use of isotopes enables the ecologist to

**Figure 34.** Nutrient solution with radioactive tracer is placed in cup to be absorbed by the osmotic action of tree trunk. Photo courtesy of Oak Ridge National Laboratory.

study the movements of radioactive substances through the environment, figure 35. From such data the ecologist hopes to be able to predict the effect of radioactive contaminants on the environment. Radioactive tracers permit the ecologist to trace the pathways of nutrients through an ecosystem. But to determine the nutrient budgets of an ecosysem, he must employ other methods. For example to work up a nutrient budget of a forest ecosystem, the ecologist must estimate the inputs of nutrients to the ecosystem, the amount retained, and the amount carried away from the ecosystem.

To estimate the quantity of nutrients being stored or retained, the investigator first has to estimate the biomass of the vegetation, just as he did in the study of energy flow. By burning samples of leaves, twigs, stems, bark, and trunk, of trees and shrubs and samples of understory vegetation and analyzing them chemically, the ecologist can estimate the quantity of nitrogen, phosphorus, calcium, magnesium and other elements contained in the biomass. By analyzing the leaves and new woody tissue separate from older woody tissue, he can determine approximately how much of the various nutrients are being fixed in current growth and what quantity will be returned to the soil through litterfall. In a similar manner, the investigator can estimate the quantity of mineral elements in the litter. By the analysis of the soil he can arrive at an estimate of the nutrients in the soil available to plants. By these methods the ecologist can assess the amount of nutrients contained in the biomass and soil.

Nutrients are being transferred from the soil to the vegetation and back again. To determine the amounts and transfer rates of nutrients, the ecologist collects litterfall periodically and analyzes it to estimate the amount of nutrients being returned to the soil by that route. He also collects samples of rain water that falls through the forest canopy and that flows down the stems to estimate the quantity of nutrients being leached from the leaves and carried to the soil. The amount of nutrients entering and leaving the rooting zone of vegetation he measures by placing a system of tension lysimeters beneath the forest litter and approximately one meter deep in the soil. And he can determine the uptake of nutrients by the vegetation by sampling the current year's growth of foliage, branches and trunk.

**Figure 35.** Infusing a nutrient with a radioactive tracer into a small plant community. Photo courtesy of Oak Ridge National Laboratory.

This leaves two aspects of the biogeochemical cycle yet to be estimated, the input from rain and dust or aerosols and the output through water and erosion. The investigator can measure the quantity of such nutrients as calcium, phosphorus and sulphur carried in by the rain by chemical analysis of rain water. The input from dust and aerosols intercepted by the vegetation and the ground can be measured by collecting particulate matter carried by the wind with special traps containing filter paper disks. These are placed at several heights on towers constructed in forest clearings. The disks are analyzed for their nutrient content. The values obtained are then expanded to relate to the number of leaves per acre, obtained by tree litter sampling, and the variation of leaf size throughout the season.

To determine the loss of nutrients through the outflow of water requires the construction of flumes and weirs to intercept the water flowing from the watershed. The weir collects organic material and sediment being carried from the site. By analyzing sediments, particulate matter and stream water and relating their nutrient concentration to the volume of stream flow, the investigator can estimate the amount of nutrients being leached from the ecosystem. By putting all this data together, the investigator can come up with an estimated nutrient budget for a given ecosystem. Studies or nutrient cycling within ecosystems and the preparation of nutrient budgets provide a better understanding of nutrient-energy relationships of terrestrial ecosystems and between terrestrial and aquatic ecosystems. Such knowledge is necessary for an understanding of nutrient depletion in terrestrial ecosystems and to the eutrophication of bodies of water. And it is basic to efficient management of agricultural, range, and forest ecosystems.

### Microcosms

Under the best of conditions ecosystems are difficult to study. Size alone is a problem. There are also the lack of definitive boundaries to an area and the many variable conditions such as temperature, moisture and light. To eliminate some of these problems, ecologists have resorted to miniature ecosystems or microcosms. They attempt to bring into the laboratory various biotic and non-biotic material from natural ecosystems and hold it under constant conditions for study. Once established in the laboratory the system which develops can be replicated many times. Any number of experimental manipulations, reproducible as well as controllable can be performed on them. Microcosms are useful in measuring such ecosystem functions as rates of metabolism, ratio of photosynthesis to respiration, the effects of pollutants on organisms, predator-prey relationships and the like. Most widely used are aquatic microcosms, developed in aquaria. These may simulate a fresh water system, a salt water or marine system, and they may be seeded with any combination or organisms typical of a spring, pond, estuary or marine environment. Within these microcosms, the investigator can determine certain functional interactions such as oxygen and carbon dioxide production and investigate interactions within and between species.

Another aquatic type is the stream microcosm. This is a relatively small constricted channel in which the flow of water and the gradient can be controlled. By changing the gradient and thus the velocity, and by varying the type of substrate such as rubble and gravel, the investigator can simulate various types of fast-running water. By enclosing at least a portion of it, he can control light and temperature. Once established the stream microcosm can be stocked with selected aquatic plants and insects and fish. In this system the aquatic biologist can study the reactions of organisms to such environmental variables as the nature of the stream botton and velocity of water. He can gain information on the transport of detritus or dissolved and suspended material in the water, and spacial succession of plant and animal communities of running water. The biologists also possess a system in which the effects of heat and waste disposal on the life of streams and the reactions of lotic organisms to pollutants can be observed.

Just as samples of aquatic ecosystems can be confined to microcosms and studied in the laboratory, so can samples of soil and litter. But for the study of soil and litter communities, microcosms established in place are more rewarding. These microcosms may be nylon net bags containing samples of litter or plastic boxes containing sections of the forest or grassland floor. Such microcosms enable the biologist to employ experimental design using such combinations as the type of forest or grassland community, species of leaves, substrate, and altitude. He can identify the effects of these variables on the soil community, observe the succession of decomposer species and the influence of selected soil organisms on the decomposition of plant and animal material.

However useful such microcosms may be, they do have severe limitations. For one they do not reflect conditions as they actually exist in nature. The complexity of the system is greatly reduced. One can question the relevancy of the conclusions of studies employing microcosms to the natural ecosystems. The investigator can never be sure that the reaction of the organisms studied or the interaction between the organism and the simulated environment in any way mirrors interactions as they might exist in natural situations. But in spite of these limitations, microcosms do provide a means of gaining at least a general insight into the functioning of ecosystems that could not be obtained otherwise.

## Methods of Studying Populations

The flow of energy and the cycling of nutrients has to operate within some structural element of the ecosystem. These elements are the populations of the various plants and animals that make up the living portion of the ecosystem. The nature of these plant and animal populations, the diversity of species, their numbers, their age structure, their distribution in time and space, the growth and change in populations, all are important parameters within the ecosystem. To obtain some idea of the structure of ecosystems, some sort of sampling of the population is necessary. Once collected the data must be analyzed by appropriate statistical procedures.

## Problems of Sampling

Sampling a population is no easy task. Before he attempts to sample a population, the investigator has a number of decisions to make. He has to decide how the area to be sampled will be subdivided and how many samples he will take in each subdivision. If the sampling is to meet certain statistical requirements, then the sample unit and the methods employed must meet certain criteria. All units of the sampling universe must have an equal chance of being selected; in other words the sampling must be random. During the period of sampling the population should be stable. The sampling unit must be convertible to unit areas, since a population to be defined must relate to some unit of space, such as per acre or per square mile. The sampling unit must be easily delimited in the field. And it must be such a size that it can be reasonably sampled with the time and money available. Also to be decided is the number of samples to be taken. The total number of samples pulled depends upon the degree of precision required. Statistical procedures are available to determine the number of sampling units needed to meet the level of error the investigator is willing to accept.

The pattern of sampling must be given consideration. To obtain an unbiased estimate of the population, the sampling data should be collected at random. The simplest method of doing this is to select sample units by the use of random numbers representing the whole area to be sampled. Tables of random numbers are found in all statistics textbooks. Often unrestricted random sampling is inefficient since the majority of samples may easily come all from one section of the area being sampled. To avoid this, the investigator may use stratified random sampling. By this technique, the area is subdivided into equal sized parts or strata and one or more samples are selected from each strata. By using this method the biologist can restrict himself to a habitat where the plants or animals being studied are most likely to occur.

Another approach is the systematic sample, in which the samples are taken at fixed intervals in space or time. While such samples present certain problems in statistical analysis they do have an advantage in the ease of operation that is missing from unrestricted randomized samples. The time that the samples are taken must relate to the life cycle and diurnal and seasonal rhythms of the plants and animals being studied. One could not adequately sample the herbaceous vegetation of a forest only in midsummer and get an adequate sample of plants present in the understory, for the spring flowering plants would be dormant or inconspicuous. Neither could one obtain an adequate sample of rabbits by restricting counts to the mid-afternoon when the animals are most active at dusk and dawn. Dispersion of the population, how plants and animals are spaced over the area, is an important parameter. Dispersion has to be considered in the selection of the sampling program to be employed. The measure of dispersion through time provides insights into fluctuations of populations, density, seasonal movements, mortality, competition, and the like. Dispersion can be analyzed by the use of such statistical procedures as the *Poisson* series and the negative binomial. These enable the investigator to describe the distribution of individuals within a population as uniform, all organisms equally spaced, as random, or as clumped often called contagious.

## Sampling Plant Populations

Methods of sampling and analyzing vegetation occupying a given site are numerous and the literature discussing them, the underlying philosophies and the statistical treatments are extensive. Except for phytoplankton, plants are relatively stationary, which reduces the problems of sampling.

One of the most popular sampling techniques is the *quadrat* method. Strictly speaking the quadrat applies to a square sample unit or plot. It may be a single sample unit or it may be divided into sub-plots. Quadrats may vary in size, shape, arrangement, and number, depending upon the nature of the vegetation and the objectives of the study. The size of the quadrat must be adapted to the characteristics of the community. The richer the flora, the larger or more numerous the quadrats must be. Within each quadrat the vegetation may be counted by species, or the actual or relative abundance and cover may be estimated.

The quadrat method is easily employed but tedious and time-consuming. If the individual organisms are randomly distributed, then the accuracy of the sample and the estimate of density depend upon the size of the sample. But since individuals are rarely dispersed randomly, the accuracy of quadrat sampling may be low, unless a great number of plots are involved.

The *transect* is a cross section of an area used as a sample for recording, mapping or studying vegetation. It may be a belt, a strip, or only a line on which the intercepts by types of vegetation are marked off. Because of its continuity through the study area, the transect can be used to relate changes in vegetation along the line or strip with changes in the environment. The *belt transect* of predetermined width and length is well adapted to estimate abundance, frequency, and distribution. The one-dimensional line transect is rapid, objective, and relatively accurate. It is well adapted for measuring changes in vegetation if the ends of the lines are well marked. Generally it is more accurate in mixed plant communities than quadrat sampling and is especially well suited for measuring low vegetation. However, the method is not well adapted for estimating frequency and abundance, since the probability of an individual being sampled is proportional to its size. Nor is it suited where vegetation types are intermingled and the boundaries indistinct.

In recent years *variable-plot* methods have become popular. All are related to the variable-radius method of tree sampling developed in Germany by Bitterlich, who used it to determine timber volume without establishing plot boundaries. The variable plot method requires no actual measurement in the field; no plots or lines are laid out on the ground; no dimension of any plant is taken. Basically the variable-plot is a means of selecting trees and shrubs to be counted on the basis of size rather than on frequency of occurrence. The usual plots or lines are replaced by a series of sampling points distributed at random over the area to be surveyed. At each sampling point the observer uses an angle gauge or prism that subtends a fixed angle of view to sight in the DBH of each tree or crown diameter of a shrub. Tree boles and shrub crowns that are close enough to the observation point to completely fill the defined angle are tallied; otherwise they are ignored. The most commonly used sighting angle is 104.81 minutes. Since this angle can be defined by a one-inch intercept on a sighting base of 33 inches, then all trees located to further than 33 times their diameter from the sampling point will be tallied. This angle gives a basal factor of 10, which means that the average number of trees per point multiplied by 10 gives the basal area in square feet per acre. A variation of the variable-plot is the *quarter method.* This method involves a number of random points along a series of line transects. The area around each point is divided into four equal parts or quadrants. The distance from the point to the nearest tree in each quadrant is measured and the species and DBH, later converted to basal area, of each tree is recorded. The mean of all distances equals the square root of the mean area. From this value the density of all species can be determined. From the basal area, one can determine dominance and from the number of trees of each species, the frequency.

Grass and other low vegetation may be sampled by the *point-quadrat* method. Pins with sharpened points are spaced at ten 1-decimeter intervals in a frame. As the pins are dropped, the vegetation first contacted by each pin is recorded. The data can be obtained either on a transect or at random points over the area. The plant ecologist is interested in at least three parameters from his sampling data: 1) relative density, the ratio of the number of an individual species to the total individuals of all species; 2) relative dominance, the ratio of the basal area of a species to the total basal area; 3) relative frequency, the ratio of the frequency value of an individual species to the total frequency values of all species. Frequency is the number of points in which a species occurs divided by the total number of points or plots sampled. By adding together relative density, relative frequency, and relative dominance the ecologist can obtain an index, *importance value,* which he can use to develop a logical arrangement of a number of stands. The index is based on the fact that most species do not attain a high level of importance in the community, but those that do serve as an index, or guiding species. Once importance values have been obtained for a species within a stand, the stands can be grouped by their leading dominants according to importance values. The groups are then placed in a logical order based on the relationships of the several predominant species.

Sampling procedures differ considerably in the aquatic ecosystems where algae are the dominant plants. Two kinds of growth are involved: plankton suspended in the water, and the periphyton growing attached to some substrate. Samples of phytoplankton can be obtained by drawing water samples from several depths. Cell counts of algae present in each sample, either normal or concentrated can be made with a Sedgwich-Rafter counting chamber and a Whipple ocular. If necessary the samples can be concentrated by centrifugation. The centrifuged samples are then diluted to a suitable volume. The cells are counted, a separate tally being kept for each species. Another method of handling phytoplankton is filtration. The organisms in the sample are fixed in formaldehyde. The sample is agitated, and a fraction is withdrawn. This sample is filtered; the filter paper is placed on a glass slide, treated with oil, and scanned under the microscope to make a quadrat count of the presence and numbers of individual species. Periphyton growing on living plants and animals can be observed in place on the organism if the substrate is thin or transparent enough to allow the transmission of light. Small leaves can be examined over the whole area. Large leaves can be sampled in strips or grids. Algae growing on turtles, mollusks, etc., and stones must be removed for study. One method employs a simple hollow square instrument with a sharpened edge which is pressed closely or driven into the substrate. This separates out a small area of given size around which the periphyton is washed away. The instrument is then raised and the periphyton remaining in the sample square is scraped into a collecting bottle. If stones can be picked from the bottom, then the periphyton can be removed with an apparatus consisting of a polyethylene bottle with the bottom removed and a brush with nylon bristles. A section of the stone is delimited by the neck of the bottle held tightly on the surface. The periphyton is scraped loose by the brush and washed into a collecting bottle with a fine-jet pipette. The periphyton can be counted in a Sedgwich-Rafter cell.

Periphyton can be grown on artificial surfaces, usually glass or transparent slides. In still water the slides can be placed on sand or stones in water. In running water they can be placed in sawcuts on boards, clipped to a rope, attached to a wooden frame, or tucked into rubber corks. In shallow water the slides are usually laid horizontally, directly on the bottom, especially if the influence of light is one of the objectives of the study. Or they can be hung vertically in the water, in which case both sides will be covered with periphyton. Slides placed horizontally collect true periphyton on the upper surface, and heterotrophic organisms on the bottom. The diversity and composition of phytoplankton and periphyton populations are indications of pollution and changes in water quality.

## Sampling Animal Populations

The study of animals is more difficult than the study of vegetation in several ways. Animals are harder to see, they move, they are subject to a higher rate of mortality through the year. For some studies they must be captured and handled temporarily, figure 36, and sampling techniques are based on some questionable assumptions. Basic to the study of animal populations is the estimation of their numbers. Methods of estimating the numbers of animals fall into three categories, a true census, sampling estimates, and indices. A *true census* is a direct count of all individuals in a given area. This is difficult to do for most wild populations. Many territorial species, easily seen and heard, can be located in their specific area. Such a census is used for birds. Direct counts can be made in areas of concentration. Herds of elk and caribou, waterfowl on wintering grounds, and breeding colonies of birds and mammals can be counted from the air or from aerial photographs.

*Sampling estimates* are derived from counts on sample plots. Estimations of populations from sampling involve two basic assumptions: 1) mortality and recruitment during the sampling period are negligible or can be accounted for; 2) all members have an equal probability of being counted, that is the animals are not trap shy or trap addicted, they do not group by sex, age or some other characteristic, and they are distributed randomly through the population if marked and released. The method of sampling must be adapted to the particular species, the time, the place, and the purpose.

Relatively immobile forms such as barnacles and mollusks can be estimated by the quadrat method, similar to that employed for plants. Foliage arthropods may be sampled by a number of strokes with a sweep net over a 10-square-meter area. The number of strokes needed to secure the sample will vary with the type of vegetation and must be predetermined. Estimates of zooplankton, obtained by pulling a plankton net through a given distance of water at several depths can be made by filtering

**Figure 36.** Canadian Geese at a bird sanctuary in Maryland. Photo courtesy of the Bureau of Sport Fisheries and Wildlife, Department of the Interior.

a known volume of sample through a funnel using a filter pump and then counting the individuals in the same manner as phytoplankton.

Widely used is the *mark-recapture* method, often called the *Lincoln Index* and *Peterson Index.* The method is based on trapping banding, or marking, and then later recapturing sample individuals. A known number of marked animals are released in the sampling area. After an appropriate interval of time, a sample of the population is taken. An estimate of the population is computed from the ratio of marked to unmarked individuals:

$$N:T :: n:t$$

or

$$N = \frac{T}{t/n} \text{ or } \frac{nT}{t}$$

where

T = number marked in precensus period

t = number of marked animals trapped in the census period

n = total animals trapped in the census period

N = the population estimate.

The confidence limits at the 95 percent level may be calculated from

$$S.E. = \cdot \frac{T^2 n\,(n-t)}{t^3}$$

There are a number of variations to this basic method. One is to accumulate the captures and recaptures. All animals captured are tagged or marked and released daily. A record is kept of the total animals caught each day, the number of recaptures, and the number of animals newly tagged. The method of calculating the population is the same as that above. However, as T becomes progressively larger, Populations may be calculated daily later in the period, or the trapping period can be divided into subperiods, and the population computed for each subperiod.

The *removal method* or *"catch curves"* is commonly employed in fishery research. It is based on the principle that as the population decreases, the rate of capture decreases. Thus capture is the function of population size. The rate of capture, the ordinate, is plotted against cumulative catch, the abscissa. Eventually the curve will approach the horizontal axis. A graphical or mathematical estimate of the intercept, the point at which the rate of capture becomes zero, will also estimate the size of the population, N.

*Indices* are estimates of the trends of populations from year to year or area to area. They are derived from counts of animal signs, calls, observations and so on.

*Call counts* are used to obtain population trends of certain game birds such as mourning doves and ringnecked pheasants. A predetermined route is established along infrequently traveled country roads. Its length should be no longer than can be covered in an hour's time. Stations are located at set intervals, depending upon the terrain and the species involved. The route is run at a time appropriate for the species concerned, early morning or late evening. At each station the observer stops, listens for a set time, records the calls heard, and goes on to the next station. Routes are run several times and an average taken. The number of calls divided by the number of stops gives a call index.

Similar to call counts are *road counts*. The number of animals observed along the route is recorded and the results divided by the number of miles.

Counts of pellet or fecal groups is used to estimate big game populations. This method involves the counting of pellet groups in sample plots or transects and relating this to the time during which the pellets accumulated. The method is based on the knowledge that deer drop approximately 13 pellets groups per day. The method works best in the northern parts of the country where the cold weather preserves the pellets over winter.

Age structure is another important parameter of the population. Changes in age structure associated with changes in numbers reflect natality, mortality, and provide an insight into the future course of the population. While information on the age structure of wild populations is not easily obtained, a number of aging techniques mostly for game and fish, have been developed over the years. Aging of fish is based on the fact that a fish scale starts as a tiny plate and grows as the fish grows. A number of microscopic ridges, the circuli, are laid down about the center of the scale each year. When the fish are growing well in summer, the ridges are widely spaced. In winter when growth slows down, the ridges are close together. The age of the fish can be determined by counting the number of areas of closed growth rings, the annuli. Since the scale continues to grow throughout the life of the fish, it also provides information on the growth rate. This is obtained by measuring the total radius of the scale, the radius to each year's growth ring, and the total body length of the fish. Then by simple proportion, the yearly growth rate can be determined, figure 37.

Aging techniques for birds and mammals are not as refined as those for fish. Among birds the color and pattern of plumage may be a clue to age, as well as shape and wear of the tail feathers, and the shape of primary wing feathers. The presence or depth of the *bursa of Fabricus*, a blind pouch lying dorsal to the caecum and opening into the cloaca, indicates juvenile birds.

In aging mammals the examination of reproductive organs is useful since most mammals do not breed until the second year. Annual growth in the roots of teeth indicate age in such mammals as deer, bear, fur seals and other canids. Growth rings are also found on the horns of mountain sheep. Deer and elk can be aged by the wear and replacement of teeth. Since the lens of the eye of most mammals grows continuously throughout life, and since there is only slight variation between individuals in lens size and growth, the measurement of the lens is a useful method for aging such mammals as rabbits, raccoon, and black bear. The technique involves the weighing of the dry lens and comparing its weight against a chart of lens weights of known age individuals. The len's growth curve permits a rather close approximation of the age of the mammal. These methods permit the aging of at least some species of mammals, figure 38.

## Remote Sensing

The methods so far mentioned involve the actual handling and measurement of some component of the ecosystem. Of growing importance is a group of

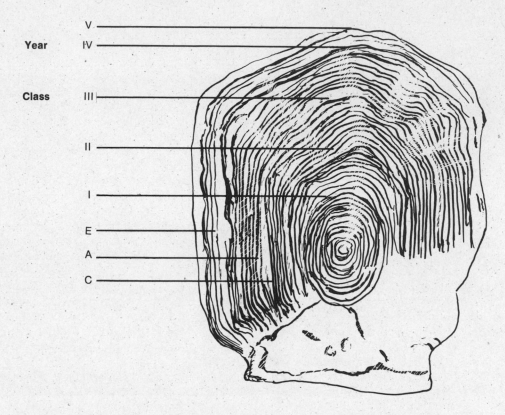

Year

Class

**Figure 37.** Fish can be aged by examining the growth rings on the scale. C is the circuli; A the annuli; E the erosion of the scale from spawning. (From Robert L. Smith, **Ecology and Field Biology,** Harper and Row, Publishers, New York, 1966, used by permission)

techniques that involve ways of learning about ecosystems and populations without direct contact. These methods, involving the use of electromagnetic and sound waves, radiant energy, and ionizing radiation, are known as remote sensing. Remote sensing in many ways is not new. One method, the use of aerial photography, has been employed for some time. Black-and-white stereoscopic photos are used to record and study vegetational and man-made changes on the land, to prepare maps and soil maps, to obtain topographic information, to determine the density and volume of timber stands, to count wild animals and to meet many other objectives, figure 39.

Color photography expands the usefulness of aerial photography, for visual contrasts are more easily interpreted. For example, color photography permits more accurate identification of tree species, the detection of diseases (figure 40) and insect damage, and the study of underwater details such as turbidity, siltation, and algal blooms.

Aerial photographs, both black-and-white and color, are more informative if used along with newer photogrammetric techniques. One is infrared photography which utilizes infrared radiation instead of visible light. Infrared photography is particularly valuable in vegetational studies, since molecules of pigments in plants do not absorb infrared wavelengths. They are either transmitted through the leaf or are strongly reflected by the cell walls. Normal cells have a different reflectivity than abnormal cell walls and the cells of one species have a different reflectivity than those of another. By comparing infrared spectral differences or variations in shading, one can separate out species and detect unhealthy plants. Thus diseases and environmental stresses in plants can be discovered by infrared photography before the symptoms can be detected visually.

Infrared photography also permits the scanning of the environment to detect differences in temperatures. Such pictures or thermographs, in which differences in temperature appear as contrasting bands of color, permit the detection of thermal pollution and other buildups of heat in the environment, figure 40. Since much more information can be derived from a series of different types

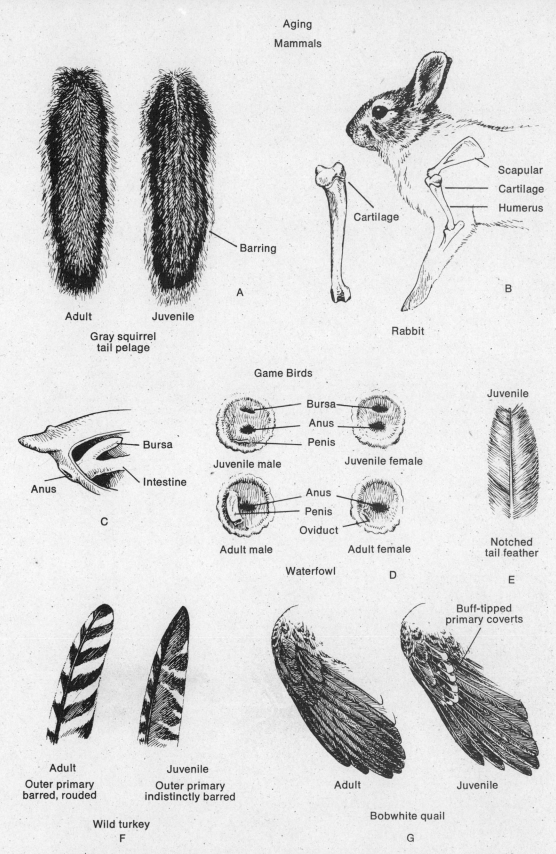

**Figure 38.** Age determination in some game birds and mammals. (a) Regular barring on the underside of the tail distinguishes the juvenile gray squirrel from the adult. (b) Presence of the epiphyseal cartilage on the humerus of the juvenile cottontail rabbit separates that age class from the adult. (c) The *Bursa of Fabricus* (enlarged). Its presence or greater depth indicates a juvenile bird. Depths vary with the species. Most useful with waterfowl and gallinaceous game birds. (d) Sexing and aging waterfowl by examining the cloaca. A bursal opening is present on juvenile waterfowl, absent on adults. (e) Notched tail feather on a juvenile waterfowl. (f) The number X (ten) primary in juvenile gallinaceous birds is sharply pointed; in adult it is rounded. In addition the juvenile wild turkey has indistinct barring on the outer primary. (g) The juvenile bobwhite quail, in addition to possessing a sharp-pointed number X primary also has buff-tipped primary coverts. (From Robert L. Smith, **Ecology and Field Biology,** Harper and Row Publishers, New York, 1966. Used by permission.)

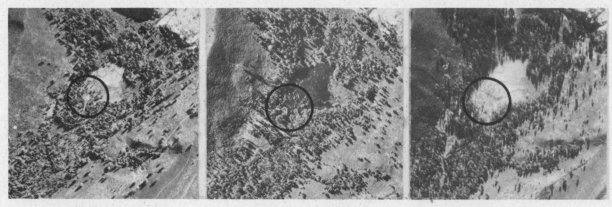

Panchromatic Photo
August 25, 1941

Panchromatic Photo
August 22, 1956

Panchromatic Photo
September 24, 1966

**Figure 39.** The sequential aerial photographs shown were taken, over a time span of 25 years, on a remote region in the Sierra Nevada Mountains. A skilled analyst can detect a stand of dead timber on the 1966 photo. Without initiating an expensive and time consuming field check, he could only speculate on the cause of this damage which might have been induced by a bark beetle infestation, pathogenic attack or lightning fire. However, with the aid of time-lapse photo coverage, he can tell that the timber was healthy in 1941 and that sometime between 1941 and 1966 the damage was inflicted. The 1956 photo shows the timber stand and adjacent meadow in a flooded state which, obviously, has caused the damage. Depending upon his experience, training and familiarity with the area, the analyst could also deduce from the sequential photography that beaver activity caused the flooding in this area and, therefore, the surrounding timber is not in jeopardy. Information of this nature is prerequisite to the more intelligent management of these resources. Thus, a sequential, pictorial record of the timber stand shown here (circled) allows for a skilled image analyst to (1) identify a degenerate timber stand, (2) determine the cause of damage, (3) determine the severity of damage and (4) predict the effect the damage will have on adjacent, healthy trees. Photo courtesy of D. T. Lauer and the Forestry Remote Sensing Laboratory, University of California, Barkeley.

Panchromatic Film
0.4-0.7 micron sensitivity

Infrared Film
0.7-1.0 micron sensitivity

**Figure 40.** Part of the mangrove trees shown in this swamp near Brisbane, Australia have been damaged by mud pumped into the basin. Unhealthy trees at the upper left cannot be distinguished from healthy trees on the panchromatic photograph; however, on the near-infrared photograph a loss in near infrared reflectance, caused by a loss in vigor, renders the unhealthy trees dark in tone as compared to the light toned healthy trees. Photo courtesy of Prof. R. N. Colwell, University of California, Berkeley.

of aerial photographs than from one type alone, a new technique called multispectral sensing, has been developed, figure 41. With this method several or many spectral bands from ultraviolet to infrared are recorded. The visual data is gathered by telemetry or recorded on film or magnetic tape. The image data is fed into a multi-channeled sensor which sorts out the various displays or images. Since all channels were recorded simultaneously, each image is in perfect register. This enables the interpreter to pick out and study an object of interest with different types of images taken at the same instant in time.

Another tool of remote sensing is *radar* which involves microwave frequencies. Conventional radar has been used to study the direction, pattern, movement, and timing of the migration of birds, as well as local concentrations of birds at roosting and feeding areas. But for ecological work the most useful is side-looking radar. Appropriate energy generated in the aircraft is directed a short distance out to the side of the aircraft rather than beneath it and the return signal is captured by the antennae. The result is a continuous strip image that looks like a continuous strip photography taken at very high altitude. Such images provide details of gross land form characteristics, figure 42.

*Biotelemetry* is the gaining and transmission of physiological and logistical information from living organisms and their environments to a remote observer. It has been made possible by miniaturized solid state transistors. The use of telemetry in ecology overlaps space and medical telemetry. The techniques have been adapted from those areas and modified for use in ecological work. Most bio-telemetry in ecology is concerned with following the movements of animals. The animal is fitted with a harness or collar holding the transmitter powered with mercury batteries, figure 43; or for certain studies, the transmitter may be implanted into the animal. The signals are picked up by a receiver with directional antenna, and by the use of triangulation the movements of the animal are plotted on a map. By using highly sophisticated equipment the signals can be fed directly into a computer for continuous monitoring. Recently satellites have been

  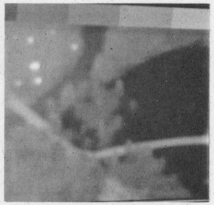

| Panchromatic Photo | Infrared Photo | Thermal Infrared Image |
|---|---|---|
| 0.4-0.7 microns | 0.7-0.9 microns | 3.5-5.5 microns |

**Figure 41.** Multispectral image techniques in aerial photography allow distinctions to be seen which are not visible to the eye or on regular film. The two different plant communities can be seen in the panchromatic photograph, the hardwood trees can be distinguished from the conifers in the near-infrared photograph and the seven small campfires among the trees can be seen in the thermal infrared photograph. Photo courtesy of D. T. Lauer and the Forestry Remote Sensing Laboratory, University of California, Berkeley. Thermal image taken with a Barnes Infrared camera.

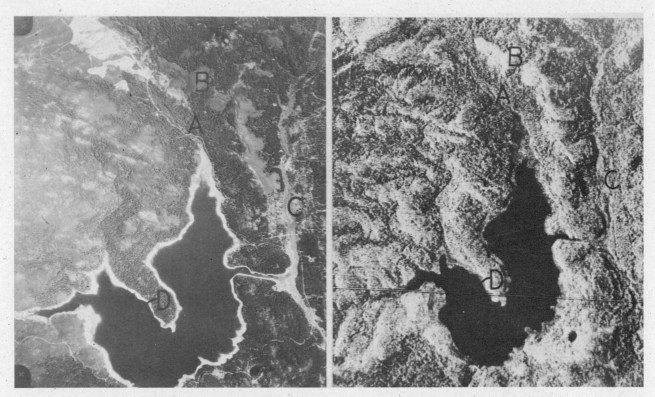

**Figure 42.** The lefthand image shown here is a panchromatic-minus blue photo (0.5-0.7 microns) and the righthand image is a microwave-side looking radar image (1.0-3.0 centimeters). Despite the effects of two limiting factors, poor resolution and influence of topography on an object's tone value seen on the image, useful information can be derived from radar imagery, especially for situations where inclement weather prevents the acquisition of conventional photography. In this example, the following major vegetation types and terrain features can be delineated: timber at "A", brush at "B", meadow at "C" and water at "D". Microwave image made by Aero Space Division of Westinghouse Corporation, Baltimore, Maryland. Photo courtesy of D. T. Lauer and the Forestry Remote Sensing Laboratory, University of California, Berkeley.

**Figure 43.** Instrumented bear with radio transmitter. Photo courtesy of Dr. John J. Craighead, Montana Cooperative Wildlife Research Unit.

employed to track the movements of large mammals such as elk. Such use of biotelemetry permits the study not only of animal movements but also of habitat selection and utilization, predator-prey relationships, mortality, and behavior. But telemetry has even wider uses. It can be used to monitor the climate of microenvironments, to obtain information on physiological parameters of animals such as the heart beat, blood pressure, and temperature of free-ranging animals.

## Systems Ecology

As studies of ecosystems and populations become more and more quantified the need for a more sophisticated approach to the analysis and study of the data has developed. Out of this need is emerging a new field, *systems ecology*. It is the application of systems analysis to ecological problems. Systems analysis is the rendering of selected physical and biological characteristics of ecosystems and populations into sets of mathematical relationships or systems called models. The model is an abstract and simplified representation of the real world situation. Models usually consist of flow diagrams and corresponding mathematical descriptions. Models consist of at least four parts. One is the components or variables of the system such as the producer, consumer, decomposer, mineral elements, or number of animals. The next part is the flow or interaction between components or compartments. Another element is the input into the system, that which affects the system but is not affected by it. And finally there are the necessary mathematical constants. Numbers representing the information are inserted into the model and the model is then subject to mathematical analysis. Widely employed in such analysis are matrix algebra and differential and linear equations. Because of the complexity of the situation in the real world and the lack of sufficiently valid data, some models have no analytical solution. In this case the systems ecologist may resort to simulation of a real world situation using analogue and digital computers. Systems analysis has its widest application in natural resource management where predictive answers are required. The variables such as population size, growth rate of both organisms and population are clear-cut, and the system is clearly defined. The model in this instance is designed to answer specific questions such as optimum yield or rate of harvest. In contrast modelling an ecosystem is difficult since the variables such as energy and nutrients, are not so clearly defined, and the inputs often cannot be measured.

This brief review of methods in ecology may give some insight into the diversity of approaches to the study of ecosystems and populations. More has been omitted than described. Nothing was said about marking and trapping animals, about ways of studying population genetics and behavior. Nothing was written about dendrochronology, the accurate dating of past events through the aging of trees; nor about the methods of studying competitive relations between populations, predator-prey interactions, and species diversity. But enough has been written to suggest the means by which ecologists seek to learn more about the environment and its problems.

Dr. Robert Leo Smith
Division of Forestry
West Virginia University
Morgantown, West Virginia

# Chapter 11

# Biomes

The *Biome* is the largest geographical subdivision of the biosphere, characterized by certain general types of plants and animals, but not by any particular species. The interaction of the regional climate, the soil and various topographic features, and the organisms themselves produce this easily recognizable major biological unit. The number of biomes varies somewhat depending on the person listing them, but the following represents those most commonly accepted: *tundra, northern coniferous forest* (or *taiga*), *temperate deciduous forest, tropical grassland* (or *savanna*). Mountains present a special problem since there is a zonation of organisms from bottom to top in which a number of biome types are represented. High mountains in the tropics may grade from tropical rain forest to tundra. See figure 44 for a map which shows the distribution of these biomes.

The concept of the biome in ecology developed gradually during the first three decades of the twentieth century. The term biome was first used by the plant ecologist, Frederic Clements, at the first meeting of the Ecological Society of America in 1916. This first usage of the term was in a more restricted sense than our present one. It was used by Clements on this occasion as a synonym for biotic community, not for the larger enity which it connotes today. The term biome did not actually come into general usage for some twenty-five years. Early in the century, plant ecologists working on problems of ecological succession and on the classification of plant associations came to think in terms of a unit called the "plant formation," or just "formation," as being the major vegetational unit. It is characterized by the climax type of plant community along with all of its successional stages and is the product of and controlled by the regional climate.

Meanwhile, animal ecologists progressed from a major emphasis on the study of individual organisms to a study of animal communities. The time was ripe in 1929 for animal ecologist Victor Shelford to make a plea at a meeting of the Ecological Society of America for a unified approach to ecology. He introduced a concept which he called bio-ecology and urged that ecologists look at a community of plants and animals as a unit, not as separate enities. Today we have little trouble appreciating this point of view, But Shelford's suggestion did not meet with wide approval immediately.

In 1939, J. R. Carpenter presented an exhaustive treatment of the biome concept at a Cold Spring Harbor Symposium dealing with plant and animal communities. In that same year, Clements and Shelford published *Bio-ecology* in which the biome was used as a major ecological entity—in essence the old plant formation of the plant ecologists plus its animal inhabitants. Since then, the concept of the biome has been generally accepted by most plant and animal ecologists. Fifteen years later, Shelford's student, Eugene Odum, was a major influence in establishing another basic ecological unit, the *ecosystem* (see his detailed explanation of the concept in Chapter Number 7), as the smallest functional unit of the biosphere, made up of plants, animals and their non-living environment. The term ecosystem had been introduced in 1935 by A. G. Tansley, but like the biome concept, it took some time for the ecosystem concept to receive wide acceptance.

As indicated above, a biome is characterized by general plant types, not particular species. In the deciduous forest biome of the Eastern United States, deciduous trees constitute the major climax type of vegetation but different regions have different dominant deciduous trees, for example, beech-maple or oak-hickory. Similarly, grasslands are composed of a number of kinds of grasses, depending on several factors—of which the annual rainfall is probably the most significant. Thus we have tall grass prairie, mixed prairie and short grass prairie.

A biome is not a region where the vegetation is completely uniform. Differences in soil and topography from place to place within the geographic boundaries of the biome result in grassy meadows in the deciduous forest biome and groves of trees in a grassland biome. In any biome, bodies of water such as lakes, ponds, and rivers are normally present.

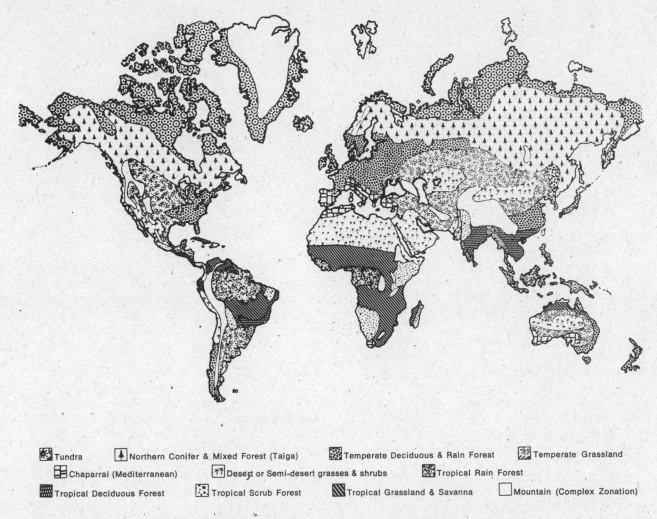

Tundra    Northern Conifer & Mixed Forest (Taiga)    Temperate Deciduous & Rain Forest    Temperate Grassland

Chaparral (Mediterranean)    Desert or Semi-desert grasses & shrubs    Tropical Rain Forest

Tropical Deciduous Forest    Tropical Scrub Forest    Tropical Grassland & Savanna    Mountain (Complex Zonation)

**Figure 44.** Map of the major biomes of the world.

A biome is therefore a composite of many diverse types of biological association (ecosystems)—with one major, easily recognizable dominant form being the most prevalent. Given enough time and ruling out the influence of man, ecological succession theoretically would eventually bring the entire region to a homogeneous state. In our modern society it is unlikely that this will occur on any large scale since man's impact has become a major disturbance factor resulting in retrogression of normal succession. Man is not likely to retire from the field and allow succession to run its normal course.

There is no sharp line of demarkation between biomes. Instead, there is normally an extensive transition zone, which constitutes an "*ecotone*." Although this term is widely used in reference to less extensive areas between communities within a biome, it may be used for any transition zone, irrespective of size.

The ecotone has interesting ecological properties since it has certain attributes of each of the adjacent biomes and in addition has some of its own. Some of the plants and animals of the ecotone come from one biome and some from the other. In addition, there is a third group which normally occur only in the ecotone. Usually the diversity of species is greater in the ecotone than in either of the biomes between which it occurs. This phenomenon is often called the *edge effect* and it can be demonstrated in many situations in which two community types meet.

A look at the map of the world with the various biomes indicated will point up some significant features of the biome concept. Only two of the biomes are even remotely continuous around the globe. The tundra and the northern coniferous forest stretch across the northern hemisphere. The flora and fauna of these two biomes in North America are much more similar to those of their counterparts in Eurasia in terms of actual species make-up than any other biome counterparts. Although temperate deciduous forest biomes are found in Eastern North America, Europe, Central

China and Southeastern Australia, and there are some similarities in plant species, the animal components are in most cases quite dissimilar.

The plants and animals of widely separated biomes of the same type are usually distinctive in terms of the actual species, yet ecologically the similarities are quite striking. The deserts of the Western hemisphere are characterized by the presence of cacti, plants whose characteristics are well-known to all. In the deserts of Africa one finds members of a completely different family, the *Euphorbiaceae,* which are amazingly like our own cacti in appearance and growth form, figure 45.

The bison and the pronghorn antelope were the large herbivores which fed on the prairie grasses of western United States. In the grassland biome of Europe and Asia their counterparts were the wild horse, wild ass and the saga antelope. In the vast savannahs of Africa a whole host of large herbivores are found, including antelope, gazelles, zebras, and even the rhinoceros. In Australia, the kangaroo is the major herbivore in grassland areas.

These plants and animals perform the same role in the ecosystem and are said to fill the same ecological niche. Often the organisms filling the same niche are called *ecological equivalents.* In essence, these ecological equivalents make it possible to divide the earth's surface into the recognizable units which we call biomes.

Although the biome was originally conceived as a terrestrial unit, it is not unreasonable to consider the entire system of oceans as the marine biome. True, there is some variation in the plants and animals from place to place, but there are few real barriers to dispersal and the interaction between the components of the ecosystem are so similar in all areas that the sea has much in common ecologically with terrestrial ecosystems.

### Tundra

The vast portions of North America and Eurasia north of the timberline make up the tundra biome. Perhaps one should speak of two tundra biomes, but the presence of many of the same (or closely related) species in both land masses, and their close proximity at the Bering Straits, make them much more alike than any other biome counterparts.

The tundra landscape is gently rolling, with lakes and bogs in the lowlands. The climate is rigorous, with but a sixty day growing season and with mean monthly temperatures varying between −30°F (−35°C) and 55°F (13°C). Much of the tundra is above the arctic circle and it receives almost continuous sunlight for approximately half of the year. At a latitude of 70° there are 46 weeks with some sunlight and 6 weeks of complete darkness. Precipitation is sparse: between 12 and 20 inches per year, decreasing northward.

During the summer months the soil thaws only to a depth of a few inches in most places. Below this is the permafrost—permanently frozen soil. When the upper part does thaw, the water cannot sink in the permafrost and thus the standing water produces great expanses of shallow ponds and bogs. Mosquitoes and other insects which undergo development in water are often unbelievably abundant during this time.

The vegetation of the tundra gives the impression of a short grass prairie, but closer examination reveals sphagum mosses, lichens (reindeer moss), sedges, rushes, grasses, low shrubs such as bilberry, and some flowering herbaceous plants. Most plants are stunted as compared with related forms in warmer climes, an adaptation which affords them some protection against the drying and abrasion to which they are subjected by blowing sand, sleet and snow. In this stern environment seeds do not readily germinate and most plants are perennials with a good capacity for vegetative reproduction.

A surprising number of birds and mammals are found in the tundra. The caribou of North America and its close relative the reindeer of northern Europe are the most abundant large mammals. The caribou migrates each year to southern portions of the tundra, often moving into the forested regions to the south. Although the caribou herds are small in the summer months, they band together when migrating into herds of several thousand. One migrating herd containing over 20,000 caribou has been reported. Carnivores include the grey wolf, arctic fox, wolverine and polar bear.

The musk-ox, formerly found throughout much of the arctic tundra, is now restricted to North America and Greenland and its numbers are dwindling.

Lemmings are the most abundant mammals and their numbers rise and fall in a very regular three-to-four-year cycle, with enormous populations developing in the peak years. Other small mammals include the arctic hare, the suslik (which is a relative of the ground squirrel), and several kinds of shrews and voles. The suslik is the only truly hibernating mammal of the arctic.

Most tundra animals are white in the winter months, many remaining so all year around. Others acquire a darker color during the short summer period. The changes in coat color, like the instinct

**Figure 45.** Euphorbia morinii, from the parched areas of East Africa filling the same ecological niche as the cactus in America, photo courtesy of Dr. A. B. Graf, **Exotica 3** (1968)

to fly south in the bird population, are triggered by changes in day-length.

Although birds are abundant in the tundra during the short summer, most of them leave before winter arrives. They take advantage of the abundance of food in the form of fish and insects, but most species are ill-adapted to withstand the rigors of an arctic winter. The ptarmigan and the snowy owl are among the few permanent residents of the tundra. Even the snowy owl, which depends largely on the lemming for food, will migrate south when the lemming population ebbs and food is in short supply. In such years, snowy owls are often reported from northern United States.

There is little if any tundra in the antarctic, most of the land surface being permanently covered by ice and snow. On the other hand, the alpine vegetation found at high elevations in many other parts of the world is quite similar to that of the arctic tundra.

### Northern Coniferous Forest or Taiga

Like, the tundra, the northern coniferous forest extends around the northern portion of the globe. It is just south of the tundra and is continuous except for the extreme western portion of Alaska and the extreme eastern part of Siberia. In addition, it extends as far south as Costa Rica at higher elevations of the great mountain chain which runs along the western edge of the continent. The following discussion will deal specifically with the coniferous forest biome of Canada and northern United States, but the general nature of the flora and fauna of the Eurasian taiga and of the forest found at high elevation in the western mountain chain is much the same.

Rainfall in the eastern portion of the biome ranges from 20 to 50 inches a year and from 10 to 20 inches in the west. Most of the precipitation comes in the summer months but there is some in all seasons. Mean monthly temperatures vary from a low of −30°F. (−34°C.) to a high of 73°F. (23°C.). There is permafrost under the forest in its northernmost portion. Because of the width of the biome, there is a great spread in the length of the growing period: from 60 days in the north to 150 days in the south.

There are few mountains and they are not of high elevation. The forest is found at elevations ranging from sea level to 1200 feet, except at the northern edge of the Appalachians where it is found up to about 2500 feet. As a result of the glaciation of much of the biome, many lakes are present.

In some regions the surface area of the lakes approximately equals that of the land.

Dominant trees of the forest are balsam fir, white spruce, black spruce, and red spruce. Along the banks of streams and rivers, tamarack, willow, paper birch and poplar trees are found. The dominant evergreens are well adapted for the stern climate in which they live. Their leaves (needles) prevent excessive loss of water during winter and dry periods, and they can also withstand freezing temperatures. Their branches are flexible and thus can hold heavy snow without breaking. The dead needles remain on the tree for some time with the result that huge crown fires, started by lightning, are common. Over fifty species of insects, including the spruce budworm, the pine sawfly and the tent caterpillar, are a serious problem, often destroying large areas of forest.

The large mammals of the biome are the moose and the woodland caribou, which was formerly more abundant than at present. Carnivores include the grey wolf, lynx, wolverine, ermine, fisher and marten. Black bear and porcupine are common, the former being an omnivore and the latter, feeding on bark, often doing serious damage by girdling trees. The snowshoe rabbit also does considerable damage to young trees. Like the lemming of the tundra, the snowshoe rabbit has a regular population cycle, this one of nine to ten years. The snowshoe rabbit is the main food of the lynx, and lynx populations follow the same cyclic pattern as the rabbit, with about a year's lag. This classic cycle has been traced back to the 18th century by a study of the number of lynx pelts purchased each year by the Hudson's Bay Company.

Other small mammals found throughout the biome are flying squirrels and red squirrels, chipmunks, shrews and mice.

Birds are abundant, with the spruce grouse, great horned owl, pine siskin, red crossbill, and black-backed three-toed woodpecker being characteristic. Many warblers breed in the forest but migrate south for the winter.

Economically, the northern coniferous forest is of great importance as the source of much of our lumber and of our wood for paper pulp as well.

### Temperate Deciduous Forest

Temperate deciduous forests originally covered eastern North America, Europe, Southern Japan, and Eastern China. Many of the same genera

**Figure 46.** Eastern Deciduous Forest Pennsylvania, West Virginia. Photo courtesy U.S. Forest Service.

(though not the same species) of both plants and animals are found in all of these widely separated regions, indicating a former geographic connection. The temperate deciduous forest biome of eastern United States will be used as an example.

Since the biome extends from the center of the Great Lakes region south to the Gulf of Mexico, there is considerable spread in the climatic features. Precipitation ranges from 30 to 60 inches a year, being highest in the Gulf States. It is rather evenly distributed over the year. Mean monthly temperatures from north to south vary from lows in January of 10° F. to 60° F. (−12° C. to 15° C.) to highs of 70° F. to 80° F. (21° C. to 27° C.). The growing season varies from 150 days in the north to 300 days in the south.

The major characteristic of the biome is its climax forest dominated by broad-leaved trees which shed their leaves each fall. Below the dense forest canopy are several understories of smaller trees and shrubs which are also deciduous. The shrubs are often sparse in the dense shade of the deep forest, but are quite evident at the forest edge and in clearings within the forest resulting from windfallen trees. In the spring, before the trees leaf out fully and cut off the light, a wide variety of flowering herbaceous plants are characteristic.

In the far south, the deciduous trees are supplanted by broad-leafed evergreens such as magnolia and live oak. Along the southern coastal plain, the southern pine forest may seem like an anomaly within the eastern deciduous forest biome. However, this forest type would be supplanted by deciduous trees if it were not for the periodic fires which prevent normal succession from occurring.

Because of the great area covered by the biome, there are important regional differences in climate as well as differences in soil and topography. As a result, there are many subdivisions of the biome, each with a distinctive climax forest type. Among the recognized types are: the beech-maple forest of the north-central region; the maple-basswood forest of Minnesota and Wisconsin; the oak-hickory forest of Missouri, Arkansas, and southern Illinois, as well as parts of the Gulf and South Atlantic states; and the rich, dense, mixed-mesophytic forest of the unglaciated Appalachian plateau, with a large number of dominant tree species including white basswood, yellow buckeye, and tulip poplar.

When the white man arrived on the scene, the eastern deciduous forest was essentially intact, covering most of the area, figure 46. In the ensuing years, man has almost completely destroyed the original forest for agricultural purposes or seriously modified it by logging. It is estimated that not over 0.1 percent of the original forested area has been immune from at least partial cutting.

Many of the original native animals are now either extinct or essentially absent from the biome. Those which are extinct include the passenger pigeon—which was present by the hundreds of millions, the colorful Carolina parakeet, and the eastern bison. Formerly present, but now existing in other areas, were the wolf, mountain lion, and the wapiti, or American elk.

The white-tailed deer and wild turkey were originally very characteristic animals throughout the biome and they, along with the black bear, are commonly present in many areas which have been allowed to return to their original forested state. Other important mammals of the biome are the gray fox, striped skunk, raccoon, bob cat, opossum, gray squirrel, southern flying squirrel, mice, moles and shrews. Many of the smaller birds such as cuckoos, flycatchers, and vireos are not permanent residents. Among those which are year-'round residents we find the tufted titmouse, several species of woodpeckers, white breasted nuthatch, two kinds of owls, and several hawks.

Reptiles of the biome include the box turtle, garter snake, timber rattlesnake, and copperhead. Amphibians are represented by several kinds of salamanders and frogs, figure 47.

Invertebrates are abundant, including land snails and slugs, countless species of insects and spiders, and a variety of millipeds.

## Temperate Grassland or Prairie

Temperate grasslands cover vast areas of North America (prairie and plains), Eurasia (steppe), and South America (pampas). In almost all of these areas man has replaced the original vegetation with cultivated grasses such as wheat and corn or has introduced domestic cattle which have greatly modified the plant cover. Until late in the last century, most of these areas still retained their native vegetation, which supported great herds of large herbivorous mammals such as the bison of North America.

All of the temperate grasslands are remarkably similar in climate, topography, plant and animal types. Perhaps it was this feature which led Frederic Clements, a student of the American grasslands, to develop the biome concept. In the following discussion the North American temperate grasslands biome will be used as the example. The descriptions are largely based on the condi-

**Figure 47**

111

tions which existed prior to the invasion of the area by the white man.

The North American Temperate Grassland biome reaches from western Indiana to California and from Texas to Alberta and Saskatchewan. In the east and central part of the grassland, this vast area is continuous—with the exception of strips of forested land in the river valleys—and the terrain is either flat or gently rolling. In the west, the great mountain ranges interrupt the continuity of the grassland, but large expanses still occur in lower elevations between the ranges.

The wide spread from north to south provides a tremendous variation in temperature. In the north, the low mean monthly temperature is 50° F. (−15° C.) and in the south, the summer temperatures often exceed 90° F. (32° C.). The growing season in the north may be as short as 100 days, while in the south, frost rarely if ever occurs.

The controlling climatic feature for the temperate grasslands, however, is precipitation. Rainfall in the eastern part of the grassland biome may be as much as 40 inches a year and decreases in a rather regular gradient moving westward, to a low of about 10 inches. In general, the high rainfall in the east would be sufficient to support forests, but the spread of trees into the grassland is prevented by high evaporation rates, periodic severe and prolonged droughts, and formerly by intermittent fires. The latter kill any tree seedlings but do little harm to the grasses. In fact, it is generally believed that periodic burning is beneficial to the maintenance of a healthy grassland.

The moisture gradient from east to west was responsible for a variance in the type of native grassland that existed prior to cultivation. In the eastern part, the grasses were tall (5 to 10 feet), grading to mid-height grasses (1.5 to 5 feet), and finally to short grasses (less than 1.5 feet) in the western part. The tall and mid-height grasses (the latter usually called mixed grasses) made up the prairie while the region of short grasses was called the plains.

The major grasses of the tall- and mixed-grass prairie were blue stem, wheat grass, wild rye, June grass, prairie grass, blue grass, needle grass and drop seed. The important short grasses were grama grass, buffalo grass and triple-awned grass. In addition to the dominant grasses, there were other herbaceous plants such as sedges.

The most characteristic animals were the bison and the pronghorn antelope, each of whose populations are estimated to have been about 45,000,000 in 1600. Although the bison was found throughout the biome, the pronghorn occurred only as far east as western Missouri. These two grazing animals were important factors in determining the nature of the vegetation and in the actual maintenance of the grassland system. Jack rabbits were also important in the mixed- and short-grass regions, as were the prairie dogs, while the cottontail rabbit was found in the long-grass portions. Pocket gophers, ground squirrels, several kinds of rats and a number of species of mice rounded out the mammalian herbivores. They were all preyed upon by the gray or buffalo wolf, coyote, swift fox, weasel, badger and ferret. Several kinds of hawks were important predators also, including the rough-legged hawk and the ferruginous hawk.

The prairie chicken, a grassland relative of the ruffed grouse, was a very characteristic bird. Others were the meadowlark, horned lark, and dickcissel. Reptiles were represented by the prairie rattlesnake, blue racer, bull snake, and garter snake. Insects were quite abundant, with many species of grasshopper, many true bugs, leaf beetles, and ants, all of which fed upon the grasses. In the winter, animals were almost completely absent above ground. The bison moved south while the pronghorn moved to the west and south, mainly to the pine forests of the mountain foothills. Most of the birds also moved south, and the reptiles and small mammals hibernated in dens or in burrows beneath the surface of the ground.

**Tropical Rain Forest**

Tropical rain forests are found at low elevations in equatorial regions with an annual rainfall greater than 80 inches. The three regions in which extensive rain forests occur are: Central and South America, where the vast, dense forests of the Amazon and Orinoco basins form the world's largest continuous rain forest; Central and West Africa, in the basins of the Zambesi, Congo, and Niger, as well as on the island of Madagascar to the east; and discontinuously on the Indian subcontinent, on the Malay Peninsula, south and east into the East Indies, the Philippines, New Guinea, and a small section in northern Australia. Although the plants and animals differ, these biomes are all quite similar.

There is little if any variation in mean temperature from month to month, and daily high temperatures are in the 85-95° F. range (30-35° C.) and daily lows are from 65-80° F. (18-27° C.). Actually, the dense forest moderates the temperatures so that within the forest itself the highs are not much above 82° F. (29° C.) and lows are about 70° F. (21° C.). Rainfall is over 80 inches a year, in some places reaching well over 400 inches. There is always some rainfall each month, but usually there will be one or two so-called dry seasons in which less than five inches of rain fall each month. Many plants grow continuously, and there is no time when there are not some trees in flower, some in fruit, and others in the process of shedding their leaves.

Although stratification is found in temperate deciduous forests and other forest types, it reaches a peak of development in the tropical rain forest, figure 48. On Barro Colorado Island, Panama Canal Zone, there are eight recognizable strata: 1. the air above the forest; 2. tall trees which emerge from the canopy, over 125 feet tall; 3. the upper forest canopy, a continuous green layer 75-100 feet above the ground; 4. the second story trees, 40-60 feet tall; 5. small trees 20-30 feet tall; 6. shrubs to 10 feet; 7. the forest floor; and 8. the subterranean stratum, figure 48.

Only minimal amounts of light penetrate the dense canopy and as a consequence there is little or no herbaceous growth on the forest floor, except in the clearings due to fallen trees and at the forest edge. Great vines or lianas twine around the trunks of the forest giants, growing upward into the light. The strangler fig starts out as such a vine, but it eventually surrounds its supporting tree, develops sturdy, woody roots, stems, and branches, and finally replaces the tree on which it had started to grow.

Large numbers of other kinds of plants called epiphytes are found on the trunks and branches of the great trees. The lovely orchid is such a plant, found in both the old- and new-world tropics. In the new world, many of the epiphytes are members of the pineapple family, the *Bromeliaceae.* Significant amounts of water are normally present in the spaces at the bases of their leaves and many small aquatic organisms live in these tiny isolated "ponds" high in the air. In at least one species of tree frog, eggs are laid and early development takes place in this unusual environment.

Probably the most unique feature of the tropical rain forest is the great diversity of species of both plants and animals. In the most diverse types of temperate deciduous forest one finds a maximum of 10 to 15 species of trees, and extensive natural stands of a single species are not uncommon. In contrast, over 500 species of trees and 800 species of smaller woody plants have been described in the rain forest of equatorial Africa.

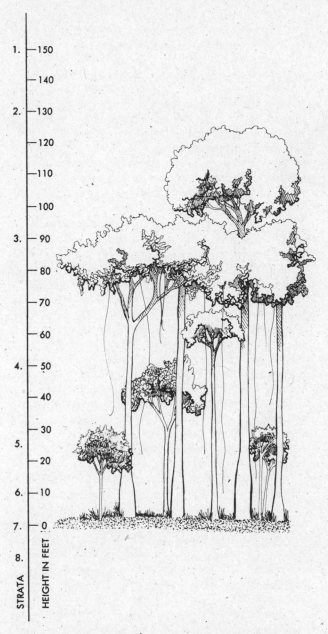

**Figure 48.** Stratification in the Panamanian rain forest.

In the rain forest there may be miles between the closest specimens of the same species of the larger trees. One can find more species of trees in a square mile than in all of Europe.

Diversity of species is also found in the fauna of the rain forest. A 6½ square mile island in the Panama Canal Zone, Barro Colorado Island, is the site of a tropical research station and its fauna is better known than that of most rain forests. Over 20,000 different species of insects have been found there, many times the number of species found in temperate forests of much larger size.

113

A visitor to the rain forest is usually impressed by a seeming absence of animal life. In large part this is due to the fact that the majority of animals are arboreal, spending much if not all of their time high in the canopy, feeding on flowers, fruits, leaves, and insects. Among the important arboreal mammals in the American rain forest are the howler monkeys, spider monkeys, capuchin monkeys, and the squirrel marmoset. Prehensile-tailed porcupines and a relative of the raccoon, the kinkajou, which also has a prehensile tail, are common. The sloth is so well adapted for aboreal life that it is quite helpless on the ground. It spends its life hanging upside down from the branches, lazily feeding on the abundant supply of leaves. Bats abound, with some fruit eaters, some insect eaters, and even one species which feeds on fish. Over 100 species of birds inhabit the canopy of Barro Colorado Island, including the fruit-eating toucan with its outlandish beak, parrots, parakeets, and trogons. Insect feeders such as woodpeckers find plenty of food in the tree tops and on the upper trunks. A number of arboreal or tree dwelling geckos and lizards are also present, including the giant iguana, whose appearance reminds one of its long-dead dinosaur ancestors. Some animals are equally at home either on the ground or in the trees. In this category are the coati, ocelot, opossum, some species of squirrels, many lizards and snakes.

The forest floor has its own assemblage of inhabitants as well. Mammals include the puma, the tapir (which also spends some of its time in the water), peccaries, agoutis, rats and mice. Ground-dwelling birds include the greater tinamou and the great curassow. Numerous snakes and lizards are also found in the American rain forest, including such poisonous species as the bushmaster, *fer-de-lance* and coral snake.

Insects are abundant at all levels of the forest, with ants and termites probably the most numerous. The army ants are very characteristic, living in enormous colonies which construct a nest of their own bodies, from which raiding parties move out over the forest floor in search of insects and other soft-bodied invertebrates upon which these carnivores feed. The leaf-cutting ants are also plentiful. Long columns of ants in single file, each carrying a freshly-cut piece of leaf many times larger than itself, are a familiar sight along the forest floor. The head of the column suddenly disappears as the ants move into their underground nest. The leaves are used as a mulch in the cultivation of fungus gardens which provide the colony with its food.

As a result of the constant high humidity, there are species of leeches and planarians, normally aquatic organisms, which lead a terrestrial existence in the rain forest. Peripatus, a worm-like animal with features of both annelid worms and arthropods is found only in the warm, humid environment of the rain forest. The relative silence within the rain forest during the day is usually noticed by those who visit it for the first time. If they remain until nightfall, which occurs quite abruptly in the tropics, there is a sudden burst of sound from howler monkeys, many birds, and above all, insects and tree frogs. The insects and frogs keep up their loud chorus for much of the night, while the birds and monkeys soon settle down to rest until just before dawn, when they join in the finale.

## Tropical Grassland or Savanna

Extensive savanna biomes are found to the north and south of the tropical rain forest biomes of South America and Africa, and there is a strip of savanna across the northern part of Australia. The most extensive savanna biome is in Africa and the following account will deal largely with that, more specifically with the savanna which straddles the equator in East Africa.

The African savanna extends north to the Sahara Desert and south almost to the southeastern tip of the continent. It is continuous from north to south by means of a strip between the great equatorial rain forest in the west and the semi-arid scrub or thorn forest which spans the equator in the east. Rainfall is from 30 to 50 inches a year, but there are prolonged dry seasons between the rainy ones. Temperatures range from highs in mid afternoon of 85-95° F. (30-35° C.) to lows at night of 65-75° F. (18-24° C.). At higher elevations, these figures are of course lower. As in the tropical rain forest biome, there is more variation from day to night than from month to month.

The savanna differs from the temperate grassland in the presence of scattered trees, either singly or in clumps, figure 49, along with the characteristic grasses and sedges. All of these plants have in common a great resistance to fire and drought. As in the temperate grasslands, periodic fires have been important factors in determining the nature of the vegetation. In the absence of fire, other types of trees and shrubs would have invaded and greatly modified the system. The grasses, some of which belong to the same genera that make up the American prairie and plains, provide the dominant vegetation of the savanna.

**Figure 49.** African Savanna, showing the grassland with scattered trees and one of the most important carnivores.

The most characteristic tree is the flat-topped acacia, but almost as prominent are the large euphorbia trees, the ecological equivalent of our American cacti. Other trees include the squat, massive-trunked baobab tree (representatives of which are the oldest known living trees on earth today), and the sausage tree—whose great fruits, three feet long and six inches in diameter, hanging by long slender stalks, prove the common name of the tree to be quite appropriate, figure 50. Several species of palm, including the stately borassus palm, also occur.

In many cases the clumps of trees grow around old termite mounds. These mounts, often four or five feet high, are scattered in a fairly regular fashion over the savanna. It seems probable that the termites, which are fungus-growers and carry large amounts of dead vegetation into their underground galleries for mulch, greatly enrich the soil in the area near the nest and thus provide a favorable place for the trees to grow.

Nowhere in the world is there anything to compare with the varied population of large herbivorous mammals which evolved in the favorable environment of the African savanna. The more common large animals which feed chiefly or entirely on grasses are the wildebeeste (or gnu), zebra, topi, African buffalo, figure 51, impala,

Grant's gazelle, Thompson's gazelle, hartebeeste, eland, wart hog, figure 52, elephant, figure 53, rhinoceros, figure 54, and hippopotamus. The latter spends its days in the water, but comes out on land at night to feed on grasses. Browsers (animals which feed on leaves and twigs) include the giraffe, figure 55, and the gerenuk.

Just as the American bison and the pronghorn antelope migrated south and west to escape the rigors of winter, so do the great herds of African herbivores migrate from place to place with changes in rainfall and resultant changes in the amount of grass available.

Associated with the great herbivore populations we find a rather large number of carnivores and scavengers in the mammal population, including the lion, figure 56, leopard, cheetah, figure 57, hyena, jackal and wild dog.

In Tanzania, east Africa, most of which is savannah, there are over 280 species of land mammals, probably more than in any other area of comparable size in the world. Included in this number are 89 small herbivores which feed on grasses and their seeds. Fifty-seven of these are rats and mice and 10 species are squirrels. Forty-five large herbivores, 37 of which are antelopes, are found. Ninety-one mammals are insect-eaters, including 56 species of bats and 19 shrews. Many

**Figure 50.** The "sausage tree" of the African Savanna.

**Figure 51.** The Cape Buffalo, one of the large herbivores of the Savanna.

**Figure 52.** The Wart Hog.

**Figure 53.** Elephant standing near a tree clump.

**Figure 54.** Rhinoceros, one of the many species threatened with extinction.

**Figure 55.** A group of Giraffe browsing on tree leaves.

**Figure 56.** The Lion, the chief carnivore of the Savanna.

**Figure 57.** The Cheetah sitting near its kill, the fastest carnivore of East Africa.

of them feed on a variety of insects, but the peculiar armor-plated pangolin feeds only on termites. Forty-one species are carnivorous, including many which scavenge upon the leavings of those which track and kill their own prey. Eleven species of bats live on fruits as do some of the ten species of monkeys. The abundant baboons are primarily flesh-eaters.

Birds are very abundant in the savanna, feeding on the plentiful seeds and insects as well as preying on other inhabitants or living as scavengers. Several species of storks, ostriches, bustards of several kinds, guinea-fowl, many species of weaver birds (which live in intricate nests of woven grasses), and the colorful sun-birds are a few examples of the non-predatory birds. Predators include several eagles, the African kite and many kinds of hawks. The white-backed vulture and a number of other species are to be found wherever there is carrion.

Reptiles are well-represented in the savanna. Many, like the crocodile and monitor lizard, are found in and near the rivers which flow through the region. Among terrestrial reptiles, the African chameleon is one of the best-adapted of animals, with several striking characteristics in addition to its well known ability to change color. The digits of its feet enable them to grasp a branch or twig quite firmly because two of them are directly opposite the other three. With its long, sticky tongue, the chameleon can pick off an insect at a distance almost as great as its entire body length. The most remarkable adaptation is in the eyes, which are located on top of conical, movable, turret-like structures. These eyes can be pointed in any direction in a full 360 degree arc, and they operate independently, making it possible for one eye to look forward while the other keeps tabs on the situation to the rear.

Other reptiles include numerous other kinds of lizards, such as the colorful agama lizard, also skinks, geckos and numerous snakes. The large reticulated python often inhabits abandoned termite nests. Poisonous snakes include the puff-adder, gaboon viper, and a number of kinds of cobras.

Of all the natural wonders of the modern biological world, the East African savanna is probably the least spoiled. Great areas have been set aside as national parks by the governments of Kenya, Uganda and Tanzania. Similar reserves are also found in other parts of Africa.

Great pressure is being exerted in some if not in all of these places to allow the introduction of domestic cattle into this rich pastureland. Previous experience proves that such a move would result in drastic changes in the vegetation and endanger the very existence of many of the magnificent native animals.

One can only hope that these pressures will be firmly resisted by the governments of the developing nations. The animals evolved along with the grassland, and together the plants and animals are able to coexist in a self-sustaining, fantastically productive ecosystem. Many ecologists believe that it is more economically feasible to harvest the native grazers such as the hartebeeste, wildebeeste and topi—or even to domesticate them than to drive them out and attempt to substitute animals which are not native to the environment.

## Chaparral or Mediterranean Biomes

Chaparral biomes on the North American continent are found inland from the coastal plain of central and southern California extending into Baja California, and in scattered areas on north to the Oregon border. The shores of the Mediterranean (except for the eastern part of the north coast of Africa), including much of Spain and Italy, are also chaparral, the local name for which is maqui. The third major chaparral biome is along the south coast of Australia, where it is called the mellee scrub. The following discussion will deal with the southern California chaparral biome.

As in the grassland, the rainfall—particularly the pattern of its fall during the year—is the most important climatic factor. Rainfall ranges from 20–30 inches a year, usually closer to 20. Most of the rain falls between November and May. For example, in one area the annual rainfall is 21.7 inches but only 0.5 to 1.5 inches fall in June, July, August, and September, with most of that falling in September. The average frost-free period is over 240 days, with many parts almost never experiencing frost. High temperatures in the nineties are common in the summer time. Along with the rainfall pattern, fires are important in determining the nature of the chaparral biome. The plants are resistant to fire damage and many are actually aided by periodic burning, through stimulation of seed germination by the fire's heat and consequent production of many new sprouts. The long, hot, dry summer season produces conditions which are conducive to devastating fires—as many residents of southern California who built homes in the chaparral have discovered to their sorrow.

Plants of the chaparral include chamise, Christmasberry, scrub oak, leather oak, interior live oak, manzanita, buckbush, mountain mahogany, and redberry. The trees and shrubs are characterized by having hard, thick, evergreen leaves. Many of the trees are quite stunted and have a shrub-like growth form. Seldom do they reach a height of over eight feet.

Mammals of the chaparral formerly included the grizzly bear and the puma. Today these are absent, but many other mammals still abound, including the mule deer, bobcat, grey fox, coyote, wood rats, and several species of skunks. Brush rabbits and chipmunks, as well as several kinds of mice, are also found. The most common birds of the dry season are the common bushtit, the rufous-sided towhee and the wren tit. The rainy season sees the appearance of several sparrows, hermit thrushes, ruby-crowned kinglets, and many warblers.

A number of lizards and snakes, including several rattlesnake species, are found in the chaparral, but none are limited to this biome. Most of them are found in adjacent biomes, particularly the more arid ones. The dry nature of the climate precludes the presence of many amphibia, but there is a tree frog which breeds in temporary ponds resulting from heavy-early-winter rains.

Insects are common, including ants, leafhoppers, grasshoppers, beetles, and flies. Millipeds and centipedes are sparse, living under rocks. Many of these invertebrates undergo a period of dormancy comparable with hibernation during the dry season—a condition termed aestivation.

Prof. Eliot C. Williams, Ph.D.
Biology Department,
Wabash College
Crawfordsville, Indiana.

# Chapter 12

# The Desert Biome

To most people, the word 'desert' conjures up visions of a barren, baking-hot expanse of sand dunes. But deserts are not always hot, are often rich in life-forms and are not necessarily sand dunes. In fact in the United States although deserts abound, dunes are rare enough so that any significant stretch of sand dunes is worthy of becoming a protected park or monument and frequently is.

What then is a desert? A desert is a place which has a sparse and undependable supply of moisture. One measure of sparseness is simple: it is the average annual rainfall. A typical area in the cool temperate east or midwestern United States might have 40 inches of precipitation per year. Deserts rarely average more than 10 inches, often considerably less. The driest place on the Earth is the Atacama desert of Chile with an average annual rainfall of less than two inches.

Compare that with the Olympic peninsula of Washington where 100 inches per year is commonplace or with Mt. Waialeale in Hawaii, the world's rainiest place, which averages 486 inches per year!

Although the general sparseness of rainfall is one cause of deserts, annual precipitation averages often give an overly optimistic picture. Much of the rain in some deserts falls as violent, heavy thundershowers. Most of such rain is useless or actually harmful to plant and animal life. Not only does it fall too quickly to soak into the ground, but in running off so rapidly it carries with it much soil and causes flash floods miles away. Were it not eroded away, the soil would help to store water and ease the severe drought conditions. The constant threat of flash floods during the thundershower season keeps the experienced desert dweller out of the small dry stream beds called arroyos, as much as possible. In the desert the two most dangerous things are water, too much or too little.

Furthermore, the average rainfall of a desert cannot be relied on for every year. Often twice as much falls and just as often there is only half as much as other years. Even larger departures from average can be expected at certain times and places. This great fickleness forces most life to be adapted to conditions far worse than found elsewhere on earth. It has also produced many highly opportunistic species which are capable of taking advantage of moist times whenever they arrive but can also retreat and wait out the more commonly protracted droughts.

Some places get very little rainfall and yet are not deserts. They are so cold that however little water does fall, it is more than sufficient to replace the even smaller amounts that evaporate. Tundra is usually like this; its soil is generally saturated with moisture. True deserts have large moisture deficits; there is enough warm sunshine so that much more water could evaporate than falls as rain, figure 58. Scanty water supplies are generally caused by the circulation of large masses of air on a global scale. In the tropics, warm air, full of moisture, rises and is carried away from the equator. As it travels it produces rain until by the time it reaches about 30° Latitude, it is dry. There it falls and leaves a desert. It then travels back along the surface picking up moisture along the way. It is no accident that the world's deserts tend to cluster along the two 30th parallels, North and South of the equator. Lack of rain may also be caused by a more local phenomenon, the rain shadow. If an area is to the lee of a large mountain range, prevailing winds will force air up and over the peaks in order to reach it. As the air is blown up the mountains, it is cooled (adiabatic cooling) and it cannot hold its moisture. Its water falls on the mountains leaving the next encountered lowlands in a dry rain shadow. The extensive great basin desert of the northwestern USA is a rain shadow desert, figure 59. It is shielded from moisture by the Sierra Nevada Mountains and the Cascade Range. The high plains east of the Rocky Mountains are also in a rain shadow and are consequently rather arid.

Deserts are also characterized by great variations in temperature. They are very hot during the day and cool off rapidly after sunset. The extreme aridity is responsible; the dry, cloudless air provides neither insulation from the sun's radiant energy nor from reradiational loss of the energy from the ground after the sun is set. Dawn and darkness come very rapidly with almost no twilight. Differences of 30° and 40°F between the daily

**Figure 58.** A mountain in the Sinai Desert not far from the southern end of the Dead Sea, one of the most desolate locations on earth.

**Figure 59.** A mountainous area forces the prevailing rain clouds to loose their moisture and renders the lee side a parched desert. Thus is formed a classic rain shadow.

maximum temperature and its minumium are commonplace; no cold front or other major change in weather is required to produce then. A strong cold front coming after a warm morning can plunge the temperature 60° or more by the following dawn. It would seem quite likely that such rigorous climatic conditions would preclude the survival of any life but incredibly, they do not. Many plants and animals have succeeded in adapting to the aridity and unpredictability of the desert. The desert biome is therefore quite rich and diverse in interesting life-forms. Most survive either by avoiding the desert's harsh times or by having special structures or chemistrys which allow them to withstand the desert's full force. Some combine both methods in one life system. The basic problem of the desert, aridity, is solved in many ways.

Plants in moisture places have thin leaves which would lose water rapidly in a desert because of their large surface area. To solve this, cacti simply lack leaves. They carry on photosynthesis in their stems, called joints. The stems are fleshy and present only a very small surface area to the sun (compared to their bulk). Creosote bushes on the other hand always have leaves, but they are coated with a waxy substance which practically eliminates water loss.

Instead of having no leaves, some plants have big fleshy leaves that achieve the proportions of cactus joints. Prime examples of this adaptation are the century plants, *Agave*, figure 60, and the stonecrops such as *Sedum*, figure 61, and the

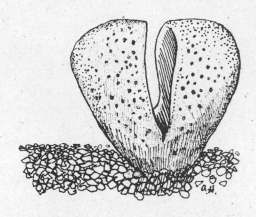

**Figure 61.** Stonecrop, a type of Sedum.

common houseplant *Crassula* (jadeplant). The leaves of *Agave* are so fleshy that in Mexico they are harvested and fermented to produce the hard liquor, tequila.

Last there are the escapees. *Ocotillo* produces leaves after every period of significant rainfall, then drops them as environmental conditions deteriorate. Magnificent little desert wildflowers appear soon after those rare periods of soaking rain. They blossom rapidly, set seed and die all within a few weeks. The appearance of their progeny may take years because each seed contains special chemicals which inhibit germination and may only be washed away by a rain of sufficient duration and intensity. Such plants are known as desert *ephemerals*, from the Greek, "only a day."

Of course some plants appear to be "dissatisfied" with only one adaptation. Palo verde (a tree related to locusts) has green stems for photosynthesis. It also grows tiny, slightly fleshy leaflets, but only after sufficient rain. Furthermore, its seeds are encased in a very hard seed coat and they do not germinate until scarified—perhaps by being ground between two rocks in a flash flood. In any case its seeds may remain dormant and viable for many years thus giving their parents many chances for at least a few seeds to win at the game of blindly picking a good time and place to germinate.

The fleshy photosynthetic structures that are so common in desert plants have evolved many times in response to this environment. Cacti whose splendid flowers, figure 62, belie their relationship to roses, are native to the Americas. (All or all but one cactus growing elsewhere was brought there by man.) But the African deserts need no cacti. There, members of the large, weedy family

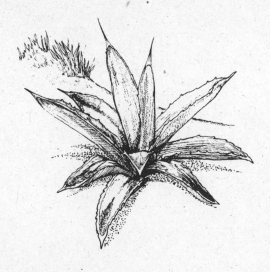

**Figure 60.** Agave, or century plant.

**Figure 62.** Although rain is scarce in the desert, when a shower does fall many types of cacti put forth very beautiful flowers.

**Figure 64.** Euphorbia valida, from the parched areas of South Africa, photo courtesy of Dr. A. B. Graf, *Exotica 3* (1968).

**Figure 63.** Euphorbias growing in the mountains of Ethiopia, and south to Cape Town, filling the same ecological niche as the cacti of the American desert.

Euphorbiaceae display their modest, weedy *Euphorbia* flowers on leafless, spiny, cactus-like stems, figure 63, figure 64. There also, *Aloes,* appearing much like *Agaves,* produce their rosettes of stout, succulent leaves. These are examples of convergent evolution: species that were once dissimilar have been subjected to similar environmental pressures and have become at least superficially, quite similar.

The desert abounds in examples of convergence because of the great pressure it exerts on its denizens. Sometimes convergences are far from "skin-deep." There is a highly useful biochemical specialty possessed by the stonecrops called crassulacean acid metabolism. These plants possess chemicals which allow them to store carbon dioxide. Some cacti are known to have the same ability. Carbon dioxide storage is another weapon in the plant's struggle to save water.

In order to photosynthesize, a plant needs sunlight, water and carbon dioxide. Thus it must produce sugar during the day when the sun is out. But if it were to open its stomata (pores) during the hot, dry day, in order to get its carbon dioxide, much water would evaporate from it. Instead it opens its stomata during the cooler, more humid night when water loss is minimal. It freshens its supply of $CO_2$ and meets the sun of the new day as a water-tight sealed unit.

It is certainly useless to conserve water unless it is first obtained. In order to obtain it some desert plants, especially those of tree form like the locust relatives (e.g. mesquite), spend much of their youth growing downward. Enormous taproots probing 30 and 40 feet into the ground seek out permanent underground streams. Humans have become deadly competitors of these plants; we sink our wells deeper than the taproots and lower the water tables below the level which the plants can reach.

Instead of deep roots, some desert plants such as most cacti have a very shallow root system which is broadcast over a wide area. Such a system takes best advantage of thundershowers. Since a thundershower's moisture falls so fast and runs off so completely, soil just a bit below the surface is often left bone-dry. The cactus keeps its roots near the surface, where they can absorb vast quantities of water. This water is transported for storage to the joints which swell perceptibly as they take on their burden.

Aerial photos of desert shrublands reveal that plants such as cresote bush are not randomly scattered throughout their appropriate habitats uniformly. Instead they grow so that each plant is not very close nor very far from its nearest neighbors. It is almost as if they had been planted by a nurseryman. The cause of this uniformity is that cresote adults inhibit the germination and growth of seeds, especially seeds of its own species. To achieve this they actually produce antibiotic chemicals which effectively declare their sole right to the water which is going to fall in their immediate vicinity.

Despite all the adaptations of plants to obtain and to conserve moisture, they cannot escape the fact that little is available in the first place. Because of this, energy flow or productivity of deserts is extremely poor. Although not often barren, plants cover only a small fraction of the surface of their habitat. In richer deserts, photosynthetic processes succeed in fixing only 100 grams of dry material (as sugar) per meter$^2$ per year. The total production (gross productivity) is about twice that. Moist biomes, except in the coldest places usually have a much greater annual energy flow. Like plants, animals exhibit adaptations which center on the problem of water scarcity. For them however, conservation of water is the most important problem since all animals obtain some water from their food. Free water (that is drinking water) is rarely depended on by desert animals. Some in fact will not even drink it when it is offered; a small pocket-mouse was kept alive and healthy in a laboratory for eight years without once having accepted a drink of water. Besides free water, animals obtain water from two sources: the water actually contained in

their food and the water released when that food is metabolized. (Using food produces energy, carbon dioxide and water.) In some cases the staple food of desert animals is seeds. The pocket mouse and other members of its family are the common granivores of the North American deserts. Since seeds are among the driest of foods, containing only 10 or 15% water, the pocket mouse must obtain almost all its water by metabolizing the seed's oils, proteins and fats.

Rodents generally manage to live on such little water for two reasons. Most avoid the extreme temperatures of the desert by constructing burrows and the soil acts as insulation and preserves a relatively even temperature throughout the day. In fact if the burrows are deep enough (some go down several feet), the rodents have relatively even temperatures and fairly high humidities throughout the year. Rodents emerge from their burrows only in the cool evening. A second feature which permits them to conserve water is physiological. Each small unit of the mammalian kidney, called a *nephron*, is responsible for drawing pure water from the urine. The part of the nephron responsible is a long tube called *Henle's loop.* The longer the Henle's loop, the more water can be drawn from the urine before it is passed. Desert rodents have the longest Henle's loops and consequently lose very little water in their urine.

Nocturnal desert rodents are not the only animals of the desert. They are hunted by various kinds of snakes and carnivores such as the desert fox (there are several species each in its own part of the world). In some deserts, especially those of Australia, lizards abound and even replace many species of mammals in environmental niches. Although there are many beautiful and unusual kinds of desert insects, spiders and their close allies (such as centipedes and scorpions), insects do not appear to be as relatively abundant in deserts as elsewhere on the earth. Perhaps insects are more subject to drying out called *dessication.* The big exceptions to this are the ants. The desert can be carpeted with them. But, ants have subterranean burrow systems too and thus they are another example of an organism that escape from desert extremes. In at least one way however ants simply endure. When caught in floods they stop all activity and appear dead. As pseudo-corpses they can endure submersion for periods of two or three days—more than enough time for the waters of desert flash floods to recede. Then after a half-hour or so in which to dry out and they are resuscitated.

Some animals are active in the desert during the day. Certain kinds of ground squirrels have bushy tails which they angle over their backs to shade themselves. Lizards prevent themselves from over-heating and regulate their body temperature by carefully adjusting their behavior and position. When they are too hot they seek shade such as a rock overhang, or a small shrub and they carry their bodies high off the ground. When they are too cool they bask in the sun and hold their bellies close to a warm rock or the hot earth's surface to give them warmth.

One of the more interesting animals active by day in the desert is surely the camel. As most people know the success of the camel is linked to its hump. But the hump is used to store heat not water. The rest of its success is physiological: it can function normally with a wider variety of body temperatures than most mammals. The camel begins the day quite cool. Instead of using a great deal of water to maintain that low temperature by perspiring, it simply allows its body temperature to rise. Its hump gives it extra bulk in which to store the incoming solar energy. Toward the end of the day the camel's temperature often reaches what would be a respectable fever for a human. At that point the camel's normal mammalian control mechanisms are called into action and no further temperature rise is allowed. After dark the camel's body simply radiates all this extra heat to the sky, much as if the camel were a hot rock. Thus it uses very little water in dissipating its heat. From the camel's extreme tolerance, one should turn to the ultimate answer to desert problems. Retreat! Turn off one's metabolism and wait for better times. It is not difficult to imagine plants (at least as seeds) doing this or *poikilotherms,* "cold-blooded" animals such as toads and snakes. But the fact is that active warm-blooded mammals do it too. Cactus mice

"turn off" during the hottest, driest part of summer; this is called *estivation* and is akin to hibernation. Several kinds of pocket mice apparently hibernate in defense against the cold and poverty of a dry desert winter. But pocket mice have an incredibly flexible metabolism. Not only do some go into a torpor regularly at certain seasons but all species can become torpid if and when food ever gets scarce. Torpid captives appear almost dead except for a lack of stiffening or *rigor mortis.* But they can be reactivated in minutes by being put next to a pile of seeds.

The desert requires a different life style from most environments, and successful life forms learn to duck its climatic jabs or roll with its punches. Human cultures that successfully deal with the water and temperature problems of the desert exist, but they offer only marginal support to their citizens. One must wonder however at the advisibility for the long term of grafting modern civilization onto the desert biome. Currently this union lacks any semblance of compatibility. Modern desert towns depend for their existence upon the underground reserves of water and they are using them faster than stores can be replenished. Every drop of Colorado River water is allocated for human use and the once magic wetlands of the Colorado River delta, that surprise area which teemed with marsh life in the midst of desert, is dry. Clearly humanity cannot continue to expend more water than it receives. Even if it can solve this problem by recycling its water, it appears that modern civilization will not exist harmoniously with the desert biome. It has instead chosen to engage itself in a kill or be killed struggle with the desert. If civilization wins, one can only hope it will also choose to preserve unconquered, vast amounts of natural desert so that the plants and animals which live with the desert on its own terms can continue to amaze us.

Prof. Michael L. Rosenzweig, Ph.D.
Department of Biology
The University of New Mexico
Albuquerque, New Mexico.

# Chapter 13

# The Arctic Biome

This chapter is divided into three major sections: (1) the Arctic biome, (2) ecology of the Arctic, and (3) Arctic ecology and man. The first two are primarily intended to provide background and frame-of-reference for the implications developed in the last.

### The Arctic Biome

The Arctic is that northern circumpolar region of the earth whose southern boundary is usually described as the Arctic Circle (Latitude 66° 30′ N). However this boundary is based on the southerly extension of the "midnight sun", those astronomical phenomena of summer days when the sun does not set and winter days when it does not rise, and fails to coincide with other geographic phenomena such as weather, vegetation, water currents, etc. Therefore we shall employ a more functional definition, considering the Arctic as the land of several phenomena including: permafrost—subsurface earth which never thaws and consequently forms a nearly impassible barrier to any excavation beyond 12 inches unless heavy machinery is used; low, severe temperatures—sometimes as low as −60°F (−75°C.) which will freeze exposed skin in a few seconds; and, low evaporation—coupled with the permafrost, this prevents underground water runoff. Also, low precipitation further combines with both of the above to yield little or no "surplus" truly making the Arctic a "desert" in the sense of available water. Therefore water in itself constitutes an interesting problem in the Arctic.

Real interest in the Arctic was greatly stimulated with the advent of WW II which emphasized the Arctic as a strategic zone for all involved. With the easing of tensions and improvements on control and detection of long range missiles, the major interest in the Arctic has shifted to the presence of natural resources. Currently interest is being directed there more and more, primarily for two reasons: declining natural resources and increasing world populations.

### The Ecology of the Arctic

Both fauna and flora of the Arctic are inhabitants of a precarious environment and both animals and plants reflect this in many ways. Species diversity (the number of different kinds of plants and/or animals in a given area) is low in the Arctic. There is not the vast array of animals and plants that one finds in more southern biomes. The reason is fairly simple and this phenomenon is characteristic of all harsh or marginal environments. Specifically, it takes a distinct sort of animal not only to withstand the extreme low temperatures common but also to survive in an area where any plants it might eat lie buried under snow nine months of the year. Plants on the other hand, must be prepared to complete their business swiftly, growing and reproducing during those few months they are uncovered and exposed to sunlight. Finally, as is common and might be expected, heavy mortality is a fact of life to both flora and fauna in the Arctic. Not only do stringent ambient conditions take their toll on different plant and animal types which may wander, drift, or be introduced there but many of the indigenous types succumb. Therefore to survive and continue to exist in the Arctic, any species, plant or animal, must rely on sheer numbers—in place and reproductivity. The Arctic is one biome that defies both individuals and their offspring to survive. Corollary and related to this is one final generalization concerning Arctic fauna and flora, i.e. that it displays remarkable uniformities.[1] One can expect to find much the same type of plant or animal there whether on one side of the globe or the other, a condition which rarely occurs in more southerly latitudes.

These circumpolar distribution patterns also result from the fact that the earth has much smaller diameter and circumference near the poles and consequently migration routes are shorter. As might be expected from their mobility, the Arctic fauna displays more homogeneity than the flora. Nevertheless each presents primarily quantitative rather than qualitative problems for study. However, specifically regarding the flora, several other points should be made.

• Only those species can survive which are able to withstand constant disturbances of the soil, buffeting by the wind and abrasion from wind-carried particles of soil and ice[3]. Therefore, vegetation in the Arctic is found in those areas which are "moderate" with regard to environmental stresses.

• Arctic soils are extremely poor in nutrients. There are several reasons for this. They are: a. short summers, b. low soil temperatures, c. sparse vegetation, and d. slow decay of organic matter.[5]

• Because of these stringent limits on life, plant-denuded areas are generally replaced by the same types originally existing there.[4]

• Further, while there is controversy on the matter, some workers feel that the factor of competition appears to play little or no role because the plants are so spaced out that any plant able to withstand the environment is virtually assured a spot in the community.[2]

The snow itself also plays a major role in the existence and survival of both Arctic flora and fauna, figure 65. Its importance is dual in nature: supplying both protection from winds and extreme temperatures.[5] Trees in the high Arctic grow only to a height of a few inches and, regarding moisture, summer snow melt may provide the entire yearly supply in many areas.

As expected from the above, the list of plants which we find in the Arctic is not very long. Hanson[4] categorized five vegetation types that can be distinguished aerially from planes: (1) forests, (2) shrubs (more than 2.5 feet tall), (3) dwarf shrub types, (4) herb-dwarf shrub and (5) herb type. The important consideration here is that this list would be quite lengthy in southern habitats.

Principle plants of the Arctic are: peat-mosses, mosses, figure 66 lichens, sedges, grasses and low herbs. Lichens most often overlie deposits of organic matter essentially not decomposing because of low temperature.[6] However, in the lower Arctic the following flowering plants from further south may be found: buttercups, dandelions, poppies and bluebells.[7]

As might be expected from our discussion above, most especially relative to the "patchiness" of the Arctic flora: The Arctic fauna in general has developed migratory habits to survive in interaction with it. Arctic animals tend to move from one vegetation zone to another quite frequently, most undergo some sort of periodic migration whether regular or irregular, and circumpolar migrations are common occurrences.[6]

Specific adaptations of Arctic animals to their environment are several: (1) protective coloration —white is the color of the vast majority, figure 67. In the high Arctic it is ubiquitous and apparent year-round as exhibited by the polar bear, snow goose and others.[5] (2) Insulation—most possess thick layers of fat or mats of hair. Superficially less obvious are the muskoxen's hollow hairs, the column of trapped air within each further acting as an insulation device. (3) Life Cycles—probably the most common adaptation is compactness of life cycles; getting about the business of growth, maturation and precocious reproduction during the short period when conditions are amenable.[5] Some animals on the other hand spend at least part of their lifespan in a "slowed-down" state, usually as eggs or very young during the most severe periods

**Figure 65.** Frozen snow frames a gravel beach in Arctic Canada.

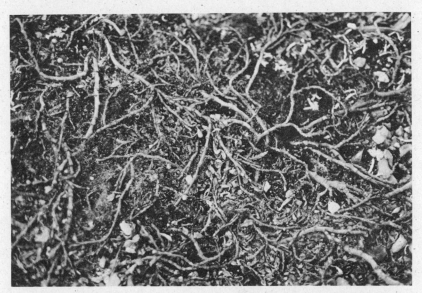

**Figure 66.** Mosses and lichens, principle plants of the Arctic.

**Figure 67.** Two polar bears run from a Canadian icebreaker in the Northwest Territory, photo courtesy of United Press International.

of the Arctic year. Once hatched or emerged however, speed is of the essence.

Reference to specific faunal (species) lists immediately makes apparent two other facts. Two major vertebrate groups are missing, both reptiles and amphibians . . . with the minor exception of some occasional frogs. The vast majority of Arctic birds are there only in the summer, nesting and then going south for the winter. Among the most prominent bird species are: phalaropes, sandpipers, plovers, snowy owls, jaegers, snowbuntings and longspurs.[8]

The mammals of the Arctic include: aquatic mammals—seal, walrus and whale; and, terrestrial mammals—fox, muskox, caribou, wolf, lemming, Arctic hare and the ground squirrel. This latter species is the only Arctic animal which "hibernates"[5].

Concerning fish, the blackfish is one found exclusively in the Arctic. It is a small fish which can tolerate being frozen—another case of adaptation. Another is the capelin, a small fish that exhibits the same sort of population numbers once characteristic of the California area sardines[8]. Other species also found here include the Arctic char, lake trout and grayling, all of which take significantly longer to mature than their southern relatives. The consequences of this slow maturation will be dealt with in the next section.

Of the invertebrates known terrestrially or aquatically, the vast majority are insects. Since the maximum water temperature in lakes of the high Arctic reaches only 4°C, many zooplankton such as the crustacean the *Daphnia* which are common in temperate lakes are not able to reproduce; consequently, there is a paucity of zooplankton. For example in Char Lake which is located on Cornwallis Island (74° 42′ N), there are only two species of zooplankton, with one, *Limnocalanus macurus* dominating throughout the year.

Summarizing Arctic life, it should be immediately apparent but nevertheless needs underscoring . . . that: The ecology of the Arctic and the interrelationships that exist there are in extremely precarious balance.

### Arctic Ecology and Man

How has man already affected the ecology of the Arctic? Put another way: How has he altered the interactions between organisms and their environments which existed in stable fashion for eons?

Man's effect seems to center around two main themes: exploitation of resources—both renewable and non-renewable, and pollution from his actions —both direct and indirect. Ideally one might wish to treat each separately but in reality this is impossible. Rather, the distinction between what constitutes exploitation and what is pollution is often nebulous and extremely semantic at best. Even historically we must take them in combination

The earliest resources exploited from the Arctic by modern or "western" man primarily centered about its wildlife. In fact the economy of the north for many years was solely based on animal resources[9]. So strong was the impulse and so rich the rewards that the initial settlement of northern (and southern) Canada was due to the fur trade[10], it being a major factor inducing war as early as the middle of the 18th century—e.g., our own French and Indian War. Settlement of the north by professional hunters began to mark a real change in the ecology of the wildlife. Previously only Eskimos and some Indians were the lone hunters and they were severly limited by their technology, i.e., weapons, mode of travel, etc., and lack of import to support greater numbers. Today the only true native hunting can be found only in those areas so remote that "southern visitors" characteristically have avoided them.

Now however technological slaughter of animals became a massive action, the beaver being one of the early victims and a prime example of what can happen. Driven to near extinction it soon became apparent that the dam-building activities of this animal were essential to certain stream-dwelling fish species requiring deep pools. Other species at one time or another in serious jeopardy have been the Arctic fox, wolf, muskox and even today, the polar bear. Only fluctuations in the market and a decline in demand, hence economic worth, saved the first three from extinction. The latest threat, especially to the polar bear, is in the form of "tourist-hunters" and their trophy hunting. Only the implementation of very stringent policing and regulations may ward it off.

Marine fisheries constitute another resource of the Arctic and are now being exploited. Man's utilization of the sea as a food source is obvious and has long been primary in these regions because yield from terrestrial type agriculture is negligible here.[1] Further, as might be expected marine fish life in Arctic water is extremely abundant: The capelin for example seeming to exhibit sardine-like densities, as mentioned above. Examples of current pressure on these Arctic fisheries are the fact that fishing constitutes the main support of the entire Greenland population

**Figure 68.** A Scientific research ship working in Arctic waters

and the populations of both northern Canada and Alaska rely heavily on harvesting large amounts of fisheries products from their surrounding waters.

Exploitation of Arctic fisheries seems more detrimental than that in more southerly regions since individuals taken require a longer time to mature. As a result, heavy pressure may tend to remove many individuals before they have reproduced thereby minimizing chances for perpetuation of the species. An interesting fact and sidelight is that the Great Lakes sturgeon was reduced to a rarity through similar circumstances[11].

In spite of the fact that one would expect Arctic fisheries to be blessed by being largely free of man-made contaminants, such is not necessarily the case. We find the waters of Alaska, the source of 85% of U.S. domestic salmon, currently threatened by such industries as pulp and paper[12] which are only now really beginning to be exploited.

Inland fisheries are another aspect that is currently being pursued. Probably the most noteworthy is that of Great Slave Lake which has been described[9] as a fine example of what can be accomplished when scientific investigation, figure 68, precedes commercial operations. However, because of economic restraints like harvesting and shipping it appears unlikely that inland Arctic fisheries will succeed to any great extent.[9]

Whales also deserve mention before closing on fisheries. Hunted to near extinction, only the introduction of international rules avoided catastrophe.[12] However it now appears that some countries are violating these rules and once again extinction of whales in the Arctic is a major problem.

Non-renewable resources are proving both boon and bane. It appears that these resources and their development constitutes the only feasible sound monetary basis for development of the region. Among the major ones in the Arctic are metals, oil, coal, natural gas and non-metallic minerals such as sulfur and asbestos.[10] Removal and the accompanying pollution associated with mining comprise yet another ecological catastrophe in the Arctic. Although it is difficult to ascertain what wealth of minerals still lies in the ground it is generally accepted that their value is sufficiently ample to offset cost of removal and transportation. One typical example is the fact that the first gold mine opened in Yellowknife is still in operation.[10]

Coal is probably one of the more uncertain of Arctic resources though peat beds are common in the low Arctic. Good[1] indicates that coal is most certainly there but equally certain is the fact that deposits vary greatly quantitatively as well as qualitatively. Regarding ores such as iron, the exploitation has thus far been retarded on purely a pragmatic and economic basis. The cost of shipping, across land or ice, is prohibitively expensive. However, geological surveys indicate mineral resources of the Canadian Arctic may be more than the rest of Canada combined.

Ecological havoc accompanying mining while not always direct or directly apparent is nevertheless very real. The landscape of the Arctic is of such a nature that what would constitute normal activity in temperate areas may cause irreversible damage in the North. For example, driving of a tracked vehicle (or even in some instances, footsteps) can change the drainage pattern of an area.

**Figure 69.** Seismic tracks filled and refrozen in the Arctic cold.

Water then flows where it has not before and even if not the track may still be clearly seen years afterward. The implications are frightening.

Seismic tracks comprise another (associated) problem. These are sites where man has bulldozed off both surface and underlying, now melting mud to yield a hard road (or ice) for heavy equipment.[13] The result is soon mud filled ditches and melting permafrost, figure 69. Even more disastrous results occur when the bulldozers expose ice wedges which soon melt causing shifts in the adjacent earth's surface. So outlandish have been the repercussions that the Canadian government has outlawed summer seismic tracks.[13]

The last, but by absolutely no means the least, resource we shall discuss is oil and the precious reserves of it that exist in the Arctic. The face of the Arctic has in some places literally been scarred beyond recognition by oil prospectors. Nevertheless spurred on by prediction such as that concerning Alaska (reserves can produce at least one-third the amount of oil produced by all other states combined) exploration throughout the Arctic continues.[14] Probably the most illustrative as well as indicative way of dealing with the oil problem in the Arctic is by describing one specific situation.

The projected Alaskan pipeline from Purdoe Bay to the Port of Valdez and the related oil find seems most appropriate. The two major ways of transporting this oil is by either tankers or pipeline. If tankers were used, spills of the nature of the Torrey Canyon and innumerous others are a very real situation. The chance of oil spills is greatly magnified by: (1) the lack of ships which can withstand ice conditions and (2) shortage of personnel skilled at handling ships under adverse conditions of this nature. If a pipeline is built, the number of imminent dangers increases. Conservationists speculate the pipeline may interfere with the migratory patterns of caribou, etc. or, it may melt the permafrost and cause earth shifts with subsequent breaks in the line. Further, the pipeline will cross several fault zones and therefore be subjected to breaks during earthquakes.[14] The fact that the pipeline will cross 350 streams and rivers suggests the effect would be on both terrestrial and aquatic ecosystems. The conclusion which must be drawn is that man is faced with a challenge. A challenge to show that he has "learned his lesson" from his experiences in more southerly latitudes. That is, man must realize the precariousness of ecosystems and think before he acts.

# REFERENCES

1. Good, D. in *Geography of the Northlands* (Good and Kimble eds.) The American Geographical Society and John Wiley and Sons. New York (1955).
2. Griggs, R. F. "The vegetation of the Katmai District," *Ecology* 17:380-417 (1936).
3. Storer, T. I. and Usinger, R. L. *General Zoology.* McGraw-Hill and Co. New York (1965).
4. Hanson, H. C. "Vegetation types in northwestern Alaska and comparison with communities in other Arctic regions," *Ecology* 34:111-140 (1953).
5. Smith, R. L. *Ecology and Field Biology.* Harper and Row Publishers Inc. New York (1966).
6. Shelford, V. E. and Twomey, A. C. "Tundra animal communities in the vicinity of Churchill, Manitoba," *Ecology* 22:47-69 (1941).
7. Mirsky, J. *To the North.* Viking Press. New York (1934).
8. Wiggins, I. L. in *Arctic Biology* (Henry P. Hansen ed.) Oregon State University Press, Oregon 307-313 (1967).
9. Rea, K. J. *The Political Economy of the Canadian North.* University of Toronto Press, Toronto (1968).
10. Loyd, T. "Canada's Arctic in the age of ecology," *Foreign Affairs* 48:726-40 (1970).
11. Claffey, F. J. Personal communications (1972).
12. Haden-Guest. in *Geograpy of the Northlands* (Good and Kimble eds.) The American Geographical Society and John Wiley and Sons. New York (1955).
13. Phillips, R. A. J., *Canada's North.* MacMillan of Canada, Toronto (1967).
14. Kornberg, W., "Concern for the Arctic environment," *Science News* 97:486-8 (1970).
15. Ellis, W. S., "Will oil and tundra mix?" *National Geographic* 140(4):485-518 (1971).

Warren B. Kirchner
and
Harold V. Kibby, Ph.D.
Department of Biological Sciences
State University of New York
at Brockport
Suny-Brockport, New York

# Chapter 14

# Society, Politics, Technology, and Pollution

No one act pollutes. No one person or group pollutes. The pollution of the environment is the result of a vast number of acts carried out over a period of time. The creation is not bound by the laws of supply and demand. It is regulated by another law which limits consumption within the boundaries of existing resources. Pollution begins the moment a living organism violates this law and consumes more than it produces or reproduces.

In the moment that man became a consumer more than a producer, he became a polluter also. It is in the nature of things that there is a restricted, limited amount of energy and material upon earth. In the simplest terms this means that there is only so much energy and material which may be exploited for the use of man, there is a limit to how much of the cosmos can be turned by digestion or procreation into human flesh.

It is now high time that we ask fundamental questions of ourselves and our leaders. Who are we? How did we get here? What is our role in the order of things? What are the legitimate levels of consumption? Where are we going?

Answers to these questions will be forthcoming when we find methods which will allow us to properly analyze reality—past, present, and future. In our search for these methods, we must look for solutions which will give us guidance and direction based on the nature of reality rather than the pragmatic need of the moment. We must cease our practice of transporting our pollution to unknown, least understood or neglected areas of the globe or universe, such actions as dumping nerve gas into the oceans; using outer space as a trash dump and the like. We must develop ears and eyes to see and hear the cosmic cry for reorientation that pollution represents. We cannot dull our senses with a "technologic fix."

This astral standpoint can be gained in three ways: firstly by looking at the course of events in millennia rather than centuries or generations; secondly by assuming a transcendental philosophical method; and thirdly, by recognizing the unique character and responsibility of mankind in the universe, in effect by affirming a new anthropology. This means the recognition that the current fascination with behavioralism must run its course. The animal cannot pollute at the level, or destroy at the whim potential to man. This awsome power makes man sensitive and at the same time responsible to the creation in his charge. Such a position in the universe demands a knowledge and a recognition of the law spheres of the cosmos which are inherent in the creation.

For the sake of understanding, the history of man, particularly western man, must be divided into three basic periods. These are divisible by the nature of the worldviews which characterized them and which came to completion within them.

### The Period of the Pre-Scientific/Pre-Theoretical Worldview

The vast expanse of human events is prehistoric, that is, it occurred at a time when only horizontal communication was available. Horizontal or face to face communication can traverse time for only one generation, from grandfather to grandson or a time frame of 60 years more or less. In such a state of affairs, myth and saga are the only remnants of the thought of the past. It is well known that in such societies dancing, singing, playing and speaking are highly formalized and are carefully preserved under the general heading of "tradition."

Methods and styles of handworking, hunting, herding, and the like also fall into stylized patterns which appear to change but slowly. The best evidence for the efforts of man during this era are his artifacts, those items which he formed from the materials of his environment. During most of man's long trek on earth, or about 99.9999% of history up to the present, he lived in small roving bands and preyed upon the herds of animals. This meant that human organisms filled their ecological niche along with other predators. When the seasons were abundant, the bands grew, when the years were lean, infant mortality would have increased proportionally and the bands diminished. It is estimated that at 15,000 B.C. there were no more than 10,000 souls in all of the British Isles. The life of man was part of the cycle of life, and he refurbished the soil with his wastes and his body as the generations rose and fell.

For want of an absolute chronology let us take the date, 10,000,000 B.C. and place the rise of man in the Garden of Eden. The progress of human society was slow during the hundreds of thousands of years that followed. Sometime between the last glaciation and 10,000 B.C. associations of human bands settled around the freshwater lakes in the southern region of Eurasia. The remains of such hunting groups, living in regular yearly campsites have been excavated from the areas of the Swiss lakes, the Black Sea, Lake Van, Galilee, the Caspian, and Baikal. By 8,000 B.C. the food-getters were setting aside their habits of pursuit and settling along the banks of streams and rivers to take up food-producing. This momentous change is usually called the "Neolithic food-producing revolution." From it the settlement of the first townships emerged in the foothills of the Zagros Mountains of Iraq.

With increased production of grains and cereals came the domestication of animals and the intrusion of man into the ecological cycle of plant succession. It was necessary for man to forcibly keep the ecosystem young to achieve increased production, growth, and quantity. The increase in information now available in the town demanded a processing and display technique and so writing appeared. Initially it was used as an aid to memory to keep records in the temple-storehouse. The

seasonal festivals and fertility rites which had arisen during the millennia of hunting were retained and reshaped and the *archaic-religious-state* was born. This took the form of a walled town with the temple as the center of the community. The leader or elder of the township was also high priest of the local cult and law giver. Each individual had his assigned place in a static and durable order and all were immediately empathetic to the general mood. It was archaic in that its beginnings are lost in antiquity, it was religious in that its basic structure was theocratic, and it was a state with all the modern connotations of that word.

From that remote beginning two notions survived until the present: 1. literacy denotes knowledge and 2. technique can ultimately overcome phenomenon. The archaic-religious-state liberated man from reliance upon nature and gave him superiority over nature.

Through township organization swampy land could be drained and raised yielding polder. In dry areas irrigation could increase the farm land. The success of these endeavors was directly dependent upon the cohesion of the archaic-religious-state.

The incredible durability of these towns has never been equaled. Places such as Jericho, figure 70, Babylon, and the Egyptian sites of the

**Figure 70.** Neolithic wall, steps and round tower at Jericho (c. 6,000 B.C.)

**Figure 71.** Roman Fish-Wharf in Upper Galilee near the ruins of Migdol. (c. A.D. 100)

Fayum supported a nearly continuous population for longer than 8,000 years. However with the increase of available food, the natural ravages of famine and the hardships of the hunt no longer kept the population stable as under the conditions of primitive forest efficiency. Slowly, the towns began to increase their size. Yet by the standards of the Nineteenth and Twentieth Centuries they were small indeed. Ancient Athens covered less than 1,000 acres, one could walk around Jerusalem in a half hour, and the Troy besieged by the Greeks would have fit into Yankee Stadium.

Pollution was well known to these ancient city dwellers and trash heaps as high as the houses have been recovered. Pestilence and disease often carried off whole populations and newcomers would rebuild the town in the next generation. The Book of Exodus makes clear that plague was nothing new to Egypt. The great herds of wild animals and the strongest of the predators were annihilated from most of the lands surrounding the Mediterranean during Greek times. Arrian the historian tells of Alexander's troops killing the herds of wild asses and bustards for food and sport. Earlier men had killed off the great mammals of the Pleistocene and burned their habitats so that most were extinct by 40,000 B.C. and the remnant severly reduced in range by 800 B.C. The habits of the irrigation cultures caused both salination and silting along the river valleys which were the "cradles of civilization." When the Roman Empire spread east and west in the First

Christian Century it carried with it the most avaricious and efficient machinery for exploitation the world had ever seen. The Greeks had brought the slowly developing notions of mathematics, geometry, and physics to the market-place. The Romans followed Alexander and brought an end to the ancient archaic-religious-states, and with them the stability and limitation of local agricultural units.

The Romans built granaries on the Danube, smelters on the Thames, and fish-works on the Jordan, figure 71, they exploited with all the energies of their armies and engineers and shipped the wealth of the ancient nations back to Rome. The private farmer who had been the backbone of the early Republic failed and was swept into the city mob by the cheap production of vast "latifundia" or slave estates. These were copied on the Carthaginian model made famous in a book on agriculture by a certain Mago. His thesis was that a large area could be made to grow vast quantities of basic agricultural products with an immense profit with the use of slave labor. Thus the idea of the plantation was born. Rome became such a stinking miasma that the upper classes fled each summer to the North shore of Africa where great resorts such as Lepcis Magna had been built.

Under Rome, hillsides were denuded of their timber, rivers were redirected over mountains, and the flat lowlands of Northern Europe pumped dry to yield the Netherlands. When Caesar went to

Gaul in 60 or so B.C. much of the land between the Rhine and the Elbe was forested. Within a mere three centuries much of the endless forest had been cut down and the land burned over.

During the long centuries of the dark ages from A.D. 400 to 1200 the land seems to have partially recovered. The depopulation brought by the bubonic plague in the middle ages destroyed as much as 90% of the towns in many areas of South Central Europe. However the archaic-religious-states were gone and the vestiges were remnants and not antecedents.

### The Rise and Expansion of the Mechanistic Worldview

The foundations of the classical physics and all the technologies which stemmed from it have been thoroughly studied by E. J. Dijksterhuis. The historian, Lynn White Jr., has traced the impact of these innovations upon the environment. He points out that the introduction of water power by A.D. 1200 at the latest had an enormous impact upon the landscape of the west. No doubt the location of the mills, smelters, refactories, dye

vats, and graneries of Europe besides the rivers from the White Sea to the Bay of Biscay created the monstrous health hazards of the Eighteenth Century towns, figure 72. The ultimate triumph of Newtonian science was not in the invention of the calculus and the explanation of the laws of motion, but in the tragic but mighty advance of practical engineering. European agriculture was given a breathing space by the introduction of new high yield crops such as potatoes, beans, and corn from the voyages of discovery. As the basis of wealth shifted away from land to economic production the republican and later the socialist fervor spread revolution across Europe. The shop-keeper mentality of the new United States, when transplanted to the bourgeoise of Europe, produced the French Revolution, the upheavals of Bonaparte and the collapse of all the kings of Europe approximately one hundred years later in the First World War. But the industrial revolution while forcing the fate of the aristocracy, caused the worst social dislocation of the centuries. The need for coal and raw materials made millions into the handmaids of the steam engine and placed

**Figure 72.** Many factories built in the late Nine-teenth and Early Twentieth Centuries have been located along river beds and streams. This manu-facturing plant astride a gravel-river-bed in central New York State pollutes the water for many miles down stream.

the heartless laws of "raw material" upon the colonies of all the continents. In the face of this massive reorganization of social classes, the ancient triumvirate of church-king-landowner succumbed to the rising corporation. The crisp neatness of Newton's theories had appealed to the enlightened despots of the early modern states and many men sought to find the same sorts of natural laws in the biological sphere.

Darwin ended the quest with his romantic biological view. It did for the Life-sciences what Newton had done for the Physical. However its instant and universal acceptance by the intellectual community of Europe was based upon its application as a justification for the status quo. Its most popular form was not biologic Darwinism, but social Darwinism. The United States with its westward expansion and extermination of the Indian, Great Britain with her Asian and African exploitations, and Germany with her scientific leadership all embraced the notion of the "survival of the fittest" as a new secular creed. Only the French and the backward legions of the Czar held back. Long after the advance of scientific experiment had left the melioristic and optimistic views of Darwin and the Huxleys behind, the Americans were still chopping down trees, the British still tapping rubber and shooting lions, and the Germans still synthesizing poisons. All devoutly believing that nature was rising to meet the challenge and newer forms of adaptation would restore what was destroyed. It was during this era that the grossest extermination and the most flagrant pollution were begun. Any problem could find a mechanical solution. The Suez Canal, the Panama Canal, the Brooklyn Bridge, the Automobile, Henry Ford and the Hindenburg, were all products of this dictum. But all of these would decay and recycle themselves into the earth. Each would be succeeded by another. Even at best air travel after World War I was very risky in storms and the Titanic had been sunk by as common a material as ice. There was still a frontier in 1905 when Albert Einstein published his theory of special relativity, America was still pushing West and pouring tons of waste into the Missouri, the Mississippi and the Columbia on the way. Russia was still pushing East toward Sakhalin and Kamchatka, and Victoria was but four years dead. The contents and meaning of that obscure paper would restructure and revolutionize the world known to all the others.

In addition it must be remembered that the immediate precursor of industrialization was an agrarian life style. In making the transition less

care than was really necessary was taken to understand the nature of the new mechanical phenomena. More often than not, the new marvels of industrialization were seen through the eyes of persons whose hearts and minds were rooted in the earth. The Eighteenth and Nineteenth century economic theory, for example, which advocated the reinvestment of funds for capital improvement and maintenance were undoubtedly drawn from the familiar agricultural process of fertilization, land retirement, and crop alternation. What the new industrialists failed to understand about the new industrialization was that capital reinvestment was not the same as replenishing the earth. The former deals in the manufacture of raw commodities—it is a consuming process more than a producing one. The latter is concerned with the production of raw commodities. Mere replacement of mechanical implements is a consumer role, no new raw materials are produced.

The early industrialist is not to blame, as has been done by Marxist historians such as J. D. Bernal, neither the manufacturer nor the farmer took the time to understand the rightful place, the law structure, call it ecosystematics if you will, that this new societal institution should have in the order of things. Consequently the proper relationship between consumption and replacement of natural resources remains a problem without solution to this present day. Pollution is the antinomy resulting from a disregard of creational laws.

### The Rise and Impact of the Thermodynamic Worldview

The seamless fabric of the classical physics so carefully woven by Newton and his followers had started to rip and tear in the mid-Nineteenth century. The devisive force was the continuing experimentation in astronomy and physical, later nuclear chemistry. Very few of the scientific community other than mathematicians recognized the problems, and even fewer cared. After Newton's discovery of the calculus a whole new array of mathematical tools were formulated and the application of these insights to physical problems, especially those associated with the molecule, the atom and their symmetries did not answer newer problems which resulted from experimentation. Increasing needs for more powerful engines brought more intense modeling of the physical processes involved. The French statesman and militarist Nicholas Carnot had published works on engineering in the first decade of the Nineteenth century, his son, Sadi became the founder of thermodynamics with its laws of heat and the conservation of energy. These became the center of modern chemical research and ultimately the basis of the new physics analyzed and proposed by Einstein and developed by his followers.

From the new technologies came two great and far-reaching innovations, electronics and synthetics. The first allowed communication and control over millions of separate operations and miles, and the latter the production of materials which would not decompose in the biologic materials cycles. Both of these have contributed to the modern degradation of the biosphere. At the same time they were the basic insights of the modern science of ecology. If energy was never destroyed or created and matter likewise, than burning did not rid Chicago of its trash, only transform it. The radioactive fallout and the radioactive tracer were sister technologies.

These then are the three great periods of human technical insight and endeavor. Certain aspects of each are still with us. Problems produced by thermodynamic technologies cannot be solved by pretheoretical committments or mechanical solutions. The all-pervasive character of the electronic atmosphere in which the twentieth century moves, covers every aspect of human communication and the wide range of radiations and signals which are not seen or detected by the unaided organism, an environment in which nature has no experience figure 73.

The popular movement in ecology suffers from the buffeting of ideas generated by all three of the modes of science. There are those who reject all exact science and seek some mystical harmony. These border on pantheism or some sort of "reverence for life" as promulgated by Gandhi or Schweitzer. Such a notion as "the great chain of being" or the "hylozoistic universe" is deeply rooted in the pretheoretical and the prescientific. Those who espouse such notions are often anti-scientific, seeing in the grosser aspects of applied science the dehumanization of man and the destruction of his environment. They are certainly correct in their insistence that industrial science has had a history of ignoring the aesthetic side of human existence. On the other hand, a return to some pristine condition of the distant past will hardly solve the basic problem. This is one of the fallacies in Lynn White's historical reconstruction. Two historiographic principles are involved:

1. The past, by definition is unobservable.
2. No one can turn history back.

The antiscientists are joined by many who feel that a new technology of environmental engineering must be implemented. The great corporations support this as it would open new product lines and new markets from which they could gain new profits. The difficulty is that the history of "Yankee ingenuity" has been one of application not particularly innovation. American corporations are very weak and slow to support basic research.

**Figure 73.** A Radar installation along the Pennsylvania Turnpike one of the increasingly more commonplace transmitters and receivers of invisible, high energy radiation.

One thing the thermodynamic thought revolution has demonstrated is that probably fewer devices rather than more are in order. Since it involves two energy loops instead of one (heating up the water, then turning the wheel) the steam car although apparently cleaner is much less efficient and less ecological than the horse.

Those who dismiss science as the ultimate solution usually return to an ancient physical conception of the universe and a prescientific world view. More often than not they prefer some vestige of the Hellenistic age. The pessimism and gnosticism of Plotinus became popular because they offered an escape for the hundreds of thousands who found themselves unwittingly liberated from the static social structure under which their forebearers had lived for a thousand years or more. The fear, frustration, and cultural shock which resulted, drove men away from accepting the rational picture of reality which their senses and logic indicated was true. This mass retreat from reality has overtaken society a number of times in recorded history and always with similar results.

Whenever the immediate results of man's activity are too painful to contemplate there is a threefold attempt to escape. The first takes the form of hedonism, drowning one's self in mood-changing, sensibility-altering situations and substances. The simplest result of this endeavor is "eat drink and be merry for tomorrow we die," a theme as old as Sumer but raised to the perfection of a philosophy by Epicurus (323–260 B.C.) in the Hellenistic Age. The modern counterpart is seen in the Holden Caufield's and "Graduates" of the popular culture. The most complex result is existentialism which is equally as suspicious of science as hedonism. Although the existentialists of Europe decry the lack of profoundity rampant in American pragmatism, they are both equally destructive of theoretical thought. The second takes the form of stoicism, the denial of the self and the effort at communal and personal triumph through dedication. While pessimism among young Americans has taken the escape of the hedonistic drug scene, it has led the Chinese youth toward the stoicism of the Red Guards. This latter escape is found in the proposals of Ayn Rand, whose self-conscious egotism turns off further sympathy at a point where it threatens the ego. This follows the American preference for "do it yourself" solutions. The third takes the form of simply turning off the whole problem. This can be done by "thinking positively" à la Dale Carnegie, Norman Vincent Peale, and the *Reader's Digest*. The incredible popularity of irrational situation comedies on American T.V. shows the popularity of this effort to get away from the problem, the insatiable desire for escape.

None of the industrial states of the modern world have effectively dealt with the problem and it is unlikely that any are likely to come to grips with it. Whenever the logical set of ideas about reality fails, a proportion of the population will seek one of the Hellenistic exits described. The sudden fads to frequent in modern society ranging from "swinging" to "sky diving" are based on escape.

Somewhat remarkable and unparalleled is the fact that of all the archaic-religious-states only Israel survived to modern times. The societal laws of ancient Israel, which was prescientific but not pretheoretical, both First and Second Commonwealth, were unique in that they demanded agricultural, economic, and sociological redemption. Through legislation and the legal structure the Jew made exploitation without restitution a crime and repayment a necessity. Guarantees against familial, organizational, and personal dominance were part of the sabbatical and jubilee years. Interestingly the Old Testament prophets are clear that the difficulties encountered by the nation in the exile and captivity were the result of the public disregard for the laws of redemption for both man and the cosmos.

The gadgetry which would reduce pollutants will of necessity reduce efficiency which will require more fuel which will only increase the pollution somewhere else in the cycle. Thus the mechanical is set over against the thermodynamic and so on. Of all of man's institutions, the church and the government are probably the least likely and the most resistant to change. Both are deeply rooted and defined in pretheoretical and prescientific terms and cannot control or comprehend science in either its older or newer determinations.

To ask a government which developed in Greece in the Sixth Century B.C. with a concept of law which is still older, to judge and order a technology innovated in the Nineteenth Century is absurd. To ask it to comprehend a theory of the Twenty-First Century is incomprehensible. Unfortunately while the oceans of pollution in smoke, water, radiation, noise and all the rest come inundating the nations of the Twentieth Century, the legal systems are bogged down in years of cases concerning patrimonies and precedents defined in Latin by Caesar Augustus. The prospects are not good since no turnabout is forthcoming.

Throughout the last two hundred years the two chief causes of pollution have ascended astronomically. They are the population of human beings and the potential to produce power. It is a clear ecological fact that the more people there are in a given area, the more synthetic the environment must become. It also follows that the more matter we turn into energy, the less matter there will be left. These two principles are on a collision course. Even if there were no poisons in the waters, no chemicals in the air, the results of the consumption of fuels would bring the burgeoning population of humans to the end of the supply and the contamination of the environment. With vastly increased population the problems and processes of one locale are immediately transferred to another. In the deserts of the Near East, the litter and junk of war is left to rust where due to dryness it will probably remain eternally, figure 74. In northern Europe polluted water swirls around the pilings of the houses and the effluents of industry wash over the fields of food plants, figure 75. In Japan, the

**Figure 74.** The Midianite Hills and Acacia Trees of the Sinai Desert with shell fragments strewn everywhere.

**Figure 75.** The Canal in Leiden, the Netherlands flowing in and around residential areas.

**Figure 76.** Fish and Seaweed farmer in the Atsumi Bay of Central Honshu.

**Figure 77.** Rice Paddy near Kyoto watered from open sewage.

fisherman must catch the fish which make up the mainstay of protein for 80 millions from water green with organic dyes and too foul to drink, figures 76, 77. Thor Heyerdahl found the junk of civilization floating in the mid-Pacific. No matter where you look on the North American continent you will find the artifacts of consumerism, the technologic innovations of post World War II, the aluminum can and the plastic bag. Few citizens however, comprehend the fact that barring any type of catastrophe, these two objects will lie where they fell without degradation forever and ever and ever    Amen.

William White Jr., Ph.D., Editor,
North American Publishing Co.
Philadelphia, Pennsylvania.

# Chapter 15

# The Popular Movement in Ecology

Since the first national "Earth Day" in April, 1970, the word "ecology" has become part of the common language. With popularization has come an explosive extension in meanings which has dismayed many professional ecologists. In this case, proliferation of language is a meaningful indication of a major conceptual change in a powerful academic idea. The popularization of ecology has its closest historical counterparts in the popularization of evolution and Freudian psychology although the details bear little similarity.

A few professional ecologists have resisted the inexorable evolution of language, with as little success as King Lear's order against the tide. The strategy of a 1971 publication was to distinguish "ecology" as a professional scientific specialty from "environment," the non-professional concern with pollution and nature. But this has been singularly ineffective. Concurrently, a more con-ciliatory definition was drawn by publicity for a new textbook—"Ecology is a science, ecology is a popular movement." The declaration may be taken as a simple statement of reality. Yet the situation is extremely complex and confused. "Ecology" has not only acquired many meanings; but has been used in ways which defy either grammatical or logical definition.

There is little justification following what has happened to the word ecology alone, but the ideological and social changes which produced the changes are of greatest importance. An analysis of the course of ecology as a popular movement may provide guide posts out of a rather chaotic period toward more effective or rational action. Since it is perilous to sit within an explosion and apply historical theory, the best course of analysis seems to select a few historical roots, follow selected issues, and identify some of the institutions which appear as peaks in the swirl of confusion.

In modern terms it may be useful to equate ecology with "natural history" and the ecologist with the naturalist. The specialty of human ecology is not contradictory, when man is approached as a natural feature, even within the urban or industrial habitat. However, the professor of natural history of the mid-19th Century may have been strictly a chemist in modern terms, "naturalist" not having been separated distinctly from "scientist." Ecology as a name and as a specialized science was indeed created in the mid-19th Century. In restropect, scientific ecology was practiced by specialists such as limnologists as well as generalists who were primarily interested in pure science (taxonomy of algae) or applied science (agriculture, fish production). Significantly, the Ecological Society of America was not established until 1907, although smaller organizations had preceded it by a few years. To this day, it remains a small academic group, whose communications are directed principally at each other. In the most restricted definition, its members are virtually the only ecologists in the nation.

The importance of the small professional ecological society should not be under estimated, although it may be completely unknown to most people. In the quiet decades of the twentieth century professional ecologists worked out the reasoned principles and the practical methodology for studying larger, more complex units of space. The quality of work may have benefited from the lack of its prestige, with the consequent pressure to produce results. Ecology has never been high on the scale of scientific specialties, measured in academic courses and research support. Within the group of professional ecologists, there were always a few who attempted popularization. Paul Sears was the first to achieve significant gains. From his dust bowl oasis at the University of Oklahoma, he wrote a widely read book, *Deserts on the March.* Moreover, his ecological view and even his language was transmitted directly to officials who initiated a bold program to restore the great plains to a self-regulation condition. In 1962, Sear's beautifully written short book, *Where There Is Life,* was published, but was not widely read. It stands alone before the mountain of paperbacks on ecology, pollution, and the environment which were to follow in less than ten years. Sears wrote of the concern needed for "inner space" at a time when "outer space" seized the public imagination and financial support. It was he who stated the space-ship concept of the earth in accurate as well as colorful terms. Although he gained limited access to decision-makers and to the public, Sears' greatest influence in the extension of ecological ideas was with his younger colleague and his students; time became receptive for the next generation.

A distinguished contemporary, Raymond Fosberg, advocated the creation of a new profession, "The Community Ecologist" in a 1957 writing, and elsewhere. Fosberg tried to give a new meaning to applied ecology. As noted previously, the practice of applied ecology predated professional organization, but the primary interest for identification was ecology for foresters, wildlife management technicians or sanitary engineers. It was not until 1971 that a Section on "Applied Ecology" was organized within the Ecological Society. The membership is drawn from the traditional biological specialties and represents only a small fraction of the specialists who practice some phase of applied ecology, such as engineers, landscape architects, oceanographers, and agriculturists just to mention a few. In the vision of Sears and Fosberg, the Community Ecologist was a scientific generalist, who would take a free hand in relating all the units of a community. In current terms, he would be problem-oriented not subject-oriented in his attention. Professional ecology did not pursue the vision successfully. Few, if any, professional ecologists occupy the positions which are being created rapidly for the practice of community ecology.

Since the ideas of professional ecology have been its chief contribution to ecology as the popular movement, it is appropriate to identify some of these ideas before examining the non-professional roots of the movement. Traditionally, ecology was commonly defined for students as the branch of life—science which studies the relationship between organism and environment. The distinction may have been added between an organism and its environment on one hand and organisms in the more collective sense (species, population) and environment in a more general sense. It is not difficult to see why Ecology was disdained by branches of science which prided themselves on preciseness and control. To this day, there remains a substantial amount of inconsistancy and overlapping in the basic vocabulary of ecology. The field gained repect, nevertheless,

because it evolved study methods which could be applied to the inconstant and overlapping real world. "Community" became an environmental concept rather than a group of humans because ecologists succeeded in describing structure and operation of woods, fields, and ponds, including the organisms. Skill was developed in measuring both biological and physical components.

The ecologists produced detailed explanations of past changes in climate, populations, and communities with their methodology and began to predict that plant communities would replace each other in a known fashion. Food was traced through series of organism from source to destruction, and group roles acquired meaning: Producers, predators, parasites, and decomposers. There was a gradual change from specific segments to general models of a system, but older illustrations still linger: Wolf or man is pictured at the end of a chain or top of a pyramid, as if the food stopped there. Specific cases pointed to the central model of the ecological system which can be expressed in two forms, either a pyramid of blocks, or a circle connecting blocks. Properly drawn, either expression transmits the two critical properties: The cycling of material and the non-cycling (capture, flow, and loss of energy). This model, a philospher has quipped, is, "the godhead of the 20th Century." Popular concern has often been at the local level, figure 78, 79.

In passing, it should be noted that ecology came to incorporate some ideas which originated well outside its circle. From engineering and physics came the second law of thermodynamics, that disorder increases in time. Thus, in 1971, the professional and popular ecologist, Barry Commoner, expressed the critical ecological idea as, "There's no such thing as a free lunch." He was saying that all growth, maintenance, production, is achieved only at the price of destroying order. Engineering also suggested the comparison between the engine, as an almost closed system, and the environments which the ecologist had been studying. Indeed, Commoner's book was entitled, *The Closing Circle.* Engineering also supplied the concept of a "steady state" which found a biological counterpart in the idea of "homeostasis" from animal physiologists. Translated into popular ecology and into political practice, it justified regulation in scientific terms. Finally, out of the philosophy of science came the idea of holism, creative evolution, and mutual aid. Simplistically, the whole is more than the parts in isolation. In union there is strength. This was a rather direct

**Figure 78.** Citizen's groups in many areas coordinated their efforts for clean-up campaigns. One result has been attractive litter receivers such as this one in Cincinnati, Ohio.

**Figure 79.** Similar efforts have been directed towards open junk yards in the U.S. and Canada.

counter to the notions of Social Darwinism with its war of all against all. In professional ecology there was a generation of rather fruitless debate as to whether a community was a group of individuals or a super organism. In the society at large, the philosophical shift in viewpoint is manifest in myriad ways, from efforts to introduce non-competitive education, to talk of "corporate responsibility." The recent call from Ralph Nader for "creative citizenship" by consumers explicitly followed his recognition of the cycle model, and the self-regulation feature of a system through mutual activity of its parts. Nader's professional background was law, but he could not have been more lucid had he been a professional ecologist. His message can be reduced to, "Learn the structure and operation of the system, then become part of its regulatory aspect."

Professional science and philosophy are not vehicles for popular education, however. Three institutions can be identified in a summary judgment of the means by which masses of people have received information. Schools, extension services, and private organizations. It is curt but realistic to dismiss the school system. Even in the third wave of the popular ecology tide, there is no comprehensive school program in effect, although many commendable piecemeal efforts have been initiated, and curricula in environmental education written, published, and placed upon shelves. Extension services include other federal and state agencies except for the U.S. Department of Agriculture Extension Service, with its integrated land-grant colleges and county agencies. However, the magnitude of this effort occupies most of the field, and it alone needs to be considered here. The Extension Service undertook a bold pioneering task of popularizing applied ecology when the science of ecology hardly existed. The elaborate institutionalization of both Agriculture and Home Economics have buried their ecological inspiration: learn about the environment, and manage it wisely. The achievements of the institution must be judged as monumental. Nevertheless, its goals became narrowed to specific production and the promotion of niceties. When the clash came between the popular ecology movement and the established patterns, the staff realized its long habit of association with the immediate goals of business and industry. It was almost automatic to side with the agro-chemical industry in defending pesticides against any criticism, and to find its efforts had been concentrated on a whiter shade of clothing in complete innocence of pollution. In spite of

the limits of professional training and the ties of dependence with "anti-ecological" institutions, the Extension Service rapidly heeded the call of the new ecology movement. Overt signs of its soul searching have been manifest in a broadening of the agricultural-home economics base. Some of the changes deserve skepticism: to retitle a College of Home Economics as College of Human Ecology should convince no serious observer. The more significant response was that some local agencies sought meaningful alliances with citizens' groups, co-sponsoring symposia on population control, and publishing independent scientific information on consumer products involved in environmental issues without asking the leave of the producing industry. It is too early to predict whether the Extension Service will have a positive or negative effect on the ecology movement, or even whether its role will be significant at all. Other institutions have rapidly become far more influential, not only in the advocacy role, but in the very research and informational role where it has demonstrated a special capacity.

The visible foundation for the prolific ecological movement is an agglomeration of private organizations. This foundation may be examined for some appreciation of the present situation when there are not only too many organizations to enumerate systematically, but their heterogeneity almost defies attempts to define the boundaries of the movement. At one time no relationship was discerned between Consumer's Union and the Sierra Club. The common concerns have been discovered in practice, where the same people have tended to gravitate to consumer and environmental groups. Naturally enough, a perceptive housewife recently concluded, "A consumer movement makes sense only in relation to ecological problems."

A few organizations can be examined in brief detail as the most important units in the foundation. They share certain features in common. All were classified as "Conservationist" until recent years. All had a primary interest which did not demand either a broad knowledge of ecology or engagement in a common cause with other organizations. All considered themselves as educational in contrast to political. Objectively, recreation was the principal activity of members. Finally, all published a magazine, which consumed a major portion of dollars in dues. In practice, they could communicate mostly with their own membership. Nevertheless, historical investigation reveals that the Audibon Society and The Sierra Club undertook broad environmental defense through available political channels well over a half century ago.

The Audibon Society was founded for a particular interest in birds. For many individual members and local chapters, bird-watching continues to be the extent of involvement, but it is a short next step to advocating or acquiring bird sanctuaries. The Society has acquired a system of 40 sanctuaries, chiefly coastal, ranging from 26,000 acres downward. Acquisition of preserved areas is the core program of another organization, Nature Conservancy. In retrospect, a progression of values may be noted, although they continue to co-exist in time. From the private preserve for the recreation of the property owner, there is extension to altruistic acquisition. At this stage society either has no community interest or no power to preserve, and the right of an owner to unrestricted use of his property is unchallenged. Audibon's national leadership recognized the need of further changes in social values, and used its supporting membership to develop an educational program. Almost alone among private organizations, it was also able to enlist the aid of professional ecologists, salaried or voluntarily. Consequently its focus on ecology was as early as the 1950s and the quality of its educational programs remains among the highest. The society has prepared an elementary school text in ecology, and aided in the creation of school nature centers. As its membership and skill have grown, Audibon has become increasingly effective in an activist role; in 1969, it distributed "Ban DDT" bumper stickers to all members. Its legal and political role will be discussed under that particular topic. In the Society's own view of itself, "Members (200,000) . . . are trying to educate their fellow citizens to understand that man is a part of the natural environment in which he lives—that man is the loser when he destroys the natural resources on which he depends for food, for materials, for recreation and for beauty, for the water he drinks and the air he breathes." (1971 news release) By any reckoning, it is a long way from bird watching.

The National Wildlife Federation was founded much later than the Audibon Society, in 1936, and it reports that it is now the largest conservation organization in the world. Many probably know the organization only as the source of wildlife picture stamps, stationery and games. In addition to adult journals, it publishes for children the *Ranger Rick's Wildlife Magazine,* which is also the vehicle of a loose membership organization. "Rick" has achieved a very high standard of scientific depth translated to simple narrative language with graphic illustrations. The Federation publishes a number of single pamphlets which forcefully present conclusions from skilled professional sources. Their view is exemplified by the abstract of "The Issue" from "By Which We Live" (1957 and 1968), "In attempting to teach the story of conservation, we are dealing with a great many adult first-graders. Comparatively few people have seen a textbook on biology, but many have felt the influence of Mr. Aesop and Mr. Disney." The Wildlife Federation regularly prepares and distributes free or at cost three significant publications. *Conservation News* is well informed on issues and militant in language. *Conservation Report* is a detailed account of federal legislation and other national activity, with sharp commentary. The three annual *EQ Index* (1969–71) "Environmental Quality," have been superb quintessences of an impossible bulk of considerations which constitute environmental quality. Most recently, "1971 EQ Index drops again . . . but an aroused public is slowing decline."

The only comparable publication to *Cconservation News* is the *CF Letter* issued by the Conservation Foundation at a cost which seems average for the format. While the coverage seems to duplicate the National Wildlife publications, there is supplementation in practice. A high water mark in the ecology movement was the appointment of the Foundation's president, Russell Train, as key environmental advisor to the President of the United States. It is an indication of the organization's integrity that it lost little time in continuing to criticize the environmenal policies and acs of the administration which had employed its leader.

The Sierra Club, only a few years ago, seemed to be a popular California hiking club. Yet, its founder (1892) was John Muir, the great 19th century naturalist, and one of its first public battles was against the construction of Hetch Hetchy Dam in the Grand Canyon of the Tuolume River. Members are still inclined to hike, and to select wilderness preservation as their preferred cause. In recent years, and conspicuously since 1970, the membership has soarded to 135,000 in 150 local chapters. Sierra came to serve as a publishing house first for books with sentimental or aesthetic appeal of wilderness, then for strongly militant literature of the popular ecology movement, e.g. *The Population Bomb,* by Paul Ehrlich, a scientific best seller. Sierra Club has become overtly political on national issues, and some local chapters have managed to engage in activist campaigns as well as educational activities.

There have been few prominent personalities in the ecology movement. One of the most intriguing is David Brower, who represented Sierra's growing influence until he left that organization to found Friends-of-the-Earth. It plunged full-tilt into the broad front of environmental issues, and foreswore tax exemption as an educational organization and declared its intention to pursue political remedies. Like Sierra, Friends-of-the-Earth became a publisher of the popular ecological literature, with somewhat more success than it enjoyed in attempting to organize a national and international political movement. Friends-of-the-Earth has retrenched severely after an early growth. Its greatest influences may well turn out to be the personality of Brower himself, described in the New Yorker Magazine as, "The Archdruid." He embodies the revolution from outdoorsman to environmentalist: Calm and tense, humorous and foreboding, a nature-lover who has learned the statistics on radiation hazard and can argue with the engineers and physicists.

Among the many conservation groups which existed when 1970 approached, several have followed the general path of the Audibon Society, National Wildlife Federation, and Sierra Club to some extent. In retrospect, it already seems remarkable that people whose principal interest is hunting and fishing, gardening, or camping (in comfortable trailers) would come so quickly to support action which is based on rather abstract and scholarly ecological concepts. The message may be muddled at times with such issues as firearms control, but it has come through rather well. Most of the new organizations are overtly environmental, although many are single issue or special interest in name and in practice. Zero Population Growth (ZPG) is one of the most successful of the new groups (40,000 members), and can be examined as a meaningful facet of the environmental movement. Its beginnings were sparked directly by Ehrlich's 1968 "Bomb," and Ehrlich has continued to be its prophet, although the organization has not been the primary vehicle by which he has become widely known, and others have been responsible for its growth. The first ZPG disciples were college students, who considered Planned Parenthood as too cowardly, complacent, or devious. Possibly the students simply did not trust an organization with so many middleaged members, or they may have learned that the founder, Margaret Sanger, had four children. Planned Parenthood did not fit the conservative

role very well, once it felt the ecological wave had struck social walls, and ZPG has no material disagreement with it. Both have promoted male sterilization, for example. ZPG was free of complexity, however, and derived its impetus from ecology,—ZPG's "Pledge of Responsible Parenthood," begins with the one-for-one reproduction . . . until (point two) "pollution is under control" . . . and (point three) "it is quite clear that there are enough resources of minerals, chemical, water, and energy to satisfy the needs of more people." Standards of heredity, health, and standard of living are also sought by ZPG. There is abundant, if non-statistical, evidence that the pledge is informally adopted where it is heard, at least among the college educated. The young promise for the future, and their elders now say openly that they would limit children to two if they had it to do over again.

The only professional scientist in addition to Paul Ehrlich and the renowned medical specialist Rene Dubos, to gain wide popular recognition has been Barry Commoner. Commoner, like Ehrlich, has worked principally within the accepted academic mold. Beginning with a solid professional reputation, he gained popular recognition by lectures and writings. Many professors followed his, Science and Survival (1966) with their own books until the number has become overwhelming. In addition, they were found in demand for public lectures, and began to institute a limitless list of academic "environmental" courses. Academic symposia, from the American Association for the Advancement of Science downward have become almost weekly events. It is difficult to trace material effects in the national decision making to this lively academic activity. Unquestionably information has freely flowed, and probably it will be reflected in action of former students.

There have been a few attempts by scientists to organize for direct access to the political machinery and to the public at large on critical issues. At the threat of nuclear warfare and weapons, testing groups took three different non-academic approaches; lobbying, moral leadership, and information in common language. The Nuclear Test Ban Treaty of 1962, and the fiasco of fallout shelter building left organizations to disband, retreat into publishing scholarly articles, or find another cause. Environment quality inherited the attention of a handful of groups which had selected the public information approach. The St. Louis Committee for Nuclear Information became the Committee for Environmental Information, and, at the specific suggestion of Barry Commoner of St. Louis, the Rochester organization became the Committee for Scientific Information and jumped into water pollution as its first target. At the national level, the Scientists Institute for Public Information was formed with participation by some most highly respected scholars: Warren Weaver of the Sloan Foundation, Margaret Mead of the American Museum, Dean Abrahamson of the University of Minnesota, Rene Dubos and Theodosius Dobzhansky of Rockefeller University, and Commoner of St. Louis. The excellence of their research and the extent of their knowledge is reflected in this Encyclopedic reference work, in which they played a major part. However, no organization or federation of local groups was ever attempted. Local groups remained independent with communication being maintained through the periodical published by the St. Louis Committee (former name *Scientist and Citizen*, presently *Environment*) and by various contacts through S.I.P.I. and with each other. *Environment* was the first, and probably the best of the many issue-centered journals which are now supported by the ecological movement, but *Environment* has never been a financial success. The Rochester Committee for Scientific Information restricted itself to duplicated bulletins, for the most part on local issues. However, it pioneered nationally in providing information on lead poisoning, the hazard of cyclamates, and both content and consequences of phosphate-based and phosphate-free detergents. Over 130 bulletins had been issued by 1972, and there had been significant statements before local, state, and national governmental hearings. The S.I.P.I. report of June 1971 listed sixteen Science Information groups, four in California, four in New York. Some have done yeoman service on one or more issues, but few have been able to sustain a multi-disciplinary information effort over a period of time. As an achievement of a small group of dedicated scientists, the record of the information groups is a proud one, but it cannot be judged significant at the present time. On a small scale, the movement served as a training ground for scientists who could work successfully with citizens and officials. It is incredible that federal and corporate funds could be given for all manner of purposes under the sun and yet rarely and insignificantly to groups such as S.I.P.I.

Most scientists who have taken an active part in the ecology movement have chosen to participate in citizens' organizations. From time to time they have organized themselves temporarily for single issues. Certainly one of the most outstanding cases was the swarming of the Cornell faculty at the ground preparation for a large nuclear power station on Cayuga Lake. A high mark in objectivity as well as skill was attained when many of the scientists came to be supported by the utilities company to study the situation and did not experience sudden changes of perspective. Some scientists have chosen to volunteer skills to Sierra Club and other established "conservation" organizations, and a small but noteworthy group have provided expert testimony for the attorneys who petition the law to apply ecology. Scientists have helped prepare cases for the Environmental Defense Fund and for the National Resources Defense Council, and in major suits joined by a number of organizations.

The burgeoning of the Environmental Law is itself an important facet of the popularization of ecology. Legal antecedents may be sought in public health laws and in the area of land use regulation through zoning. No great number of people were involved in these matters directly or were even concerned with them. Before ecology became popular, the legal evolution had only slightly eroded the concept of absolute rights of property owners and absolute right to use land, air, and water even when the latter went on into another's domain. In the extreme view, which is still held by many individuals, there are no environmental issues, but only issues of property and injury between individuals. In addition, the legal strategies of the civil rights movement have provided patterns to use on environmental issues. Suddenly, in the space of a few years, an enormous number and variety of laws and administrative regulations have been added at all governmental levels. The overall effect is to test the limit of the police power of government, inevitably to extend the limits. As in the case of civil rights, and other fields, the body of judges does change its collective mind.

Acceptance of environmental law has been rapid and far reaching on the time scale of legal history. The ecological education of individual judges and lawyers cannot account for their widened perspective. While law schools have contributed to the large pool of environmental courses, an insignificant number of students or professionals study ecology in their academic work, and only isolated individuals have undertaken deliberate self-education. Like the American medical education an inordinate amount of student time is taken up with unnecessary and irrelevant curriculum. It is only a rare individual who can qualify as an inter-disciplinary environmentalist with legal ability. On the face of the evidence, the conclusion is forced that the legal profession has responded to the popular ecology movement, and have interpreted it as a valid change in the social values which govern their reining of the law.

In the space of three years, there have been enough court cases and interest to support at least two journals devoted to environmental law and a reporting service which totals over 1000 pages a year of abstracted material. It is too soon to discern trends from the bulk of judicial decisions. There have been comparatively few critical cases, chiefly on national issues. These have been initiated for the most part by the citizens organizations. At first their own individual lawyers were employed, but more and more, cases are undertaken by public interest law firms in collaboration with citizen's groups. Ralph Nader must be credited with founding or inspiring most of the active public relations law firms. They are yet few in number, chiefly located in Washington, D.C. and the New York area. In one experiment (State University of New York at Buffalo) staff and law students have voluntarily served in the capacity of a legal research group. No great number of individuals has yet responded to this movement, and its early success may not be repeated when opponents recover from their surprise to have found that some lawyers and judges have actually sided with bird-watchers and their ilk.

Even a brief discourse on the place of environmental law would be incomplete without recognition of the court battle against the Amchitka nuclear test in 1971. It was undertaken by a coalition of citizens and legal groups, and went to the Supreme Court of the United States. The decision was lost by a narrow margin, not so much numerically, but in terms of the issues of contention. The coalition won not only the right to use the legal steps, but it won the ecological argument. In the end, ecology was simply overridden by another value system—national defense. The second case of particular significance is the extended struggle between "Scenic Hudson" (which is the rallying point for many individuals and groups) and the Consolidated Edison utilities company over the construction of a pumped storage plant in Storm King Mountain. The record is by now a rich and bitter one. Powerful Con Ed was indeed delayed, and the inadequacy of regulatory mechanisms to deal intelligently with such a complex environmental problem was displayed for the curious. Indeed, a powerful stimulus was provided to improve the machinery rapidly. The cost to the contestants has been enormous, over a half million dollars by Scenic Hudson and far more by the utilities company. Quite clearly, the courts are much too expensive places in which to study and implement ecology.

Outside the court system, an intricate labyrinth of legal routes is available for extending ecological ideas into public decision making. Among them, the public hearing is the most important. At all levels of government in, both legislative and executive functions, public hearings have attracted a flood of both individuals and citizens groups. Such participation was scarcely imagined only a few years ago. The ecology movement is not alone in using the mechanism, but environmental issues now dominate many agendas. The idea behind the public hearing is an almost pure faith in democracy, free flow of information, and the judgment of the common man. The customary alternatives are uninformed authoritarianism, private advice of staff or outside specialists, and the influence of paid lobbyists. The public hearing is generally a far more effective, if not necessarily more influential, information exchange device than letter writing. Although its use is uneven throughout the country in keeping with the general level of participative democracy, a vital issue brings out the citizens to question the ecological wisdom of a road across the Great Smoky Mountains National Park in Tennessee and a chemical plant in South Carolina.

The views of the ecology movement have been surprisingly heeded in hearings in spite of the heavy odds arrayed against them. Hearing officers (officials, legislators) lack a background in biological or ecological education, most citizens lack training and experience in investigation and reporting, and the skilled ecologists who will leave ivy towers for the noisy marketplace are few. Time and again the unusual concerned and capable professor has gone to the hearing on his own time as unpaid expert to face a battery of hired and well supported corporation scientists. The professor can publish conclusions only when his other duties are met, then to a limited audience, while the corporate view can be promulgated in full page ads and mailed to the entire membership of the Ecological Society. The balance has not been redressed with the aid of scientists from governmental staffs; these are rarely fighters for ecological reasoning, and they are susceptible to muzzling. Moreover, few government agencies, or coalitions of agencies can match the staff available to any major polluter.

"Pollution Politics" has quickly developed into a field big enough to be analysed in both scholarly and popular volumes. Professional ecologists and other scientists with claims on special knowledge about environmental matters have entered the lists in a variety of roles. There always have been "company men" among scientists, but the entry of scientists into the lists of environmental defenders has given many more an opportunity to earn consultant fees. Even among scientists, few bite the hand which pays them. A marked change has occurred in the access of scientists to officials. Several who have been vociferous associates of the popular ecology movement have been asked to advise officials and legislators. At the national level there has been for some time a precedent of referring occasional matters to the National Academy of Sciences and an annual seminar has mixed scientists and legislators. Nevertheless, it is at state and local level that the contacts have sprouted. It would have been unlikely five years ago that an ecologist would be on speaking terms with state commissioners and the heads of important legislative committees. One state has recently established a full time scientific staff for its legislature, and a second had proposed a similar approach. The inter-disciplinary contact has been a productive one. A frequent experience of the ecologist has been to discover a deep environmental concern in the politician, whom he had never really trusted or respected. The learning experience has often been mutual. To a surprising extent, officials have stood ground with the ecological movement against a tremendous barrage of paid testimony, e.g., Erie County, New York, and Dade County, Florida, held their plans to prohibit detergent phosphate. Nevertheless, the old politics of lobbying is still effective, and may prove decisive. The *Wall Street Journal* both named and described the influence of the soap and detergent industry's chief lobbyist at the top level of the national administration.

Such models of success led an increasing number of environmental activists to the view that direct political action was necessary. "Game and Fish" lobbies had for some years actively represented the interests of sportsman's groups, whose name became "Conservation" and interests broadened somewhat. Most remain suspended somewhere between defenders of a "Daniel Boone" dream and allies in the ecological movement. Where they have joined with the new lobbyists, their experience and numbers have been valuable. The new lobbyists are products of the ecological movement, and they encompass the broadest range of environmental issues. By far the most conspicuous success has been achieved in the Northwest where the advanced environmental regulations of Oregon can be credited to an effective coalition of many different organizations, the Oregon Environmental Council. A similar state-level effort was recently organized in New York. Compared with most serious political efforts, the environmental lobbies are small: a single paid director for the Oregon Council. Nevertheless, the respect of both officials and opponents already has been won in both the west and the east. The power of the ecological political arm has arisen from the fire of volunteers and the catalytic atmosphere of the public. The great difficulty will be the need for the popular movements to sustain membership and interest over the many years necessary to get corporate turn-arounds.

Political involvement is being accepted not only by citizens who would otherwise disdain politics, but by scientists. A billboard symposium announcement at the 1971 meeting of the American Institute of Biological Sciences declared, "AIBS explores Biopolitics." A few individuals have already made the jump into elected office at local level. Nevertheless, it is difficult to find a single case in which environmental issues have dominated a campaign at any level. In 1970 the League of Conservative Voters circulated a rating sheet on congressional voting records. A more widely known tactic of this sort, by the League of Women Voters for example, might begin to influence the political system. The political system has moved. It was no liberal or conservationist president who devoted a third of his State of the Union message to the environment in 1970. Action has followed words, if not as closely as might be desired. The activity has been so prolific, heterogeneous, and rapid that it may not be comprehensible. However, three distinct acts stand out. First, there has been a marked consolidation of ecological functions at federal and state levels, and a few local governments are beginning to follow the trend. Second, the Environmental Impact mechanism, instituted for federal agencies already seems to have keyed the opening of a new ecological era. It is an astonishingly direct manifesto that ecological values are to be placed high in priority. Finally, government has rapidly become the ally of the popular ecological movement, not the cat's paw of exploitation or the benign bureaucrat who does not interfere in family quarrels. In the resurrection of the 1899 Refuse Act, the citizen was asked to report polluters, and the Army Corps of Engineers was, in effect, assigned to help him.

Both elected and administrative officials have not only identified themselves with the ecology movement, but they have been among its leaders. Senator Gaylord Nelson and Representative Henry Reuss, both of Wisconsin, have been mighty in the vanguard. Stuart Udall, Secretary of the Interior before 1969, appointed a distinguished academic ecologist, Stanley A. Cain, as an under-secretary. After his retirement from office, Udall became an active evangelist among citizen's groups and students and developed a personal interest in the capabilities of public interest law practice. The finest drama on the scene, however, was the transfiguration of Walter J. Hickel, who succeeded Udall as Secretary of the Interior. Despairing defenders fully expected Hickel to sell Alaska to the highest bidding exploiter. Indeed, the defeated promoters of the Smoky Mountains high-ways were almost waiting for an appointment before he entered his office. Yet, he had told his senate examiners firmly that he was for the policy of upholding environmental quality, and he soon demonstrated that he meant what he had said. When Hickel resigned, there were reasonable explanations, but almost nobody in the ecology movement believes anything except that he had proved to be too honestly committed to the ecological ideal. Hickel implored the President in a letter, to listen to the young people of the country; he had listened to his own children. In disbelief, professional and amateur ecologists have seen "the enemy" turn; the contractor leave the swamp, the hunter put down his gun, and the corporate executive place environment ahead of profit and power. These are a few dramatic moments, but they point to a scarcely rational general conclusion.

Could it really be that the ecological revolution must be credited to the non-professional and politically powerless gathering of youths and adults? These organized the first Earth Day in 1970, and have been the soldiers for recycling cam-paigns, Nader studies, and environmental educa-tion. Their skills have been scarce; their failures manifold; and most quickly drifted off to other kicks. Yet the world was moved. This writing has examined the various other facets, and most con-clude that none of the rest are so responsible for the phenomenon which has transpired.

The future course of the popular ecology move-ment is likely to be exciting, but not predictable. To some extent ecological thought has been built into the social system, and it will not be rooted out. It may be that the broader area of participative citizenship will be given a life of its own with the nurture provided. On the other hand profiteers have quickly assembled their wagons to be loaded: Gouging contractors in pollution abatement pro-grams, toy makers, consulting scientists, lawyers, and the rest. The sale of organic vitamins, "natural" foods, and earthworms has soared. The threat of environmental degradation has been abused to maintain privacy or gain business advantage with no material community interest. In such cases, the large corporation can easily defend itself, but the small businessman suffers. In all too many cases the millions spent by massive corporations on "environment" went for movies, television and radio media to tell the public how pure were their motives and techniques.

Reaction is always a danger, although new forms could scarcely be more venomous than the opposition which the scouts of the ecological movement had encountered for almost a century. Clever perversion against the movement has already been adapted to a fine art: It is demon-strably convincing that we should pollute more in order to pay for abatement. The current scarcity of money does much to probe the firmness of the movement, and the results are not encouraging. As a headline summarized the national accomplish-ment, "1971 was not a year of conclusive action," (*BioScience*, December 1971).

## Environment Impact Statement and Review: Has Tomorrow Dawned?

In the recent flurry of concern and activity created by people all trying to do something about the environment one feature stands out in elegant clarity. The federal National Environmental Policy Act of 1969 declared a policy which was followed by directives to all agencies of the Federal Gov-ernment with specific instructions for the prepara-tion of Environmental Impact Statements. The requirement for citizen review of such statements opened the door to citizen participation wide.

Even if the thrust of the policy act is blunted in adversity or dissipated, the record of the first two years must stand as the initial success if not the high mark of achievement in society's recognition and acceptance of ecological ideas. The language of the policy statement speaks for itself:

"The Congress, recognizing the profound impact of man's activity on the interrelations of all components of the natural environment, particularly the profound influences of popula-tion growth, high-density urbanization, industrial expansion, resources exploitation, and new and expanding technological advances and recog-nizing further the critical importance of all wel-fare and development of man, declares that it is the continuing policy of the Federal Government, in cooperation with State and local governments, and other concerned public and private organizations, to use practicable means and measures, including financial and technical assistance in a manner calculated to foster and promote the general welfare, to create and maintain conditions under which man and nature can exist in productive harmony, and fulfill the social, economic, and other requirements of present and future generations of Americans."

It is encouraging that the Congress has stated such sentiments, but statement is far easier than implementation. The NEPA proceeded with these directives:

Sec. 102 "The Congress authorizes and directs that, to the fullest extent possible: (1) the policies, regulations, and public laws of the United States shall be interpreted and administered in accordance with the policies set forth in this Act, and (2) all agencies of the Federal Government shall—

(A) utilize a systematic, interdisciplinary approach which will insure the integrated use of the natural and social sciences and environmental design arts in planning and in decision-making which may have an impact on man's environment:

(B) identify and develop methods and procedures in consultation with the Council on Environmental Quality established by title II of this Act, which will insure that presently unquantified environmental amenities and values may be given appropriate consideration in decision-making along with economic and technical considerations;

(C) include in every recommendation or report on proposals for legislation and other major Federal actions significantly affecting the quality of the human environment, a detailed statement by the responsible official on:

(i) the environmental impact of the proposed action,

(ii) any adverse environmental effects which cannot be avoided should the proposal be implemented,

(iii) alternatives to the proposed action,

(iv) the relationship between local short-term uses of man's environment and the maintenance and enhancement of long-term productivity, and

(v) any irreversible and irretrievable commitments of resource which would be involved in the proposed action should it be implemented.

These five points crystalize the NEPA. Their role is thus not unlike the 10 commandments of Moses.

Nevertheless, it is the review features which places the jewel in the hands of the democracy. Section (C) continues:

"Prior to making any detailed statement, the responsible Federal official shall consult with and obtain the comments of any Federal agency which has jurisdiction by law or special expertise with respect to any environmental impact involved. Copies of such statement and the comments and views of the appropriate Federal State and local agencies which are authorized to develop and enforce environmental standards, shall be available to the President, the Council on Environmental Quality and to the public as provided in section 552 of title 5 United States Code, and shall accompany the proposal through the existing agency review process;"

The following sections of the Act, D through H, aim at providing the information needed to answer the questions posed in section C. Typical of the provisions are:

"study, develop, and describe appropriate alternatives. recognize the worldwide and long-range character of environmental problems . . . preventing a decline in the quality of mankind's world environment. make available to states, counties . . . . individuals . . . advice and information useful in restoring, maintaining and enhancing the quality of the environment. initiate and utilize ecological information. assist the Council on Environmental Quality."

The language of the act implies a much more comprehensive idea of "environment" than most people share at this time. Although it specifies the "natural environment" quite clearly urban environment and the social fabric of the community is included. The most important implications are:

1. Planning and decision making must involve the "integrated use" of a wide range of disciplines including the natural and social sciences and the "environmental design arts."

2. Unquantifiable values must be given greater weight in the planning and decision making process.

3. Attention must be given to describing the environmental impact of proposed actions. Short-range consequences must be distinguished from long-range ones."

The NEPA began to be operational through a series of directives and responses among Federal agencies, which filled in the details of applying general principals to specific cases within the given governmental structure. These details require no further discussion here. Typically, significant consequences of the implementation process were not planned. The relative power and attitude of Federal agencies began to form clear and colorful patterns. The Department of Transportation and the Army Corps of Engineers were first and second in number of statements submitted. The Atomic Energy Commission was "failed" in its first offer to meet the requirements of the act, and instructed to return with better procedures.

As an elegant idea, the mechanism of Environmental Impact and Review was taken up quickly below the federal level. California, Delaware, Montana, Washington, Puerto Rico, and more recently New York have passed legislation modeled after provisions of NEPA. At least one County (Monroe County, New York) has implemented the spirit of the act, and used its structure informally, but rather extensively, in the environmental decision making process involving both public agencies and private development. The County Environmental Management Council participated actively in the review of Federal Environmental Impact Statements, which are routed through the regional planning board.

The significance of this implementation at county level, is that it has demonstrated the potential effectiveness of a local group, composed of county officials and public members, to question the work of operating agencies with highly skilled, but not necessarily ecologically informed, staffs. Strategies were developed, including use of available public resources: academic specialists, the Cooperative Extension Service, citizens organizations. It was learned that the adequacy of the study behind Environmental Impact Statements, not simply the statements themselves were the most vulnerable points for question. For example: when the statement for a proposed new town asserted that no damage would be caused to wild life, the review asked details on the professional qualifications of the investigator who provided this conclusion, and on the amount of time which was spent on the problem. Similarly, it was pointed out that no social scientist had been engaged to examine the short or long range consequences which a new town may have on a nearby old town.

Nevertheless, the effectiveness of local review depends strongly on the attitude of governmental agencies and private developers which not only have the advantage of skilled staffs, but the advantage of time to devote. A devastating counter strategy against citizen involvement is simply to wait it out. The early practice of Environmental

Statement was a farce, and considered to be nothing but a nuisance by the preparators; reviews were superficial if they in fact occurred. Rapidly however, serious reviews have emerged, and the NEPA has been "read" by reviewers to the operating agencies. As reviews have become effective, the reaction against NEPA has become an active campaign rather than passive disdain. There undoubtedly will be more and more powerful attempts to destroy the meaningful mechanism of statement and review. These may succeed.

With this uncertain future, it is well to record for history the heights attained by an agency in taking the spirit of NEPA as its own philosophy. The Buffalo, N.Y. district Engineer, U. S. Army Corps of Engineers, responded to a thorogoing thrashing which had been given to a preliminary environmental impact statement of the Corps as follows:

"Hopefully, your Environmental Management Council and our staff ecologists will soon have an opportunity to meet and discuss not only the Irondequoit Bay project but even a broader spectrum of topics, for we fully concur with Eugene P. Odum in that, '. . . man's success in manipulating his environment is proportional to the degree to which the whole environment is considered.' "

However one army colonel provided ample evidence that he could act as well as write. He secured and organized a three man group to prepare Environmental Impact Statements, consisting of a wildlife biologist, a plant ecologist, and an environmental engineer. If the scope of a particular problem exceeded the breadth of this staff, academic consultants were secured. The brief record of events is enough to demonstrate that the colonel expected his staff and consultants to deliver environmental analysis—not to defend projects, and he emphatically shifted the onus of decision to "go ahead" to local officials and citizens.

Perhaps it is unrealistic to emphasize this example of what one man wrought. Nevertheless, the social machinery moves or stops at the command of individuals who control its critical levers. The historical importance of NEPA and the Environmental Statement and Review process is that they have provided both the inspiration and the leverage for men of good will and environmental consciousness.

Prof. Herman S. Forest, Ph.D.
Department of Biology
State University of New York,
Geneseo, New York.

# Chapter 16

# Social Problem Solving and Pollution Control

Pollution as a component of the general question of environmental quality is a problem that has increasingly occupied the interests and resources of societies. Intense social concern is relatively recent and efforts in effectively dealing with pollution have not yet proven generally successful. The reason for this lies at least in part with the special qualities of this problem for social institutions. These special qualities in turn have an impact on the process of social problem solving and the social sciences.

This brief essay will examine these qualities and their impact on social problem solving and the social sciences and then explore some of these concepts in the context of a case study of pollution of the Great Lakes. While Great Lakes pollution is a problem that it far from being solved in the sense of a successful accommodation between man and his environment, it does represent some of the most intensive efforts made to date in dealing with a critical pollution problem in an area of high concern for two national governments.

There are three qualities of water pollution problems which make solutions particularly difficult for social institutions. First of all, the boundaries of pollution problems are not congruent with the political boundaries of governments which are the agencies of social problem solving. Municipalities, provinces, and nations have in fact often been boundaried along rather than within environmental systems. Rivers for example, most often are boundaries between political entities while being in themselves a focal point for environmental systems based upon the concept of river basins or river systems. As a result, actions to control pollution must frequently call for joint action by political units rather than action by a single unit. Thus jurisdictional responsibilities are divided and conflicts are not easily resolved over issues of how decisions should be made, which decisions should be made, and what resources should be used to implement those decisions made.

A second quality of the pollution problem is the fact that man's social institutions have tended to break down control of environmental systems into man-use functions. The environmental system has been divided and has been considered separately according to the various uses man has made of the resource itself. Administration and regulation of waterways has been handled within a single political unit by a multitude of agencies to include the perspectives of navigation, commercial fishing, recreation, public health, etc. By no means have those agencies had compatible requirements for water usage; as a result, to confront a political unit with a general question of water quality is to involve a plethora of governmental organizations, each with a real though limited interest in the resource. Due to historical availability of water, man has taken a man-use perspective and has compartmentalized his requirements for regulation. This traditional organization has created conflicts in social institutions for two reasons. First, the resource has become quantitatively limited so that the man-use interests are now competing for the same volume of water. Second, water quality requirements differ markedly for different uses, industry versus recreation for example, and so there is again conflict between interests for the use of water even over the same volume at different times. Industrial use upstream of proposed recreational use may be unacceptable in terms of qualitative and quantitative requirements imposed if pollution is of a certain magnitude. In essence, because of the nature of water as a resource and the high level of its use by man, society must address water use in the broad context of the total environmental relationships between water, the remainder of the natural environment, and the man-made environment. This broad orientation must be reflected in the social institutions created to deal with water quality.

A third quality of the pollution problem is the fact that it has generated broad social interest and demands for remedial action within a relatively short time frame. It is this pressure which has spurred governmental activity at such a rapid pace. The basis for this social pressure, while ambiguous and sometimes ill defined, rests on changes within the society itself. The magnitude of water pollution has increased in visibility as an issue to the general populace. Instrumental in this increased visibility is greater scientific knowledge which has confirmed and demonstrated environmental relationships heretofore unknown. The adverse effects of water pollution on human health have been demonstrated as well as causal connections with total environmental systems. An obvious reason for this pressure is increasing pollution from industrial growth without pollution control measures. Beyond this, industrialized societies have developed greater leisure and recreational demands on the resources have increased accordingly so that not only are the objective water quality conditions worsened, but societies have raised the standards of what is acceptable water quality.

Together these three qualities of the problem of water pollution have made difficult any attempts at solution of what has been agreed upon generally in industrialized nations as a pressing social problem.

Industrialized societies are today confronted with many complex problems. Perspectives on the scope and nature of these problems have changed largely in response to increased knowledge in societies drawn more closely together in the relationships between men and between man and his environment. This situation is characterized by broad social awareness of these large scale problems and popular perception of them in generalized terms such as the "pollution problem" and the "poverty problem." In fact, these large problems are a cluster of lesser problems each with major impact themselves upon society. This clustering of problems into one aggregate problem has developed what has been described as "metaproblems."[1] Environmental quality is clearly one of these areas of complexity for modern societies. Pollution as a popularly labeled social problem is a symbol for the broader questions of the man-environment relationship. It involves not the question of whether or not to pollute but rather how much pollution, under what conditions, and with what effects. Also, of course, it involves a whole set of relationships between man and the environment and within the physical environment itself to include land, air, and water.

While the scope of social problems has broadened, the basic tasks of social problem solving remain essentially the same. At the outset, there must be the development of goals relating to the task addressed. In a pollution context this means the development on a society wide basis of the definition of a desired man-environment relationship. The development of operational goals

for water pollution control can only be developed when there is broad consensus on this definition. A second task is the development of knowledge of the man-environment relationship sufficient to allow construction or policy alternatives which will support the definition developed. This is still a critical area for many aspects of the pollution problem. Rational policy must be based upon a solid foundation of scientific and technical knowledge. The ambivalence of the United States government on the question of banning phosphate detergents stems at least in part from a lack of common agreement among scientific experts on the nature of the relationship between phosphates and eutrophication or lake aging. The third and fourth tasks are most closely associated with what is described as the tasks of government. Here one is speaking of selecting and implementing policy alternatives. To be successful the policies selected must be the outgrowth of a commonly agreed upon social definition of the man-environment relationship and knowledge of how such a definition can be implemented. For example, water pollution legislation to include a ban on phosphate detergents will not be effective in this sense unless it reflects broad social agreement leading to voluntary compliance with the law and unless it is demonstrated that banning phosphate detergents will achieve desired results in water pollution control.

These four tasks must be considered with two additional tasks, that of linkage and feedback. Both deal with connecting processes between the steps discussed. Linkages include connecting the forward processes between steps to arrive at policy. It is critical that social definition, scientific expertise, and government policy formulation and implementation be connected. This involves a high level of government responsibility to encourage social definition and exert initiative in funding scientific studies as well as creating channels for communication of such knowledge to flow from experts to government decisionmaker's and society at large. This is particularly important when the problem is new and society's general knowledge of the problem is low. The sixth critical task is that of feedback. This is the backward connection of tasks, primarily from the implementation of policy alternatives back to the initial task of goals and the definition of the desired man-environment relationship. Only in the broadest sense can the desired man-environment definition remain static since it is descriptive of a dynamic situation which will reflect changing social values. Goals will change within the broader framework of social values changing and this must be continually reflected in the policies pursued by a society. Thus societies must consciously think of recycling ideas with scrutiny of social policies not only within the government but within the society itself. Because of the magnitude of change within industrial and post-industrial societies, this process must be consciously directed if choices made are not to be obsolete before they are implemented. Feedback must flow not only in the lines of sequence discussed but back from each task of the process. This in itself poses a difficult task for government institutions, yet it is crucial for handling the complex problems society faces in a rapidly changing human environment.[2]

The social problem solving task with its sequential steps as complicated by the special qualities of the man-environment problem has implications for all of the social sciences. To the extent that this new dimension has become an integral part of social considerations, it must be incorporated into social science endeavors.

The importance of such an incorporation stems from two trends in modern societies. First of all, man's relationship to his environment is now perceived as a closed system including relationships between man, his man made environment, and his natural environment. These relationships have been perceived as man has begun to test the real limits of the natural environment. In a direct sense this means, for example, that the worker in an industrialized concentration can no longer always find good recreational water use within a day's drive of his home. In terms of the use of the natural environment the system is closed and the awareness of this relationship forces man to decide upon and take action with respect to this closed system which leads to the second trend; modern man's increasing capacity for choice.

In the twentieth century man's knowledge of his physical environment and capacity to alter it have increased dramatically. Man is no longer a prisoner of his natural environment; he had chosen to modify it greatly, and he must continue to develop the capacity to make such choices wisely. Perhaps the central problem of modern man is how to deal with the problem of choice, for his physical and mental well being are and will continue to be based upon the choices he makes.[3] With respect to the natural environment, he must perform the social problem solving task and the social sciences play a vital role in enhancing this capacity for choice. This orientation toward social problem solving at

the very minimum demands inter-disciplinary efforts in the social sciences and forces each of the disciplines into a wider focus of interest.

To be sure, each of the disciplines can retain its identity but only in a modified form. To the extent one addresses broad social problems one must integrate a variety of considerations which cross the boundaries of the traditional disciplines. While certain disciplines can be associated with particular tasks in social problem solving, each discipline is a part of each task; and all of the tasks, as has been are neither purely sequential nor mutually exclusive. It is within this context that disciplines can be discussed in their relationships to the problem of man and his environment.[4]

In the psychological and sociological dimensions of the social sciences it is important to realize that man's behavior toward his natural and man-made environment is the outcome of individual and social perceptions of what that environment is and what he wishes it to be. The values underlying man's actions have not remained static, they have changed through time and have varied between cultures.[5] To examine and understand these attitudes is critical since they lie at the root of the first task in social problem solving; the definition of a desired man/environment relationship. While man has not always consciously defined this relationship, his attitudes and actions taken together have implicitly defined the relationship. This relationship coupled with man-man relationships and the psychological qualities within the individual form and triad upon which, in part, individual and social identities are based. Previously, the man-environment relationship has been neglected, but it must be understood if social choices are to be made with full knowledge of their implications. The decision of a society to preserve aspects of the natural environment at the cost of some industrial growth has broad implications for what that society stands for in the sense of encompassing broad values and attitudes within a society. Efforts to preserve aspects of the natural environment will be no more successful than the extent to which the values underlying such efforts are internalized within the society.[6]

Economic perspectives of society have been greatly altered by considerations of the man-environment relationship. Most important is the end of the historical assumption of air and water as free goods.[7] This assumption of air and water as available in unlimited supply to all is no longer valid because of the demands made upon limited available resources. Thus all economic considerations with respect to air and water now have a new factor of cost. This factor is a social problem solving context is critical to the selection and implementation of policy alternatives. It deals particularly with the costs and benefits of varying levels of environmental quality as well as who is to bear or enjoy which of the costs and benefits. The examination of these problems is essential to wise social policy and these economic aspects must develop in a climate where general social goals have been established and where there are processes and institutions for realizing the goals developed which encompass the political and legal dimension of social science considerations.

Political science is critically important to the solution of environmental problems because the nature of these problems in social and political, in essence, is the process of social choice. In the same vein law is social choice which has been institutionalized. That these problems are not resolvable on an individual basis is evident not only because of the pollution problem itself but because so much of modern man's existence is bound up in social relationships. This interdependence itself necessarily broadens the consideration of the problem to boundaries within which social choices are made. This presents a definite difficulty since, as discussed previously, environmental problems generally cross political boundaries. There are two levels of consideration; the first level deals with environmental system boundaries within nations. Problems at this level can generally be resolved when there is common definition of the desired man-environment relationship and institutions and processes are compatible between municipalities, provinces, and national governments. Within the political framework of societies an effort must be made to develop institutions and processes which enhance dealing with environmental problems. Institutions must be developed which can address the problem with coordination between traditional man-use organizational relationships. Also, processes must be developed to enhance linkages between problem solving steps and feedback between tasks.[8] A much more severe problem exists in the second level when the area of environmental concern is between two societies or nations. In this instance there is the distinct possibility that there may not be agreement even on the first step of social problem solving, that of defining goals. Also, institutions and processes to address the problem may be quite different. Political and legal institutions are to a large extent outgrowths of

societies and cultures, and compatibilities in addressing environmental problems are more likely to be the exception than the rule. Efforts must be made between nations to develop international agreements and coordinating institutions for concerted environmental action. Political science addresses these broad areas in terms of focusing on understanding the process of social problem solving itself. The demands of "meta-problems" are such that governments must become broadly and deeply involved in problem solution not only in selecting and implementing policy alternatives but also in encouraging social goal definition and the development of knowledge related to the problem.

All of the social sciences have a vital role to play in addressing the questions of the environment, yet this is not to understate the role of science and technology. Scientific knowledge is essential to the development of a definition of the desired man-environment relationship, and technology is critical to any efforts to arrive at the goals established. It is not the purpose of this essay to discuss the role of science and technology not only for the use of government decisionmakers but also by all members of the society. Thus one essential to the development of a definition of the social sciences in discussing the environment but also realize the need for a greater understanding between social scientists and the scientific and technical communities.

The foregoing discussion has been abstract and general in order to cover the scope of the problem. In order to further examine some of the concepts discussed, pollution of the Great Lakes basin presents an interesting case study of social problem solving

The Great Lakes are an important water resource for both Canada and the United States in the world (20%) in a total land and water basin containing the largest concentration of fresh water of 300,000 square miles in both nations. The basin falls entirely within the province of Ontario on the Canadian side but include eight states on the American side (Illinois, Indiana, Michigan, Minnesota, New York, Ohio, Pennsylvania, and Wisconsin). The geographic area alone does not sufficiently describe the importance of the basin for it is a center for population and industry in both countries.

Of particular significance is the lower Great Lakes basin which includes Lakes Ontario and Erie. In this area alone there was a population of 10.4 million Americans and 1.4 million Canadians in 1966.[9] The population concentration is co-located with highly industrialized areas such that this region produces one-fifth of the United States and one-half the Canadian gross national products.[10] The Great Lakes waters are used for a variety of purposes by both nations to include domestic water supply, sanitary purposes, navigation, industry, recreation, fish and wildlife, and agriculture. These purposes, as discussed earlier, are not mutually compatible in the same place at the same time, and as a result there is competition for water use on a qualitative and quantitative basis.

In recent years there has been increasing public awareness of the fact that the Great Lakes were becoming polluted. Of course there were great variations within and between individual lakes, but in general the quality of water in the Great Lakes has been deteriorating.[11]

The social problem was not simply one of ending pollution but rather coming to terms with a resource that was no longer a free good. Because of the magnitude of use, water in the Great Lakes was ending as a free good forcing the question of what priorities should be attached to the above uses, how the priorities could be made operational, and, now that the good was no longer free, who was to pay and in what proportions for water use.

The social institutions were not prepared to deal with these questions at the time they were raised, and even now the questions are still in the process of being answered. One should also add that the question of water quality was, for both the United States and Canada, a question of national scope rather than purely a Great Lakes centered preoccupation.

If there is a chronological point at which to begin consideration of Great Lakes water pollution efforts it should be in the period following World War II, for it was then that the first major international considerations were made of the problem and definitive steps were begun at national levels.

From this period to the present there are certain threads of continuity in both United States and Canadian policy in approaching water pollution control. First, there has been a trend within both federal systems toward the development of federal level initiative if not authority in water pollution control at the expense of provincial or state autonomy. This has been in response to the realization of the necessity to deal as much as possible with environmental systems as units. Also, such centralization of authority was recognized as a necessary prerequisite to any effective cross-

national water pollution control efforts. A second thread has been the development of government institutions within each nation which more broadly encompass water quality concerns in order to minimize difficulties in coordination which existed when water regulation was divided into the various agencies of man-use functions. A third thread, although less continuous than the others, is the recognition by both governments of the necessity for a high level of cooperation between the United States and Canada in order to effectively address the problem. In this sense there has been an effort toward better United States—Canadian cooperation made possible only by the development of the previous two threads. This development has resulted in use of the International Joint Commission to address the question of water pollution among other boundary water issues. The threads can be best examined by discussion of United States, Canadian, and joint water pollution control efforts respectively.[12]

The first federal legislation in the United States aimed specifically at water pollution control was passed in 1948. By that time responsibility on this issue had already gravitated to a state level from municipal concerns in all of the states in recognition of the inability of municipal government to adequately deal with the problem.[13] This set the stage for the legislation in 1948 which, although in many ways weak and of temporary duration, did establish broad federal interest in the problem.[14] The law was extended in 1953 and in 1956 another water pollution act was passed which expanded the federal role further yet did not completely answer the major question of federal enforcement authority as once again Congress demonstrated a sensitivity to the continuing debate over state versus federal jurisdiction on enforcement of water pollution control legislation.[15]

The decade of the 1960's was a period of intense water pollution legislation and activity. Water pollution control acts in 1961 and 1965 were of major importance in developing federal enforcement authority and creating a separate federal agency to deal with water pollution matters. In 1966 still another act was passed which extended federal grants on water pollution research projects. Of importance for Great Lakes considerations, the 1966 act established a federal—state linkage of procedures in the event of international pollution to include establishment of a conference with foreign nations invited to participate through the Department of State.[16] Legislation in 1970 specifically authorized funds for research on Great Lakes pollution and also established the Environmental Protection Agency which absorbed many of the water pollution control activities previously scattered throughout the federal establishment.[17]

It is still too early to fully evaluate these programs and enforcement actions taken by the United States federal government based on this legislation. Nevertheless, it is important that the United States has, over the period addressed, moved progressively toward a more comprehensive approach to water quality management at the federal level, and the program has had a direct effect in dealing with the Great Lakes basin.

In addressing the Canadian side of the problem only the province of Ontario is involved. Federal provincial relations are important however since Canadian federalism leaves much more autonomy at the provincial levels than does the federalism practiced in the United States. As a result, the drive toward federal initiative and authority has been much slower than in the United States and there have been long constitutional arguments over the role of the Canadian federal government in water pollution control. The efforts of Ontario are important not only because of this relationship but also because the province has been a leader in developing a progressive approach to the problems of water quality. In 1956 the Ontario Water Resources Commission (OWRC) was created with comprehensive powers for water quality control. The OWRC has been successful in most of its endeavors but problems have remained in effectively coordinating with United States counterparts due to constitutional problems.

The responsibility of the Canadian federal government has been severely limited and authority was lodged in traditional water pollution concerns through the protection of navigable waters and fisheries legislation with a plethora of concerned agencies. Aware of actions in the United States, concerned with developing compatible institutions and standards, and directed toward meeting the need for a national water policy, the Canada Water Act was passed in 1970 after lengthy consideration by the Canadian parliament.[18] While the law does not assert federal authority over the provinces, it does establish federal initiative and make possible the effective implementation of international agreements between the United States and Canada. In the same year the Government Organization Act created a Department of the Environment to address general questions of environmental quality and coordinate the programs of other departments.

Thus by 1970 there had been created the institutional and legal basis for international agreements and, more important, implementation of those agreements in order to limit Great Lakes pollution. Aside from national efforts leading toward the capability for joint consideration of the problem, there has been a U.S.-Canadian agency dealing with international aspects of this problem, the International Joint Commission (IJC).

In 1909 the United States signed a treaty with Great Britain (acting for the Dominion of Canada) governing boundary waters questions. At the core of the treaty is the creation of an international commission to deal with boundary waters problems. Article IV of the treaty states that boundary waters "shall not be polluted on either side to the injury of health or property on the other."[19] The IJC has the capacity to address pollution of the Great Lakes but only on referral of questions from both nations, and the recommendations of the IJC in response to a referral are not binding on either government.

There have been three referrals on Great Lakes pollution to the IJC, in 1912, 1948, and 1964. In response to these referrals the IJC created technical boards with experts from both nations to examine the problem.

In the first report the IJC examined bacteriological problems from a public health perspective and found that pollution did in fact exist in violation of the treaty; both nations took corrective but limited measures. The second referral was answered by a 1951 report which again prescribed corrective action; both governments agreed to act on the recommendations but in terms of examining the respective national efforts, it is clear that neither nation had the capacity to take effective action.[17]

The 1964 referral addressed the entire lower great Lakes Basin (Lakes Erie and Ontario), considerably more of the basin than had been addressed in previous referrals, and came at a time of broad public concern. The 1970 report of the IJC presents a major technical study of the area and both technical and political recommendations for water pollution control.[21] The IJC recommended water quality objectives for the basin and, of major significance, it made two types of political recommendations. First, it made specific recommendations for joint action by the United States and Canada to meet proposed water quality objectives. Second, it requested authorization from both countries for the IJC to develop the capability for the continuous monitoring and surveillance of boundary waters for the purpose of water pollution control. These two recommendations, taken together, can establish a framework for a truly cross-national basin wide approach to water pollution control within the limits imposed by the reality of national political boundaries.

These recommendations have been tentatively accepted by the United States and Canada to the extent that public agreement was announced in June 1971 for a joint program to eliminate pollution in the Great Lakes.[22] Confirmation of this agreement was made in a treaty signed by both nations in April 1972.

The IJC report and subsequent agreements between the United States and Canada cannot be isolated from previous actions in Great Lakes pollution control efforts. They are, rather, the culmination of national and international efforts over an extended period of time which have sought to deal with this broad problem. The problem of water pollution in the Great Lakes is far from solved. Yet the example of efforts to date in the Great Lakes basin can serve as a case study from which to draw knowledge in solving the social problem of water pollution control within and between nations.

Clearly, progress has been slow on water pollution control due to the peculiar qualities of environmental problems. Time has been short and this has forced simultaneous efforts in social problem solving tasks. Technical knowledge has had to be developed to understand the nature of the problem and help propose alternative solutions. Social institutions have had to be developed to use information and develop policy alternatives for society at large. Lastly, societies have been forced for the first time to define a desired man-environment relationship. Simultaneous solution of these tasks has been difficult yet unavoidable, and the role of the social sciences has been and will continue to be crucial in these endeavors. In modern complex societies in an ever changing world there can be no expectation that these problems will ever finally be answered. What is important is that the process be understood as dynamic and that both scientific and social science experts and all of society must internalize the idea that dealing with man's relationship to his environment is a continuing social task. The stakes are high and may include the very survival of the species and the biosphere, and there is no alternative for man except to mate scientific knowledge and social values to make the best possible decisions on "spaceship earth."

# REFERENCES

1. Michel Chevalier, *Special Science and Water Management: A planning Strategy,* Department of Energy, Mines and Resources (Ottawa: Queen's Printer, p. 4 (1970).
2. See Alvin Teffler, "The Structure of Social Futurism," Chapter 20 in *Future Shock* New York: Random House-Bantam, pp. 446-487 (1970).
3. See David Apter, *The Politcs of Modernization* (Chicago: University of Chicago Press, 1965) p. 11, and Erich Fromm, "Freedom—A Psychological Problem?" Chapter 1 in *Escape From Freedom* New York: Avon, pp. 17-39 (1941).
4. For a discussion of some of these ideas see Lynton Caldwell, "Policy" Part I, in *Environment: A Challenge of Modern Society* Garden City: Doubleday-Anchor, pp. 3-55 (1970).
5. For a particularly interesting study of the relationship between environment and culture see Peter Freuchen, *The Book of Eskimos* New York: World-Fawcett, (1961).
6. For further discussion see Clarence J. Glacken, "Men Against Nature: An Outmoded Concept," in *The Environmental Crisis,* Harold W. Helfrich Jr., ed., (New Haven: Yale, 1970): and Earl Finbar Murphy, "The Origins of Environmental Control" Chapter I in *Governing Nature* (Chicago: Quadrangle, 1967): and F. Fraser Darling and John P. Milton, eds., "Social and Cultural Purposes," Chapter IV in *Future Environments of North America* Garden City: Natural History Press, (1968).
7. Allen V. Kneese and Blair T. Bower, *Managing Water Quality: Economics, Technology, Institutions* (Baltimore: Johns Hopkins, 1968).
8. Aspects of these concepts are discussed in Karl Deutsch, *Politics and Government: How People Decide Their Fate* (Boston: Houghton-Mifflin). See Chapter 7 "Self-Government: How a Political System Steer's Itself," pp. 145-162 (1970).
9. International Joint Commission, Canada and the United States, *Pollution of Lake Erie, Lake Ontario, and the International Section of the St. Lawrence River* Washington: U.S. Government Printing Office (1970).
10. *Ibid.,* p. 18.
11. For a general discussion see Gladwin Hill, "The Great and Dirty Lakes," in *Controlling Pollution: The Economics of a Cleaner America* Marshall I. Goldman, ed., New York: Prentice Hall, pp. 43-48 (1967).
12. For a full discussion of these questions see William E. Wilson, "Environment as a Cross National Policy Problem: The Great Lakes 1950-1970," unpublished doctoral dissertation, Fletcher School of Law and Diplomacy, Tufts University (1972).
13. J. Clarence Davies, *The Politics of Pollution* New York: Pegasus, p. 38 (1970).
14. U.S., *Public Law No. 845,* 80th Cong., 2d sess. (June 30, 1948) "Water Pollution Control Act."
15. U.S., *Public Law No. 660,* 84th Cong., 2d sess. (July 9, 1956), "Water Pollution Control Act Amendments of 1956."
16. U.S., *Public Law No. 753,* 89th Cong., 1st sess. (November 4, 1966), "Clean Water Restoration Act of 1966," Section 206, Sect 10(d) para. 2.
17. U.S., *Public Law No. 224,* 91st Cong., 2d sess. (April 3, 1970) "Federal Water Pollution Control Act Amendments," and *Government Reorganization Plan No. 3,* 91st Cong., 2d sess., (December 2, 1970).

18. *Canada Water Act,* Chapter 5, 1st Supple., Ottawa: Queen's Printer, (1970).
19. Treaty Series. Treaty Between the United States and Great Britain Relating to Boundary Waters and Questions Arising Between the United States and Canada: 36 Stat 2448; TS 548; IIII Redmond 2607, British Treaty Series 1910, No. 23, Treaty and Agreements Affecting Canada in force between His Majesty and the United States of America, 312 Ottawa: King's Printer (1927).
20. International Joint Commission, United States and Canada, *Report of the International Joint Commission on the Pollution of Boundary Waters,* Washington: U.S. Government Printing Office (1951).
21. International Joint Commission, United States and Canada, *Pollution of Lake Erie, Lake Ontario, and the International Section of the St. Lawrence River,* Washington: U.S. Government Printing Office, Chapter XVI, "Recommendations." pp. 149-160 (1970).
22. *The New York Times,* p. 11 (June 11, 1971).

**William E. Wilson, Ph.D.**
**Department of Political Science**
**Saint Michael's College**
**Winooski, Vermont**

# Chapter 17

# The New Ecology

Over the last two decades, beginning in the early fifties, there has grown up in the United States a dynamic and far-reaching concern with the environment. Starting with the radio-active fall-out issue (ambient radiation levels 20 years ago were some ten times what they are now), this concern gradually broadened to include all aspects of man's physical and biological environment. Now, we find ourselves troubled about air and water pollution, soil problems, toxic chemicals in the food we eat and liquid we drink, and so on. Concern has also begun to spread to aspects of the man-made environment; so that environmentalists are concerning themselves with such problems as lead poisoning afflicting children living in substandard homes and eating chips of flaking wall paint containing lead compounds.

There is, no doubt, a serious, growing problem of environmental deterioration in the United States, and more and more people are becoming aware of it. Indeed, in some places citizens are becoming so aware of the problem that it is causing a crisis of confidence in government. A recent *Wall Street Journal* article reports that growing numbers of persons in Los Angeles do not believe the city government when it says the air pollution problem is getting better; their smarting eyes tell them it is not. People are also realizing that the environmental crisis, although expressed and felt in local terms, is not just a local concern, or even a national concern, it is in fact a global concern. It is one that affects every citizen on the planet. The calling of the United Nations Conference on the Human Environment, held in Stockholm in June of 1972, was a signal to the world of the global importance of the situation, but the divisions among the viewpoints demonstrated how far the community of nations must go to reach a global understanding.

However, the environmental problem does not affect all the peoples of the world in the same way. In the industrialized nations the environmental problem is characterized by increasingly dirty air, the fouling of rivers used as sewers, ever-mounting piles of solid waste materials and a general deterioration of the ecosystem. In the developing nations, however, the primary characteristic is depletion of the natural resource base—these resources primarily being exploited the benefit and profit of the industrialized one third of the world.

In an analogous way, the environmental problem in the United States affects poor and working class people in different ways than it does the more affluent middle and upper classes. The most serious environmental problems (those directly related to the health of human beings) affect poor and working class people most often and most severely because of the areas in which they live and work, as for example, air pollution problems, which are much more serious in inner city areas than they are in the suburbs. These problems are made more severe because of the lack of adequate medical facilities to which the poor have access. It is literally true that the natural resource represented by clean air is being depleted in inner city areas for the benefit, primarily, of those who live outside the cities—those fortunate enough to be able to live in suburban areas. Industrial complexes which produce noxious compounds are placed in the urban environment. In addition to this already double burden, many of the solutions to environmental problems currently being proposed (additional taxes on electric power, strict automobile exhaust emissions codes with no concomitant improvement in the mass transit system), add yet another burden on the poor and working class citizen, because they hit him in the place where he is least able to absorb the blow—his pocketbook. The real problem is whether or not a society can render health and freedom from detrimental environment an economic commodity to be purchased by the consumer.

There has been considerable debate in the last few years over the root cause of the environmental problem. Some examples of the more serious situations on environmental degradation will serve to give us some evidence for answering this question. Probably the type of environmental degradation most noticed by citizens of the United States, at least by those who live in cities, is air pollution. Day after day, millions of people inhale countless particles, most of which are in some way detrimental to human health, in ways which we are just now beginning to understand. Poisonous gases in varying amounts also fill the air, and are the daily intake of these same millions of people. We now know that nitrogen dioxide, a principal pollutant, is one of the causes of, or at least exacerbates the dread disease emphysema; while sulfur dioxide, particularly in moist air where it combines with the water to form sulfuric and sulfurous acid, is a serious health hazard and another major pollutant. People are literally dying of air pollution. Recent studies have shown direct correlations between the incidence of respiratory diseases and air pollution levels.

Much, although by no means all, of the urban air pollution problem is caused by the automobile. An enormous amount of effort has been expended to make the automobile faster and more powerful but this same effort has resulted in making the automobile a more serious menace to human health. The reason is very simple; as the compression ratio of an automobile engine goes up, which is necessary to increase the power, the temperature inside the combustion chamber rises. With this increase in temperature comes an exponential increase in the amount of nitrogen dioxide formed. Nitrogen (forming about 79% of the air) comes in as an unwanted element, and the higher the temperature, the more oxygen combines with the nitrogen, forming nitrogen dioxide. Thus, as automobiles were made faster and more powerful, they also became more efficient producers of nitrogen dioxide.

This situation is intensified because people are driving more and more. In the years 1946 to 1967, for example, vehicle miles per capita increased 70%. This is due to a number of factors, probably the most serious among them being a massive marketing effort to sell and encourage cars with a concomitant decline of bus passenger-miles during the same period. These two factors resulted in the deterioration of the mass-transit system. Once the spiral starts, it is difficult to unwind; the deterioration of bus and other forms of mass transit creates a demand for bigger cars and more and better highways, which causes a further deterioration of mass transit service, and so forth.

Another major cause of air pollution is electric power production. Nitrogen dioxide and sulfur dioxide are the main contaminants; there are often other trace contaminants, such as mercury in an unknown form, which do not as yet pose a real threat as an air pollution hazard, but do pose a threat to human health when they are washed out by rainfall and get into the ecosystem. The current

doubling time for electric power production is seven to ten years. This is clearly a situation which, if allowed to continue, will result in serious environmental deterioration.

It is instructive to look at some of the reasons for this precipitous increase in electric power production in the United States, since neither the rate of increase of the population, nor the rate of increase of the Gross National Product, nor the two increases combined equal the rate of increase of electric power production.

Aside from increased use of appliances such as television, there has also been an increase in the manufacture of materials that require relatively large amounts of power to produce, at the cost of those things which require relatively small amounts of electric power. Thus, for example, although the per capita consumption of bottled and canned beverages has remained about the same, the use of aluminum as a container for this purpose has increased dramatically. It takes almost seven times as much electricity to produce an aluminum can as it does a steel one; the increase in electric power for this purpose alone is therefore significant.

There are a number of other power consumptive articles, such as plastics, which have displaced other manufactured articles, such as natural fibers, which are much less power consumptive and which therefore, from that standpoint alone, are less demanding upon environmental quality. As will be discussed later, these products also contribute to environmental degradation in other ways. In short, as our mode of industrial production is shifting from relatively power poor practices to relatively rich power practices, there are very serious implications for the environmental future and we must raise serious questions about the system by which we produce our goods and services.

There are other examples of serious flaws in our productive technology which raise interesting and serious questions. Take, for example, the production of agricultural goods in the United States. Data show that, although productivity per acre has increased, the total farm acreage has decreased, leaving per capita production of agricultural goods virtually unchanged. The technological causes of this increase in agricultural productivity are well known; the use of herbicides, insecticides, etc., and especially the use of inorganic fertilizer have caused dramatic increases in the yield per acre of some of the most important agricultural crops grown in the United States. However, these increased yields have not been bought without cost. Heavily fertilized areas are producing as an unwanted side effect high nitrate levels in drinking water in excess of the limit set by the United States Public Health Service. Most of the well water in such areas has for some time been marked by high nitrates; the present worry is that many municipal drinking supplies are also developing high nitrate levels. As just one example of the ill-effects of high nitrate intake, particularly in infants, it is apt to lead to a disease called *methemoglobinemia*, which results in a reduced oxygen-carrying capacity of the blood.

The excessive use of pesticides such as DDT has caused the by now well-known effect of thinning egg shells in certain birds; what high intake levels of DDT and other pesticides have done to human health is less well known, although it is certain that it is not good. Perhaps even more importantly, the application of pesticides such as DDT have been carried out in such a wholesale fashion that not only are soils in many areas poisoned, so that very few micro-organisms, necessary for long-term soil fertility, will grow, but also resistant strains of pests are developing. This of course means that more, different and stronger pesticides must be used, with the vicious cycle continuing unbroken.

In other words, here again is an example of flawed technology, a technology which because of the demands of the ecosystem, finds it increasingly difficult to meet its own demands. On top of that lies the serious problem of deleterious side-effects which result from the application of this technology. There are grave questions about the ability of this technology, as it has developed in the last twenty-five years, to meet the basic human needs it was designed to meet without causing untold damage to the human environment. It is doubtful whether it can succeed measured on its own terms.

In response to these questions, we must go to the heart of the issue of how we organize ourselves to produce the goods and services necessary to meet the human needs of a growing number of people. Many investigators, primarily those rooted in what is known as the "science information movement," have developed a set of ideas which place environmental problems in a new perspective. In their simplest form they state that environmental concerns give us a window through which we can look and examine the workings of our society. In other words, environmental problems, both as a whole and taken one at a time, give us tools with which we can unravel the very complex workings of society, not only the one we happen to live in, but others as well. That is to say that there is

no such thing as an "environmental movement" which exists in isolation from the rest of society, but there is an environmental arm of the general movement for social justice, and the increased awareness of greater and greater numbers of common citizens.

Environmental concerns are a particularly clear window into society, and also a very real tool to accomplish other ends as well. However, just as with any tool, they can be employed for retrogressive ends as well. Hence the concern with the needs and desires of poor and working class people, on a national scale, and with those of developing countries on a global scale. For it is true that unless these concerns are taken into consideration at the very beginning of the development of environmental solutions, the poor are apt to be the only ones to suffer the consequences of inept solutions to environmental problems. Thus we begin to define the outlines and dimensions of the "new ecology."

An example will serve to make the point. In St. Louis, in 1970, an ordinance was passed which banned the burning of coal for heating purposes. On the surface this seemed like a reasonable rule, since coal burning had in the past contributed greatly to the problem of air pollution, and it was well known that many of the components of coal dust caused or aggravated many diseases, including lung cancer and emphysema. Most residences and buildings had been converted to other forms of heat anyway, and it was not felt that the ordinance would visit a hardship on anyone. However, this analysis did not reckon with a population of poor people in the city, who had not been able to convert their homes to another form of residential heating, and still depended on coal as a source of heat. These residents now found themselves unable to use coal, and had therefore to turn to the only other thing their furnaces could use—wood. A bucket of wood cost the same as a bucket of coal, only it lasted for just a day, whereas the coal lasted a week. Not only was it more expensive, but it was very difficult to get enough wood and raised the amount of wood consumed. Typically, residents would be able to get enough wood to heat their homes for about one day a week. No attention was paid to this problem in the original ordinance. This, then, impresses upon us the necessity of avoiding simplistic solutions to complex problems, which end up aggravating the very situation they were intended to improve. It also impresses upon us the necessity of taking into account the people whom we propose to help, and in particular to look at environmental solutions through the eyes of the people who are apt to gain or lose the most from a change in the way society produces goods and services. These will usually prove to be the poor and working people of the world.

When the environmental problem and its proposed solutions are looked at from this point of view, some interesting facts and relationships develop. For example, in the United States there has been no net increase in the total per capita amount of fiber (synthetic plus non-synthetic) produced in the years 1945–1970, but there has been a displacement of natural fibers (cotton and wool) by synthetics. In other words, the basic consumption per person has not changed, but the mode of production has. When one realizes the environmental degradation that results from production of most plastics, the seriousness of this displacement becomes obvious. The same kinds of displacements have developed in other areas as well; aluminum replacing steel is a good example. The electric power required to produce a container from aluminum is many times that required to produce a steel container, and electric power production is one of the most seriously polluting activities in the industrialized world.

The "new ecology" is interested in viewing the environmental problems from an economic point of view, and one that tells us something about the needs of people and the modes of production. This is not standard economics, for that field has had to reduce the environmental complexities to a system of what it calls "externalities." Rather, it is an attempt at understanding the workings of a system from the point of view of one symptomatic attribute of that system, the environmental problem. But, in order for the points to be elucidated sharply, the environmental problem must be viewed from the point of view of the lower socio-economic classes.

Thus, we find the "new ecology" *not* concerned with pollution, per se, but rather with pollution as an indication of some flaw in the structure of society. After all, if one looks at just the pollution aspect of the environmental problem, it is possible to say that the only thing needed is more sophisticated pollution control equipment, to turn back the clock on environmental degradation. Both the fact that this does not seem to be being accomplished very fast and the tremendous cost of such equipment make it seem unlikely that this will be done. This is not to say that more sophisticated techniques are not necessary and even essential, but simply that they are only part of the story.

Looked at from a global point of view, and realizing that the situation in the Third World is vastly different from that in the industrialized portion, the environmental problem also gives insight into social structures and relationships among nations. It is clear, for example, that developing nations are suffering from the aftermath of a vicious social system, specifically colonialism. The current situation gives us some insight into what aspects of that system remain. The developed and industrialized portion of the world is characterized in its environment by large amounts of pollution. Irrespective of the country, the degraded environment is commonplace, it is marked by industrial poisons in air and water, excessive noise in urban regions, and diminished space. The situation is aggravated by the fact that the over production of durable goods and packaging produces masses of solid waste while the decrease in open and uncommitted land leaves no space for disposal. The plight the underdeveloped countries face is quite the reverse with depleted resources, little return benefits, and booming populations. The imbalance between the developed and underdeveloped nations must be solved if either environmental or social sanity are to triumph.

To give just one simple illustration, the manufacture and use of plastics in the United States is producing severe environmental consequences and has the potential for even worse ones. Might not the solution be to replace as much as is practical of the synthetic fibers with natural cotton and wool? Now it turns out that neither cotton nor other natural plant fibers are feasibly grown in the United States. The cotton industry survives in the United States only because of an enormous government subsidy. On the other hand, it is quite feasible to grow cotton in tropical, underdeveloped countries and it is practical for the citizens of those countries to manufacture clothing and other articles from the natural fibers. Such a trade-off not only relieves the unnatural environmental pressure upon the American ecosystem but yields economic and social benefits to the Third World states giving them the economic independence they must have to solve their environmental problems.

With little imagination one can visualize other ecological and industrial practices which would work to the worldwide benefit of mankind. Many illustrations come to mind; cotton growing clothing manufacture in Brazil and Guatemala; rubber growing and tire manufacture in Malaysia; oil production and gasoline and other petrochemical compound manufacture in the Middle East and so on. The very difficult and complex question that has to be answered is—how do we get there? What is the path by which we can travel to unwind the spiral of environmental degradation and industrial inefficiency in which we now live?

It is perfectly clear that it is not up to the industrialized part of the world to dictate to the underdeveloped part what to do. That is, we can imagine a scenario, as was done above, but fulfillment of that imaginary drama demands that people all over the world have the right to determine their own futures. The failure of European and American aid programs to bring this about is good evidence of the nature of those aid programs. In like manner, it is not for the "haves" in the United States (or any other country, for that matter) to tell the "have nots" what to do; rather, they must be allowed to come into full sharing of the richness of American life. When there are no slum houses, there will be no more lead poisoning; when there are adequate facilities for garbage pickup and disposal, there will be no more harborage and food for rats. When there are adequate jobs for all, people will be able to provide a clean and healthful environment for themselves.

Seen in this light, the "new ecology" demands a fresh look at relationships among people within a country, and among the countries themselves. For the problem of environmental degradation goes hand in hand with other problems: those of social justice, feeding the hungry, and the like; until they are solved together, they cannot be solved individually.

Prof. Alan McGowan, Ph.D.
Scientific Administrator,
Center for the Biology of Natural Systems,
Washington University,
St. Louis, Mo.

# Chapter 18

# Air Pollution, the Case of Air-borne Lead

The subject of air pollution and its effects is a complex one. Perhaps one of the most typical, complex, (and controversial) topics in this general area is air-borne lead, and whether (or how) it should be controlled. The purpose of this article is to provide an overview of this controversy, and place the problem in proper perspective. The history of lead as a pollutant, lead releases to the air, the biological effects of lead, other potential effects of lead, and current proposals for reducing air-borne lead emissions will all be discussed.

## History of lead as a pollutant

Lead has been mined and used by man for many centuries in part because it is relatively easily obtained from its ores and in part because, being a soft metal, it is fairly easy to fabricate into useful objects. The ease of fabrication of lead has other, less fortunate implications. Lead, while easily melted, is also easily volatilized; lead poisoning caused by the inhalation of lead fumes is a well recognized occupational disease. Similarly, when lead is used for the construction of cookware or of containers for the storage of food or drink, there is the possibility that the food or liquid in such containers may become contaminated with hazardous amounts of lead. This may occur both when the containers are made of lead or coated with a lead-containing substance (such as certain pigments or pottery glazes).

One particularly interesting historical example clearly demonstrates the possibility of poisoning from such sources of lead, and indicates that lead contamination may indeed, have changed the course of world history. During the time of the Roman Empire, wealthy families used lead or lead-based substances extensively for cookware, storage containers (for wine, etc.) and plumbing. Poorer families did not have access to these amenities. As time passed, there was an increasing amount of serious disease—including, apparently, mental retardation—in the ruling class. It has been suggested that lead poisoning was the cause of this; this suggestion has been strengthened by studies that show a significantly higher lead content in the bones of Roman nobles than in the bones of the slaves of the same period. Thus lead may have contributed significantly to the decline and fall of the Roman Empire. It is very likely that similar events occurred in other major civilizations, both in the Old and New World. (One still finds examples of harmful lead exposure from improperly glazed pottery.)

Lead has found its way into humans and other living things in other circuitous ways, as well. Before the advent of synthetic pesticides such as DDT several years ago, one pesticide commonly used on tobacco fields was lead arsenate. Some of the lead ended up in the tobacco leaves, and thus in cigarettes. As a result, cigarette smokers had measurably higher levels of lead in their blood than non-smokers. Even though lead arsenate use has been phased out, the residual lead in the soil still causes contamination of the tobacco, and even today the difference in lead levels between the blood of smokers and non-smokers persists.

Two other sources occasionally come to attention in modern times. One is lead-contaminated "moonshine" whiskey, which is sometimes produced when the bootleggers use as the condenser an automobile radiator containing lead solder. The alcohol solution dissolves the lead, which can reach sufficiently high concentrations to be dangerous. The second is lead poisoning in young children who eat chips of peeling, lead-based paint. Until roughly twenty years ago, essentially all paints used lead as a major ingredient; the lead often comprised 20% or more (by weight) of the paint. Thus any building over two decades old is very likely to have layers of lead-based paint on its interior surfaces. The growth of slums has, among other things meant run-down buildings, with the accompaniment of peeling paint and crumbling plaster. A certain fraction of all young children (below about the age of six) exhibit the phenomenon of "pica", the ingestion of non-food objects. If a child with this propensity happens to live in a ghetto building, he may eat lead-containing paint chips rather than other objects, and thus develop high blood concentrations of lead. This may cause frank acute lead poisoning with convulsions, coma, permanent mental retardation and even death. Estimates of the number of American children affected by lead from this source range as high as 400,000 per year.

### Lead Releases to the Air

Lead is a metallic element and one of the natural constituents of the earth's crust; thus it is theoretically possible that some air-borne lead emissions occurred even before man appeared on the face of the earth, by such mechanisms as the wind blowing lead-containing dust into the air. Such emissions were (and still are) both unavoidable and minimal. There is one good record that traces, in broad outline, the overall lead emissions to the air over geologic time. These are the polar ice caps, which are deposited every year and trap whatever is brought down by the snow. By digging a deep core into the ice, and measuring the lead content in it, one can obtain a profile over time of the relative amounts of lead in the air at different times. This work was pioneered by C. C. Patterson of California Institute of Technology.

If the sampling is done far from major cities the results give a rough profile of the "average" lead concentration in the air from all sources in the hemisphere in which the sampling is done. A schematic diagram of the results of such measurements in the snow cap of Greenland is shown in Figure 80. The lead concentration in the snow, and by implication the emissions of lead into the air in the northern hemisphere, was extremely low until 800 B.C., then began an upward trend. A significant upswing occurred at the time of the Industrial Revolution, when a greatly increased amount of lead was used per year in an ever larger number of industrial activities. Some current uses of lead include:

- Batteries (the familiar lead-acid automobile battery)
- Shielding (for protection against radiation)
- Piping
- Component of certain metal alloys
- As an element in organic compounds
- Ammunition (e.g. lead pellets in shotgun shells)
- Base of exterior paints.

This is of course not a complete listing. The last upturn, which has resulted in an extremely large increase in the amount of lead in polar snow, occurred shortly after the time that organic lead additives were added to gasoline. Without question the use of lead additives in gasoline accounts for by far the largest contribution to currently observable levels of air-borne lead collected in the Greenland snow. Lead is not a natural component of gasoline; it is added for specific purposes late in the refining process, and in fact, in the early days of the automobile, was not used at all.

However, as the power of automobile engines increased (particularly as the combustion ratio became greater), certain derivatives of lead were found to protect against engine "knock"; in

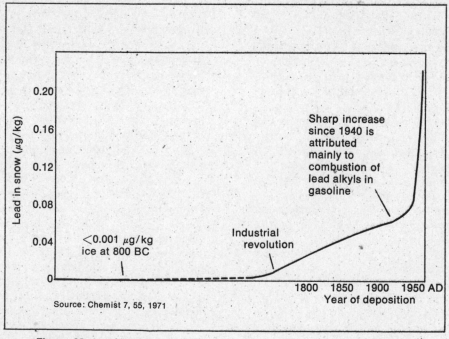

Source: Chemist 7, 55, 1971

**Figure 80.** Lead content of Greenland snow layers continues to increase.

addition, the addition of lead derivatives also raised the general octane rating of a gasoline by several points.

Not all lead compounds have this ability, only so-called "alkyl lead compounds." The original one was *tetra-ethyl* lead; ethyl referring to the $C_2H_5$-grouping. This addition gave rise to the old nickname of "ethyl" for high-octane premium gasoline. More recently *tetra-methyl* lead, methyl referring to the $CH_3$-group, or a combination of the two has been used. These organo-lead or alkyl-lead compounds are themselves quite volatile, and some of the air-borne lead is undoubtedly due to direct evaporation from storage containers, spilled gasoline in service stations, evaporation from fuel tanks and lines, and so forth. Most of the lead additives, however, pass into the combustion chambers of the car engine where they are converted into very small particles of lead chloride and/or bromide. Special substances containing chlorine and bromine, called "scavengers," are also added to gasoline for this purose.

These fine particles of lead leave the engine and pass into the exhaust system. A large but variable fraction of them escape through the tail pipe of the car. At low engine speeds, perhaps only one-fourth of the lead is emitted into the air; at higher speeds much more becomes air-borne, perhaps as much as 60–80% of the total. Overall, according to recent government figures, about 70% of the lead put into a gas tank eventually becomes air-borne.

The sizes of the lead particles vary. Larger ones may be removed from the air rather quickly. Dust swept from the side of a highway, for example, contains relatively large concentrations of lead compared to most other dusts. Similarly vegetation growing near a major road contains higher concentrations of lead on its surface than the same vegetation growing some distance away. However, many of the particles are small enough to remain air-borne, and are removed from the air only after traveling a long distance. Extremely long distance travel is possible; this is the mechanism by which the lead appears in the ice of Greenland. Shorter distances of travel can also be detected; measurements of the lead in rainfall show the transport of air-borne lead tens or hundreds of miles downwind of urban areas. Finally, local transport of air-borne lead is also detectable, from a highway to adjacent areas, for instance, as described above in dust and vegetation.

The quantities of lead released into the air by this means are very large, even though there are only two to four grams of lead per gallon of gasoline. Recent federal documents indicate that while less than 20% of the 1968 consumption of lead went into gasoline additives (approximately 262,000 tons of a total consumption of 1,330,000), 181,000 tons of lead were released into the air by the burning of gasoline and only about 3,500 tons by all other sources *combined.* Thus in 1968 gasoline combustion accounted for over 98% of the country's emissions of air-borne lead.

It is for this obvious reason that much of the concern over air-borne lead and its possible effects has been focused on the automobile. It should always be kept in mind, however, that releases of lead into the air from isolated industrial facilities (e.g., lead smelters, manufacturing of alkyl lead compounds) may also cause serious local problems due to "hot spots" of very high air-borne lead concentrations.

## Biological Effects of Lead

As indicated above, elevated intake of lead, no matter what the source, is known to be extremely hazardous. Though there are differences between organic lead compounds that is, tetraethyl and tetramethyl lead, and other lead compounds, certain general phenomena are the same.

Lead is chemically classified as a heavy metal, as are such other well-known toxic substances as mercury and cadmium. All heavy metals are poisonous in some forms; often they affect the kidneys and the nervous system as well as other important parts of the body. Very often such substances, in their pure metallic form, are relatively non-toxic. A lump of lead, for instance, or a drop of mercury are not the most biologically harmful varieties of these two substances. But when the metal is converted into an ionic form, it chemically resembles more closely certain other necessary ions (calcium, for instance) and thus is more easily absorbed through the lungs or intestine into the organism. Also, in such ionic forms, the heavy metal atoms may find their way throughout the body by passing through the circulation as though they were the useful ions. Eventually the metal ions may find themselves in a cell where they can disrupt one (or several) normal metabolic functions, leading to disorder and/or disease.

Organic derivatives of lead (tetra-ethyl lead or tetra-methyl lead) are much more dangerous than inorganic forms, such as lead chloride. This is in part because they can enter the body not only through the intestine and lungs, but even through the skin, by direct absorption. This means that very high concentrations of lead may very quickly enter the blood, thus reaching the other tissues of the body and quickly result in serious lead poisoning.

In fact, when alkyl lead compounds first came into large-scale use as gasoline additives in 1923, there was an outbreak of lead poisoning from this source, in some cases resulting in death primarily due to damage to the central nervous system, in workers exposed to these substances in the course of their occupation. Within a few years, a committee established by the Surgeon General proposed regulations to guard against this occupational hazard; implementation of these guidelines greatly reduced the incidence of this occupational disease, though constant vigilance of the protective equipment and other preventive measures are required.

Pre-dating lead poisoning due to alkyl lead compounds is lead poisoning resulting from exposure to lead vapor and inorganic lead compounds. This disease dates back many centuries; some of the sources of the lead have been described above.

The consequences of lead poisoning from inorganic lead compounds differ somewhat from the effects of alkyl lead compounds. Serious acute lead poisoning can develop from a single, massive exposure to lead-containing vapor or dusts, with much the same consequences as that described above. However, exposure to much lower levels of inorganic lead compounds for longer periods of time (days to months) can result in chronic lead poisoning. This disease is the result of the slow accumulation of lead compounds in various tissues, and has different symptoms depending on the age of the person exposed, his general state of health, the rate of absorption of excessive quantities of lead, and other factors.

Lead ions chemically resemble calcium ions, and since the body stores a considerable quantity of calcium in bones and teeth it is not surprising that lead is also found stored in these tissues. It is generally thought that the lead stored in the bones and teeth is not, in itself, particularly harmful. However, if some other factor (another disease or a hormone imbalance) causes increased release of calcium from the storage locations, lead may also be released into the blood and thus reach other tissues where it may cause acute symptoms in the nervous system. The actual storage of excessive quantities of lead may cause lead deposits, (particularly in long bones) that are visible in X-rays and the "lead line," a distinctly visible band of lead salts on the teeth near the gums.

Lead compounds can also penetrate into the bone marrow, where the lead ions are able to interfere with several enzymes involved in the biochemical process leading to the production of hemoglobin, the red, oxygen-carrying pigment in red blood cells. Thus one common symptom of lead poisoning is anemia. Interference with these orderly enzymatic processes cause certain intermediate substances, the "building blocks" of hemoglobin, as it were, to build up in concentration in the bone marrow. These substances, two of them are *delta-aminolevulinic acid*, or ALA, and *coproporphyrin*, then leak into the bloodstream and may be eliminated by the kidneys. Certain current and potential diagnostic procedures rely on detecting one or another of the biochemical consequences of the presence of lead in bone marrow by measuring the level of one or another of these substances in blood or urine.

The effect of lead on the enzymatic process of hemoglobin synthesis has been thoroughly studied. Many other biochemical processes in other tissues are such that lead could also interfere with the normal metabolic functioning of other synthetic pathways. There has, unfortunately, been relatively little research into these other likely possibilities at this time.

Calcium plays a very important role in the normal functioning of the nervous system. It is therefore not surprising to learn that excessive body burdens of lead also interfere with normal functioning of the nervous system. The consequences in any individual are variable and often not very specific; they may include headache and irritability, progressing in serious cases to convulsions and coma, and even death, which may be due to paralysis of the nerves controlling breathing. Children suffering from lead poisoning severe enough to cause these symptoms run an additional risk of suffering such permanent nervous system damage as chronic epilepsy or permanent mental retardation; these are much less frequent in adults, for reasons that are not clear.

So far, the symptoms and effects of exposure to relatively high levels of lead have been discussed. The next obvious question is: what are the biological effects of exposure to lower levels of lead, particularly exposure to the levels currently found in major metropolitan areas? This is simultaneously both a difficult and controversial question to answer. Several parts of this question are generally agreed on; these will be examined first.

Historically, the extensive release of lead into the air from automobiles has been a topic of recurrent concern since shortly after the addition of alkyl lead compounds to gasoline in 1923. It is by now clear, of course, that past and current releases do not cause the acute lead poisoning described above, nor do current exposure levels clearly result in severe chronic lead poisoning, also described above.

On the other hand, the rapid growth in the automobile population and in gasoline consumption, coupled with an increase in 1958 in the amount of alkyl lead compounds allowable in each gallon of gas, has caused a very great increase in lead emissions in recent years. Federal figures indicate the quantity of lead used per year as a gasoline additive has increased more than four-fold since 1943, and has increased 65% just within the past decade.

One might expect that this has resulted in higher average lead concentrations in the air of urban areas. Although the monitoring data does not uniformly demonstrate this for all urban areas of the country, it does indicate it is clearly true for certain cities. An extensive study was carried out by the federal government in 1962 to measure average air-borne lead levels in Cincinnati, Los Angeles, and Philadelphia; it was repeated in 1969. Several sampling sites showed statistically significant increases between 1962 and 1969; other sites did not demonstrate such increases. A more recent study by other investigators indicates that the concentration of air-borne lead in San Diego has been increasing by about 5% per year.

Simultaneously, of course, urban centers have been growing, so that more Americans, both in absolute terms and on a percentage basis, are exposed to urban air-borne lead levels now than were so exposed two or three decades ago.

Thus there is general agreement on the following conclusions:

- Overall lead emissions to the atmosphere, for the country at large, have been increasing rapidly for several years.
- Some parts of cities have higher average air-borne lead levels than existed several years ago; other parts of the same cities have unchanged levels.
- More citizens are exposed to urban air-borne lead levels than were exposed to such levels several years ago.

Two other observations have been often verified. First, in general, urban residents have higher average blood lead levels than rural residents. Second, urban workers who are exposed to automobile exhaust either for somewhat longer periods than normal, or at concentrations somewhat higher than normal for urban areas, such as tunnel workers, garage mechanics, etc., have, on the average, even higher lead levels in their blood, table 3.

**Table 3.** Summary of Concentrations of Lead in Blood of Selected Groups of Males

| Mean, mg/100g | No. of Subjects | Identity of Group |
|---|---|---|
| 0.011 | 9 | Suburban nonsmokers, Philadelphia |
| 0.012 | 16 | Residents of rural California county |
| 0.013 | 10 | Commuter nonsmokers, Philadelphia |
| 0.015 | 14 | Suburban smokers, Philadelphia |
| 0.019 | 291 | Aircraft employees, Los Angeles |
| 0.019 | 88 | City employees, Pasadena |
| 0.021 | 33 | Commuter smokers, Philadelphia |
| 0.021 | 36 | City Health Dept. employees, Cincinnati |
| 0.021 | 155 | Policemen, Los Angeles |
| 0.022 | 11 | Live and work downtown, nonsmokers, Philadephia |
| 0.023 | 140 | Post-office employees, Cincinnati |
| 0.024 | 30 | Policemen, nonsmokers, Philadelphia |
| 0.025 | 191 | Firemen, Cincinnati |
| 0.025 | 123 | All policemen, Cincinnati |
| 0.025 | 55 | Live and work downtown, smokers, Philadelphia |
| 0.026 | 83 | Police, smokers, Philadelphia |
| 0.027 | 86 | Refinery handlers of gasoline, Cincinnati (1956) |
| 0.028 | 130 | Service station attendants, Cincinnati (1956) |
| 0.030 | 40 | Traffic police, Cincinnati |
| 0.030 | 60 | Tunnel employees, Boston |
| 0.031 | 17 | Traffic police, Cincinnati (1956) |
| 0.031 | 14 | Drivers of cars, Cincinnati |
| 0.033 | 45 | Drivers of cars, Cincinnati (1956) |
| 0.034 | 48 | Parking lot attendants, Cincinnati (1956) |
| 0.038 | 152 | Garage mechanics, Cincinnati (1956) |
| | 1877 | Total |

Table courtesy of Public Health Service, U.S. Department of Health, Education, and Welfare. Publ. #999-AP-12, January 1965.

From this point on, there is no general agreement; this, of course, is where the controversy lies. The blood lead levels referred to above are measured in units of micrograms of lead per 100 grams of blood; this will be abbreviated as "micrograms" for convenience from now on. Frank, serious clinical lead poisoning in adults has been observed when blood levels are 80 micrograms or higher. This is a clearly hazardous level. Even lower levels in adults may be associated with some difficulties; a recent federal report, referring to a document published by the National Academy of Sciences, states that "symptoms and signs compatible with mild lead poisoning have been associated with whole blood concentrations of 50–80 micrograms per 100 grams." In addition, there is a possibility that any given lead level may be more hazardous to a child than to an adult.

Certain average blood lead levels in various population groups are shown in the accompanying table. No group has an average as high as 50–80 micrograms. Some would argue, however that the levels in certain groups are uncomfortably close to the levels at which poisoning is known to occur.

Obviously the presence or absence of "poisoning" has a serious limitation as the only tool for determining the effects of low-level lead exposures. Lead poisoning is a serious disease with obvious clinical symptoms, e.g., convulsions. Since there is no evidence that exposure to excessive lead is beneficial, one might expect that lead exposure not sufficiently intense to result in "poisoning" might result, instead, in measurable, potentially serious changes of a less obvious nature. Few studies looking for this possibility have been done, but the few that have been carried out are quite interesting.

One study, carried out in Finland, showed that humans with "normal" blood lead levels showed some loss of activity in an enzyme which is similar to one of the enzymes in bone marrow described above. The higher the blood lead level, the less well the enzyme performed, even though the lead levels were all in the normal range. This suggests that even quite low lead levels may have disruptive effects on biochemical processes. Another study demonstrated that rats exposed for rather short periods to air-borne levels similar to those in heavily-trafficked areas had a lower number of the so-called "scavenger cells" in their lungs than unexposed rats. These cells are considered an important line of defense against foreign bacteria and other matter in the lungs, and help to protect against such diseases as pneumonia. This raises the possibility that lead may have indirect as well as direct effects on bodily processes at rather low levels.

Obviously more such studies are needed. Unfortunately they are costly and time-consuming. Other facets of the issue may force changes in patterns of air-borne lead emissions before such studies can be carried out.

It is obvious from material presented earlier that the air is not the only source of lead in humans. Food, drinking water, peeling paint, and occupational environment may contribute to the overall lead intake of any given individual. In general, about 5–10% of the lead entering the intestine (through food, drink, dust, paint chips, etc.) is absorbed into the body tissue, while about one-third of the lead entering the lungs is absorbed. The total lead burden in any individual person is a result of his particular experience, variety, and the amounts of exposure. We have seen above the upward trends in the concentration of air-borne lead in some urban areas and in the air over Greenland. Insofar as air-borne lead is a pervasive contaminant, it is a contributor to everyone's exposure. Also, since it is "man-made," it can be controlled by man, where the "natural" lead contamination in food and water cannot. Other "man-made" sources which affect discrete population groups, such as children in run-down housing, workers in lead industries, and drinkers of moonshine whiskey, can also, in principle, be controlled by man.

The growing knowledge about the biological effects of lead and in particular the recent studies suggesting that quite low lead levels (by earlier standards) may be associated with adverse effects have, in part, caused the re-kindling of the lead-in-air controversy that has flared up, off and on, since tetra-ethyl lead was first added to gasoline about half a century ago.

## Other Potential Effects of Lead

Curiously enough, one of the most important factors in the current controversy over air-borne lead as a pollutant is not due to the *direct* effects of lead on people, but rather to the effects of lead on *other air pollutants*. It is generally recognized that air pollution in general, especially in urban areas, is at unacceptable levels, and that much of the air pollution is caused by cars. Space does not allow a detailed description of all common air pollutants and their relative effects on people, vegetation, objects and visibility. Suffice it to say that of this "pea soup" of pollutants, only two additional ones concern us here: carbon monoxide and hydrocarbons.

The toxic properties of carbon monoxide are well understood. At high levels, it can kill. At much lower levels, including those often found in urban areas, it can interfere with a wide range of physiological functions. Hydrocarbons, essentially unburned gasoline, are not harmful in and of themselves, but they are a necessary ingredient of the chemical reaction which produces "photochemical smog." This type of air pollution results from the action of sunlight (*photo*-) on two components of contaminated air ,hydrocarbons and the oxides of nitrogen (-*chemical*). First recognized in Los Angeles, this variety of air pollution can now be detected in most major cities, especially in the summer.

Various steps are underway to reduce both of these pollutants. Far and away the major source of both is the internal combustion engine, primarily the private car. Very promising devices exist for the potential control of these pollutants. These devices, employing catalysts, have the ability to very effectively cause these pollutants to react on the catalyst's surface and be turned into non-pollutant gases such as carbon dioxide and water vapor.

There is only one problem. Most, if not all, of the catalysts would have their effectiveness quickly destroyed after contact with air-borne lead. This is called "poisoning" the catalyst. Thus as long as lead is added to gasoline, these promising devices cannot be used to solve other air pollution problems. If lead were removed from gasoline, it might very well facilitate the solution to these other two pollutants. And, in fact, it is this situation which has also added to the controversy concerning air-borne lead. In this contex, then, one potential though indirect effect of air-borne lead is that it is blocking an effective attack on other air pollutants.

There are two other potential effects of air-borne lead which, at this time, must be regarded as speculative and in need of additional study. These are, first, the possibility that air-borne lead changes precipitation patterns and, second, that air-borne lead causes another common air pollutant, sulfur dioxide, to be converted into a more toxic substance, sulfur trioxide. A few words on each of these will suffice.

Certain lead compounds, in particular such materials as lead chloride or lead bromide, behave, in the laboratory, as very good substances for inducing the formation of water droplets in moist air; technically such substances are often called nucleating agents. Raw automobile exhaust from an engine using lead-containing gasoline shows this same ability. This is not surprising, since chemically similar substances have been used for cloud seeding for some time. Whether these laboratory observations have any implications for the outer environment is not certain, but there have been certain suggestive observations which clearly raise this possibility. It is well-known that the center of urban areas receive greater amounts of rainfall than the surrounding suburbs and rural areas, and that rural areas downwind of cities receive greater rainfall than rural areas located in other directions from the cities. If the lead compounds play a role in cloud formation and rainfall, this is, in fact, the distribution of rain one would expect. However, urban air contains a potpourri of pollutants; among them are many other small particles which would also serve as nucleating agents. At this point one cannot be certain whether lead compounds alone, other materials alone, or (as is more likely) a combination of the two are responsible for the observed differences in rainfall.

The second possibility is derived from other laboratory investigations which demonstrates that, under controlled conditions, the presence of lead causes another very common urban air pollutant, sulfur dioxide, to be more quickly converted into sulfur trioxide. Sulfur dioxide is produced by the combustion of sulfur-containing coal or oil, the manufacturer of sulfuric acid, and in other industrial processes. In itself it is a hazardous material, and is thought to be, at least in part, responsible for damage to human health, vegetation, buildings, etc., in urban areas. On combination with water, sulfur dioxide is converted into sulfurous acid.

Sulfur trioxide, on the other hand, is not found at as high a level as sulfur dioxide, though it is present in urban air. Sulfur trioxide, when combined with water, forms sulfuric acid, a much more strongly acid substance than sulfurous acid. Thus any material which tends to convert sulfur dioxide to sulfur trioxide must be looked on with suspicion. Lead compounds are known to speed up this conversion in the laboratory; whether they also do so in polluted urban air is not yet known.

## Current Proposals for Reducing Lead Emissions

In view of the multi-faceted controversy about air-borne lead emissions, it is not surprising to learn that, for some time, federal air pollution officials have been considering restrictions on them. These considerations became public on February 23, 1972, when William D. Ruckelshaus, Administrator of the federal Environmental Protection Agency, published his agency's proposals for regulating lead content in gasoline. In brief, his proposals would require one lead-free gasoline to be available throughout the country by July 1, 1974, and would require a gradual reduction in the lead content of other grades of gasoline between 1974 and 1977. In this announcement, it was predicted that a reduction of 60-65% in emissions of lead to the air would be achieved by mid-1977. In general, the proposed schedule would allow no lead in gasolines with a 91 octane rating after 1974; a reduction from present levels to no more than 2.0 grams of lead per gallon in both 94 octane (regular) and 100 octane (premium) gasolines; and reductions from 2.0 to 1.25 grams in gradual steps until 1977 for both 94 and 100 octane gasoline.

Federal procedures allow that interested people or individuals be given 90 days to comment on these proposed regulations.

Two reasons were cited by Ruckelshaus for his proposals. First the presence of lead in gasoline "will impair to a significant degree the performance of emission control systems that include catalytic converters which motor vehicle manufacturers are developing." Second, "air-borne lead levels exceeding 2 micrograms per cubic meter . . . are associated with a sufficient risk of adverse physiologic effects to constitute endangerment of public health." Since many urban areas have levels as high as 5 micrograms, a considerable reduction in emissions is required, and since, as noted above, the main source of air-borne lead is the automobile, restrictions on the lead content in gasoline were chosen to achieve protection of the public health.

If the proposed regulations go into effect unchanged, the current phase of the lead in air controversy will, for the first time, result in significant reductions of this particular pollutant, a significant and important step to restoring clean air.

Glenn Paulson, Ph.D.
Staff Scientist,
Natural Resources
Defense Council, Inc.
New York City

# Chapter 19

# Ozone: Another Threat to Plant Life

Ozone, a gas occurring naturally in our atmosphere, and restricted almost entirely to the upper atmosphere, plays the very important role of screening harmful ultraviolet rays from the incoming sunlight. In the relatively unpolluted areas of the lower atmosphere, ozone occurs in very minute quantities of 1 to 2 pphm (parts per hundred million =0.000001%). Ozone is produced as a result of electrical discharges, which occur with lightning during thunderstorms or with the sparking of electrical appliances. It may also be brought down from the upper atmosphere by vertical movement.

In very minute concentrations ozone has been thought to be beneficial and has popularly been associated with pure air. Its pungent odor has been associated with the cleansing action of thunderstorms. Since microorganisms are very susceptible to death from ozone, the gas has been used as a germicidal agent and air purifier. However a small increase in concentration can raise it to levels that are harmful to life. These levels are very close to where other organisms, including man can suffer harmful effects.

Ozone levels in the lower atmosphere have been increasing at an alarming rate during the last 20 to 30 years. High levels can be measured in the lower atmosphere of every metropolitan centre. Ozone is not a direct or primary emission of either our major industries or automobiles, but is rather the by-product of a series of reactions that occur in polluted air in the presence of sunlight. Where photochemical reactions are prevalent, producing a number of highly irritable and toxic substances, the term photochemical smog is used. It is the product of the tons of unburned hydrocarbons and nitrogen oxides that are spewed into the atmosphere daily from a multitude of combustion processes particularly the inefficient internal combustion engines of automobiles.

The nitric oxide (NO) emitted is oxidized by atmospheric oxygen to nitrogen dioxide ($NO_2$), however sunlight splits the nitrogen dioxide producing nitric oxide (NO) and liberating atomic oxygen (O) which combines with atmospheric oxygen to produce ozone ($O_3$), the triply bonded form of oxygen molecule.

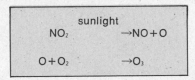

$$NO_2 \xrightarrow{\text{sunlight}} NO + O$$
$$O + O_2 \rightarrow O_3$$

Normally the ozone would react again with the nitric oxide to give nitrogen dioxide. That is, the overall reaction is reversible with the equilibrium towards $NO_2$ so that little ozone accumulates.

$$NO_2 + O_2 \xrightarrow{\text{sunlight}} NO + O_3$$

However, in the presence of hydrocarbons in the atmosphere nitric oxide reacts with these compounds stopping the back reaction and ozone accumulates. At night the splitting of nitrogen dioxide does not occur and the highly reactive ozone disappears from the atmosphere. Thus, a diurnal cycle is established with very low levels at night, building up in the morning, reaching a peak during early mid-afternoon and declining again towards evening.

Normally, the highest levels of ozone occur over large metropolitan areas but the smog may drift downwind a hundred miles or more. In many industrial cities the whole region may become blanketed with smog. Under certain conditions the whole eastern United States may have elevated levels of ozone and other pollutants.

Ozone is not the only chemical produced under the conditions mentioned; other substances with similar oxidizing properties are also produced. Because these oxidants can be very corrosive to substances such as rubber, the term *oxidant smog* has been coined. This smog is also very damaging to plant life. Damage to plants from ozone may be expected if the concentration exceeds roughly 5 pphm. Concentrations have exceeded 15 pphm in nearly every major city where measurements have been taken, and in areas where atmospheric inversions occur, concentrations have built up as high as 100 pphm (or 1 ppm).

Certain disorders on plants had been noticed for many years but had gone unidentified until the mid to late 1950's when ozone became established as a major *phytotoxicant,* destroyer of

**Figure 81.** A soybean leaf that has been exposed to ozone showing the typical flecking type of injury.

plant life. Since that time damage has been described on almost every type of ornamental and agronomic crop; leafy vegetables; cereal, forage and textile crops; shrubs and trees of all descriptions; all have shown extensive damage.

Although plants vary greatly in their susceptibility to damage from ozone, all display more or less the same type of injury. The upper surfaces of leaves are affected first. The ozone penetrates through the stomata and affects the membranes and metabolism of the cells inside the leaf. Individual cells become disrupted and die and if enough cells in a given area in the leaf are destroyed, a local lesion occurs. As a result, the leaf may take on a pitted or flecked appearance, figure 81, or the whole upper surface of the leaf may be injured giving the leaf a bronzed or glazed appearance, figure 82. The expression of injury may be dependent upon other environmental conditions or may be influenced by the presence of other pollutants. Ozone injury is not always readily distinguishable from injury produced by other pollutants, by disease organisms, or adverse environmental conditions and care must be taken in identification of the causal factor.

Plant damage can be severe enough to cause whole leaves to die or age and die prematurely. In susceptible crop species, yield may be suppressed to the point where it is no longer profitable to grow the crop. In other cases such as with spinach or green onions the product may be blemished or unsightly and therefore unmarketable. Crop losses from ozone injury are difficult to estimate particularly if yield reduction is small, but it is estimated that in the United States the damage to crops alone runs in the billions of dollars.

Until the pollution levels are brought down to a tolerable level, other steps must be taken to reduce crop injury from ozone. In many cases more tolerant crops can be substituted for sensitive ones. It has been found that not only are there differences among species in the degree of sensitivity but also among varieties or cultivars within a crop species. Thus varieties can be selected or bred for increased tolerance to ozone. Atttempts are also being made to find substances or chemicals which can be sprayed on crops at critical times to protect them from ozone injury. A number of antioxidants have been found which

**Figure 82.** A soybean leaf exposed to ozone and other air pollutants in the field displaying the characteristic "bronzing" type of damage.

**Figure 83.** Ponderosa pine tree suffering from exposure to oxidant smog in the San Bernandino mountains above the Los Angeles basin (Also note differences in suceptibility of different trees).

can be used to protect plants sufficiently to reduce injury and economic losses. These are of course only stop-gap measures which can be used until the real problem of reducing the cause of this damage can be solved.

Although it may be feasible to protect crop plants it is often not possible to do the same with ornamental trees and shrubs let alone our native forest trees, many of which are also very sensitive to ozone poisoning. Many trees and shrubs become unsightly during the summer months following episodes of high ozone. In many cities the trees are showing the effects of repeated exposures year after year and many will probably disappear from metropolitan areas. Injury to forest species is most conspicuous in the forest adjacent to the Los Angeles basin, figure 83. Where species such as the *Ponderosa* pine show extensive damage causing a serious decline in a major species which may disrupt a whole ecosystem covering hundreds of square miles. Similar but less conspicuous injury, is no doubt occurring in many other areas, reducing the health and vigor of forests and thereby also reducing the fitness of our biosphere.

Prof. Gerald Hofstra, Ph.D.
Ontario Agriculture College
Department of Environmental Biology
University of Guelph
Ontario, Canada

# Chapter 20

# Water Pollution

From earliest times man has made his home around abundant supplies of fresh waters—rivers, streams and lakes. Historically these large bodies of water, besides supplying people with drinking water were also used in commerce, as a source of food, for recreation, and also as a convenient place to discharge wastes. Usually the wastes discharged were mostly human sewage and surface runoff with its accompanying debris. Although there were occasional nuisance problems associated with water pollution, such conditions were generally limited to the immediate area adjoining the crowded cities. Extensive pollution, covering large areas and of a broad general impact such as we have today did not exist. Such pollution is a product of modern man and can be traced directly to the great industrial revolution of the last century, the concentration of people in cities, population growth, the development of more efficient agricultural techniques, but especially the development and growth of factories and industry.

Most industry relies heavily on the availability of water supplies. Thus, most industrial operations have conveniently located on the banks of rivers and lakes. Here fresh water can be withdrawn as needed, used by the industry and the used wastewater returned to the clean water source. As long as there were only a few population and industrial centers, this practice did not severely harm the receiving water environment outside of a small restricted area. Anyone, in fact, offended by the resulting environmental upset did not have to travel great distances to avoid the intrusion. Up until a short time ago this was still possible. Now, however, there is practically no large river or lake in the world that is free from the polluting hand of man. Fouled water, however, when it is returned to the environment and the hydrologic cycle, does not remain discrete or separate from any other water since eventually the same water used and discarded by industry will also be used by man for drinking water purposes. It was only a question of time until much of the water found in our rivers and lakes and even underground water supplies showed signs of pollution. Much of the available surface as well as ground water supplies, in fact, are now unfit for drinking without extensive treatment.

Of some alarm is the fact that the normal treatment methods used in the purification of drinking waters do not remove many of the materials that affect our health and which originate principally from wastewater discharges. This includes many viruses as well as many heavy metals (such as mercury, arsenic, and lead), but also includes compounds such as nitrates and organic herbicides and pesticides which have their origin in agricultural operations. The presence of these chemicals is not surprising when one considers that there are an estimated 12,000 potentially toxic chemicals used by industry, and that an estimated 500 new chemicals are synthesized or manufactured each year. Pollution, however, besides posing a health hazard, can simply lower the quality of our receiving water and make it unfit for use. For example, many bodies of water are not even fit for recreational activities such as swimming or fishing as a result of pollution.

More recently this has resulted in an awareness by the average man of the great extent of our water pollution problem. Thus, we are now living in an era where people are slowly awakening to the fact that to a large extent the deterioration of the quality of our life has paralled "progress," i.e., the development of industry—and that if we are to stop this deterioration we must practically alter our life style. This we are not so eager to undertake. We do, however, have the basic ingredients to deal with this problem. This would include an awereness that the problem exists, an urge to do something about it, and a better understanding of the natural forces involved as well as the technical and financial resources required. Gradually we are learning to respect nature and are slowly coming to the realization that we are really in a position to control these forces of deterioration and that we can improve the quality of our life.

## Water Use

It is clear that our primary concern with water pollution really has to do with protecting our available fresh water supplies. At still another level we are alarmed that polluting chemicals have been found in our food chains. In light of the steady increase in population and the continued growth of industry it is unthinkable that we will continue to recklessly foul these supplies.

While some 75% of the planet is made of water and it might appear that there is an abundance of water on the surface of the globe, most of this water (i.e. 99.5%) is not in the form that is either usable or readily available to man. This is because much of this water is found either in the ocean (as a salt), or bound in the polar ice caps, or located too deep within the earth to retrieve readily (below 2500 feet). figure 84 represents our water supplies on earth as a "cube" of water and will give some idea of the relative amount of water that is available to man, figure 84. It is apparent that

most of the available water (i.e., 99.5%) cannot be used by man. The water that can be used by man—i.e., the remaining 0.5% portion has two general sources; ground water (wells) and surface water (rivers, lakes, etc.).

In terms of use, water is used for three general purposes: in domestic households (or municipal), by industry and in agriculture, table 4. The

**Table 4.** Water Use—U.S.A.
(Billion gals/day)

|  | 1900 | 1950 | 1960 | 1970 | 1980 | 2000 |
|---|---|---|---|---|---|---|
| Municipal | 3 | 14 | 21 | 25 | 29 | 42 |
| Industrial | 15 | 77 | 140 | 250 | 363 | 662 |
| Agricultural (irrigation) | 22 | 82.6 | 113.6 | 143 | 167 | 184 |
| Total | 40 | 173.6 | 274.6 | 418 | 559 | 888 |

quantities of water used for these purposes and projected use through the year 2000 is shown as figure 85. While large quantities of water are

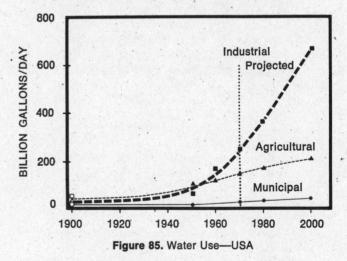

**Figure 85.** Water Use—USA

used for agricultural (i.e., irrigation) and human activities (sewage and household wastewaters), since 1900 agricultural and industrial activities have steadily accounted for more than 90% of our water usage. At present the figure stands at around 94%. Industry alone, it is also apparent, is a very large consumer of water. In 1970, for example, industry used 10 times more water than municipal consumers. It is also clear from these figures that since the turn of the century the amount of water used by industry has steadily increased from 37% to more than 50%. A projection of future water consumption is more clearly shown in figure 85. By 1980, two thirds of our water will probably go to supply the needs of industry and by the year 2000, industry may consume fifteen

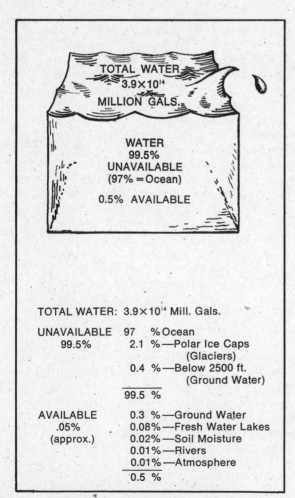

TOTAL WATER: $3.9 \times 10^{14}$ Mill. Gals.

| | | |
|---|---|---|
| UNAVAILABLE 99.5% | 97 % | Ocean |
| | 2.1 % | —Polar Ice Caps (Glaciers) |
| | 0.4 % | —Below 2500 ft. (Ground Water) |
| | 99.5 % | |
| AVAILABLE .05% (approx.) | 0.3 % | —Ground Water |
| | 0.08% | —Fresh Water Lakes |
| | 0.02% | —Soil Moisture |
| | 0.01% | —Rivers |
| | 0.01% | —Atmosphere |
| | 0.5 % | |

**Figure 84.** Approximate Water Distribution

times more water than municipalities. Probably the industries consuming the greatest quantities of water are those dealing with pulp and paper and primary metals (mostly steel), however, the chemical and refining industry and the electric power industry are also huge consumers.

### Water Pollution Sources

As shown in figure 85, in the U.S. water pollution sources have three rather general sources—municipal, agricultural and industrial. Municipal (or domestic) wastes consist essentially of wastewater discharges originating from private households—i.e., homes and apartments. Such waters consist of sewage (body wastes such as urine and feces, and kitchen and bath wastewaters as well as street washings and storm water runoff as shown in table 5. In many instances domestic

**Table 5.** Sources of Municipal Wastes

Human wastes
  urine
  feces
  oral

Storm flows

Street washings
  sand, grit, etc.
  animal wastes

Industrial wastes
  manufacturing processes
  business operations
  cooling waters

Household wastes
  laundry
  bathing
  kitchen

Groundwater
infiltration
  leaky pipes

sewage will also have added to it wastewaters originating from industrial sources. Most cities, for example, will have usually some industry connected to the city sewer lines. An example would be a car wash, filling station, laundry, or a bottling company, and, in most instances heavy industry.

Agricultural wastes originate from farm and ranch activities and consist of runoff waters from fertilized or chemically treated soils which carry fertilizing chemicals together with pesticides and herbicides as well as sediment from the erosion of land. Another large source of agricultural waste, however, is from operations related to raising and fattening cattle and hogs on small plots of land (feedlot wastes) where large amounts of animal

body wastes are generated. It has been estimated that the overall waste production from farm animals in the U.S. is some twenty times that originating from humans. In the western part of the U.S., the principal water pollution problem is caused by agricultural operations as opposed to the East where the principal problems are related to industrial operations.

Industrial sources of pollution include wastewater discharges from industrial plants, manufacturing operations, factories, and activities such as those carried out by chemical plants, paper mills, oil refineries or petrochemical plants, metal plating shops, steel mills, meat packing operations, etc. Since industrial pollution represents such a large and important source of pollution in the U.S., a separate section will consider this category in greater detail.

In addition, in the U.S., huge amounts of water are used for cooling purposes. Hot water discharges (thermal pollutants) are considered to be pollutants since they will reduce the available dissolved oxygen stores in a receiving water. *Thermal pollution* has been defined as man caused deleterious changes in the normal temperature of water. Since almost one-half of all the water used in the United States is actually used for cooling and condensing by the power and manufacturing industries, thermal pollution from these industrial sources is a major problem. A list of the more important industries using water for cooling heat generating processes (cooling waters) and discharging these hot waters into the receiving waters is shown as table 6. It can be seen that the electric power generating industry is a very large source of thermal pollutants.

**Table 6.** Use of Cooling Water by U.S. Industry, 1964

| Industry | Cooling Water Intake (billions of gallons) | % of Total |
|---|---|---|
| Electric Power | 40,680 | 81.3 |
| Primary Metals | 3,387 | 6.8 |
| Chemical and Allied Products | 3,120 | 6.2 |
| Petroleum and Coal Products | 1,212 | 2.4 |
| Paper and Allied Products | 607 | 1.2 |
| Food and Kindred Products | 392 | 0.8 |
| Machinery | 164 | 0.3 |
| Rubber and Plastics | 128 | 0.3 |
| Transportation Equipment | 102 | 0.2 |
| All Other | 273 | 0.5 |
| TOTAL | 50,065 | 100.0 |

Industry alone, however, does not account for all temperature changes in water. Many of the activities of man, in fact, can change water temperatures in more subtle ways. For example, temperature changes can be brought about by altering the environment of a water course through road building or logging or by creating impoundments or through stream diversions.

For the future, however, thermal pollution must remain one of our primary concerns. This becomes apparent when one considers that power generation has approximately doubled each ten years during this century, and that this rate is increasing. Of greater significance is the fact that the power industry alone accounts for 80% of the cooling waters used.

It should be recognized that the categories noted above might not include all of the sources of pollution in the U.S. There are, of course, other sources of pollution which do not fit the categories listed but which contribute considerably to our water pollution problem. This would include such activities as mine drainage and wastewater originating from discharges of boats and ships. For example, it is estimated that acid mine drainage, mostly from coal mines, pollutes an estimated 10,000 miles of streams in the U.S. It should also be noted that there are more than 8 million watercraft in the U.S. which discharge pollutants in the form of sanitary wastes, oils, litter, and ballast and bilge waters to our coastal and inland waters.

There are, as shown on table 7, ways of

**Table 7.** Pollutant Categories

Oxygen Demanding Wastes
   mostly organic wastes
     (sewage or industrial wastes)
   some inorganic compounds
Infectious Agents
   viruses (hepatitis, polio, etc.)
   amoebic dysentery
   bacteria (Typhoid, cholera, etc.)
Toxic Chemicals
   heavy metals (mercury, arsenic, etc.)
   pesticides, herbicides
   organic compounds
Plant Nutrients
Heat from Industry
Radioactive Substances
Sediments from Land Erosion

classifying pollutant discharged to our receiving waters. table 7 probably reflects our concern with toxic agents, pathogenic microorganisms and environmental damage resulting from the discharge of pollutants. While some of these materials merely cause a deterioration of water quality, many are potentially toxic to man or to the food chain.

Oxygen demanding wastes, as will be explained later in greater detail, are wastes which will ultimately consume oxygen from the receiving water. The bulk of our pollutant discharges are in this category. While such pollutants are predominantly organic in nature (i.e., from sewage or industrial sources), many inorganic compounds can consume dissolved oxygen in receiving waters.

The most serious infectious agent at present which is carried in sewage and which cannot be readily removed from water by current purification practices is the hepatitis virus. Plant nutrients are essentially any nutrient such as nitrogen or phosphorus compounds that will promote the growth of green plants. While nitrogen and phosphorus are the most common "eutrophic" agents, other minerals such as carbon, silica, etc., might also serve to stimulate microbial growth and thus enrich receiving waters resulting in a process of over-enrichment called *eutrophication*. Table 7 also lists other processes which might actually occur naturally such as erosion or the accumulation of sediments. In many instances, however, such deteriorative processes have as their origin the activities of man. Perhaps the most insidious pollutant since we cannot immediately see its effects, is radioactivity. As has been noted, hot water discharges are pollutants of a very serious nature.

**Water Pollution Quantities**

For comparative purposes and quantitative comparison, water pollutants can be categorized into two categories—domestic (i.e., household) and industrial wastes. While sewage wastes are often the most noticeable and the focus of much public attention, in the U.S. industrial wastewater pollutants comprise the bulk of the water pollution problem. An idea of the quantities involved in the terms most commonly used by engineers and scientists; volumes (gallons), pounds of solids and pounds of BOD (*Biochemical Oxygen Demand*), is shown as table 8. The term BOD is an estimation of the pollution burden to a receiving water in terms of the dissolved oxygen required to stabilize a waste aerobically (i.e., biologically in the presence of oxygen) over a period of five days. It is apparent from these figures that industrial pollutants exceed domestic (sewered) sources by at least two-to-one and at times three-to-one.

At still another level, individual consumers foul relatively large quantities of water compared to the quantities that are required for biological needs. For subsistence, an average person will require as little as 1.5 liters (about a quart and a half) to satisfy his biological demands. The water that we consume is far in excess of this figure. For example, an average household of four people will produce approximately 100 gallons of sewage per person per day (400 gallons per living unit per day!). This is not surprising when we consider that the flushing of a toilet can consume from four to eight gallons of water and that a five minute shower can consume as much as 50 gallons. Needless to say when the water consumption of industry is considered as a per capita average the 100 gallons per day per person figure becomes two to three times this figure.

## Industrial Pollution

It is well established that industry contributes a large quantity of our pollution. It has been estimated, for example, that in the United States there are more than 300,000 industries that produce wastes. The wastes discharged by these industries exceed the quantities of oxygen demanding wastes produced by all of the sewered population in this country by some three to four fold. While there has never been a detailed inventory of the exact quantities of industrial wastes discharged in the United States, the Federal Government has made some reasonable estimates as shown in table 8.

The major portion of our industrial wastes have their origin from only a few of the major industries: paper, organic chemicals, petroleum and steel. table 8 gives a more detailed idea of the contribution by particular industries. Although a wide spectrum of activities are included in each industrial category (i.e., under food and kindred products is included meat, dairy, canned and frozen food, etc.), the figures are quite plain. In terms of BOD, at the top of the list is the chemical industry, followed by the paper and food manufacturing operations and the primary metal industry with all of its allied activities.

Perhaps a more specific idea of the industries involved in industrial pollution can be seen from an examination of the list on table 9 of "Critical Industrial Groups" which have been classified by the U.S. Corps of Engineers. The food, chemical, metal and paper industries are well represented, although many diverse manufacturing activities are listed in addition to some in agriculture and mining.

Because of the size of the industry and the nature of its processes, probably the largest polluter of water in the U.S. is the steel industry. This is because large quantities of water are used in the steel industry amounting to 4 trillion gallons per year! This water is used to cool and condense hot metals as well as to scrub gases which are used for fuel. In addition, large volumes of water are used to remove waste scale on steel surfaces.

Industrial discharges as well as many agricultural pollutants are of particular concern to

**Table 8.** Estimated Volumes of Industrial and Domestic Wastes before Treatment (1963)

| Industry | Wastewater (billion gals.) | Standard BOD (million lbs.) | Settleable and Suspended Solids (million lbs.) |
|---|---|---|---|
| Food and kindred products | 690 | 4,300 | 6,600 |
| Textile mill products | 140 | 890 | n.a. |
| Paper & allied products | 1,900 | 5,900 | 3,000 |
| Chemical & allied products | 3,700 | 9,700 | 1,900 |
| Petroleum and coal | 1,300 | 500 | 460 |
| Rubber and plastics | 160 | 40 | 50 |
| Primary metals | 4,300 | 480 | 4,700 |
| Machinery | 150 | 60 | 50 |
| Electrical machinery | 91 | 70 | 20 |
| Transportation equipment | 240 | 120 | n.a. |
| All other manufacturing | 450 | 390 | 930 |
| **All Manufacturing** | 13,100 | 22,000 | 18,000 |
| **Domestic** | | | |
| Served by sewers (120 million people) | 5,300 | 7,300 | 8,800 |
| **Ratio** (Industry/domestic) | 2.5 | 2 | 3 |

Ref: The Cost of Clean Water, Vol. 1, Summary Report, U.S. Dept. of the Interior, FWPCA, U.S. Government Printing Office, Washington, D.C., January, 1968.

**Table 9.** Some Critical Industrial Groups

Organic and Inorganic Chemical Industry
  Plastics, resins, rubber, paints, explosives,
  drugs, pharmaceuticals, detergents cosmetics,
  etc.
Paper and Allied Products
  Sawmills, paper mills, pulping, plywood, etc.
Food Manufacturing
Fertilizers and Agricultural Activities
  Pesticides, agricultural chemicals
Textile Mill Activities
Blast Furnaces and Steel Activities
  Steel mills, foundries, ore refining
Petroleum Refining Activities
Tires and Tubes
Stone, Clay, Glass and Concrete Products
Transportation
  Ship building and repairing
Electrical Companies
Mining

environmental scientists. This is for a number of reasons, one of which has already been noted: the relatively large quantities of waste water that are generated by industry. The other reasons have to do with the fact that many industrial discharges are either toxic or biologically resistant to breakdown. For example, many industrial compounds that are discharged are toxic either directly or · indirectly to biological components of the receiving waters. For example, the toxicity of inorganic poisons such as cyanide or phenol is well known. Certain other compounds which are produced by industrial activities can be concentrated on up the food chain by microorganisms and ultimately they can reach sufficient concentrations to become potentially harmful to humans. Examples of such indirect toxic agents are herbicides, insecticides and heavy metals such as mercury.

There is still another class of compounds which are produced in large quantities in industrial as well as agricultural operations which are not degraded readily by normal biological means or processes. These are called non-degradable or biologically resistant compounds. Such resistant compounds tend to accumulate in nature and at times can be the source of toxicity or simple nuisance problems. Examples of such compounds would be plastics, herbicides and insecticides, and until a few years ago, detergents.

## Wastewater Treatment

The practical objectives of wastewater treatment are BOD (Biochemical Oxygen Demand—the term expresses wastewater strength in terms of oxygen required or an oxygen demand for the biological oxidation of sewage substrate under defined conditions) and suspended solids removal. Wastewater treatment processes can be classified by a number of ways. A useful way of categorizing treatment methods consists of dividing the processes used into physical, chemical and biological. Examples of these processes are shown as table 10. The examples of the processes listed is not exhaustive and includes only the most commonly employed methods. While treatment may consist very simply of one step (e.g., sedimentation), quite often several of the process steps listed on table 10 are used. For example, a

**Table 10.** Methods Used in Wastewater Treatment

Physical Processes
  Sedimentation
  Absorption
  Cooling
Chemical Processes
  Coagulation
  Neutralization
  Chemical Oxidation or Reduction
Biological Processes
  Aerobic
    Activated Sludge Process
    Trickling Filters
  Anaerobic
    Anaerobic Digestion
    Septics/Cess Pools
  Oxidation Ponds or Lagoons

chemical plant may settle a waste (e.g., sedimentation) neutralize and treat biologically by either of the methods listed.

Sedimentation is probably the most common unit operation employed in wastewater treatment. Usually the process consists of the use of a very simple sedimentation basin where the wastewater is held quiescently for a short period of time (15 minutes or so) during which time the heavier particles can settle to the bottom of the tank where they are removed. Many times when only sedimentation is carried out in domestic wastewater treatment, this is called "primary" treatment.

Chemical wastewater treatment processes are not often used in the treatment of domestic wastewater discharges. Usually the use of such chemical operations as coagulation or neutralization are carried out in industrial wastewater treatment processes. Coagulation, however, has been used as a part of "tertiary" treatment methods (which will be discussed) in connection with the treatment of domestic wastes.

Biological methods consist of using living microorganisms in treatment. Since these methods are used to a great extent in the U.S. they will be discussed separately.

Very often wastewater treatment processes are classed by still another system of nomenclature as "primary" and "secondary." In primary treatment, biological activity is of negligible importance in the improvement of the wastewater quality, and, generally, settleable or suspended solids are removed from the sewage by the physical process of sedimentation in settling tanks. Secondary treatment depends upon the activities of aerobic (oxygen requiring) microorganisms—generally the bacteria—for the biochemical decomposition of the sewage substrate into a "stable" end product. As would be expected, primary and secondary treatment processes also differ in effluent quality or degree of treatment. For example, primary treatment processes will only effect a 40 to 60 percent removal of suspended solids and a 20 to 25 percent removal of BOD plus the removal of floating materials.

*Primary treatment methods* consist of the use of septic tanks, *Imhoff* tanks, as well as various settling devices such as sedimentation basins, upflow clarifiers, etc., and chemical treatment methods.

*Secondary treatment* methods are broadly classed as activated sludge, continuously mixed, actively growing microbial populations, trickling filtration involving the use of microorganisms grown on a fixed medium such as stones, and the like, intermittent sand filtration, and stabilization ponds.

As noted on table 10, biological wastewater treatment can be classified as either *aerobic,* meaning in the presence of dissolved oxygen, or *anaerobic* where oxygen is toxic to the process and therefore excluded. By far the most common method of treating domestic wastewaters, as well as many industrial waste discharges or effluents, is by aerobic biological means ,specifically the use of either the activated sludge process or trickling filtration. No attempt will be made to fully explain the biological metabolic sequence in aerobic metabolism. However, the general steps used in sewage treatment are shown as figure 86. Generally aerobic end products are considered to be "stable"—e.g., no further change will occur on standing. In treatment oxygen is supplied as part of atmospheric air which is introduced into the mixture by forced aeration—e.g., bubble aeration as in the activated sludge process or by atmospheric film renewal (ventilation) as a result of flowing sewage over a microbial slime growth as in the trickling filter process. In either case, the microorganisms are predominantly aerobic and dissolved oxygen must be continuously supplied.

In both the activated sludge and the trickling filtration processes the overall pattern of treatment is similar—primary sedimentation precedes the use of either the aeration tank or the trickling filter. This is followed by a final settling sedimentation or clarification of the "sludge" (cells) and discharge of the "treated" wastewater as a clear effluent of low organic content. Recycle of sludge cells is commonly employed in both processes.

Since both the activated sludge and the trickling filter processes are quite commonly used in the treatment of domestic wastes in the U.S. and Europe they should be described in greater detail. Quite often in conjunction with the treatment of sludges (settled residues from the process), anaerobic digesters are used. Depending on the situation, oxidation ponds and holding lagoons are also used, usually more so in conjunction with the treatment of industrial waste discharges.

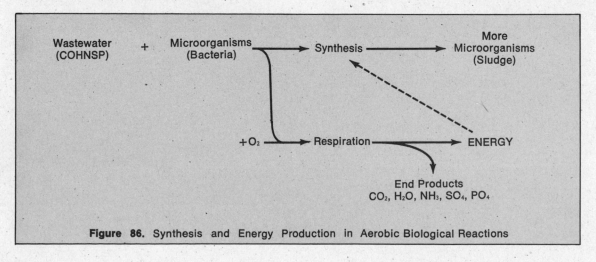

**Figure 86.** Synthesis and Energy Production in Aerobic Biological Reactions

**Figure 87.** Primary sedimentation tank.

**Figure 88.** Aeration in a tank.

**Figure 89.** Final sedimentation tank.

### The Activated Sludge Process

In wastewater treatment there are many variations that can be used in carrying out the activated sludge process. Accordingly, the process goes under a variety of names. The characteristics of the activated sludge process are that there is a continuous, intimate contact, mixing and aeration between a freely moving slurry or flocculent microbial population (mostly bacteria) and the sewage substrate, figure 87. During the period the mixture is held in the aeration tank (usually a few hours), the bacteria grow actively, utilizing the sewage as food and essentially degrading it to oxidized end products carbon dioxide and water and converting another portion of the waste to more microbial cells, termed sludge. Conditions in the aeration tank are maintained such that the microbial cells will settle as well as actively metabolize the wastewater, figure 88. The bacterial cells formed (called secondary sludges) are then removed to a final sedimentation tank where they are removed, figure 89. Usually, as shown in figure 90 A, B, C, a portion of these sludge cells are recycled back to the aeration tank. The

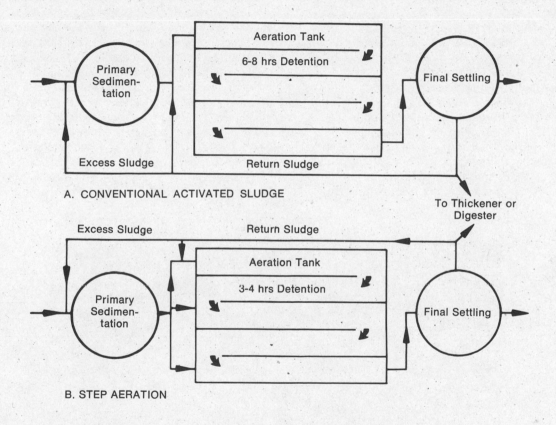

A. CONVENTIONAL ACTIVATED SLUDGE

B. STEP AERATION

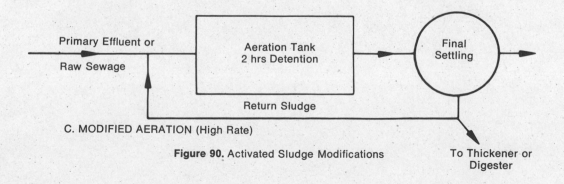

C. MODIFIED AERATION (High Rate)

**Figure 90.** Activated Sludge Modifications

remaining sludge cells must be further processed.

In general, the activated sludge process can be reduced to the following five functional steps:

1. Mixing—The "activated sludge" (return sludge) is mixed with the sewage to be treated.
2. Aeration (and agitation)—The mixture of sewage and activated sludge is aerated for a period of time.
3. Separation—The activated sludge is separated from the "treated" mixture.
4. Return—Recycling of settled activated sludge to the process.
5. Disposal—The sludge mass must ultimately be removed.

Mixing is generally achieved by adding the return sludge, figure 90 A, B, C, to the settled sewage, e.g., as primary sedimentation effluent, at the inlet end of the aeration tank, figure 91. Aeration supplies the necessary oxygen for biological oxidation as well as provides agitation for mixing the sewage and the return sludge sufficient to maintain the sludge in suspension. Aeration is carried out normally in excess of one hour, depending on the process, but can be for as much as eight or ten hours. Separation of the activated sludge mass from the mixture must be achieved prior to discharging the treated wastewater and is accomplished in a final sedimentation tank. Gravity settling using controlled flow is normally employed and is carried out for short periods of time (an hour or so). Return sludge is generally reintroduced into the process to provide microorganisms as well as to enhance BOD removal through surface adsorption. The volume of return sludge may vary from 10 to 50 percent of the volume of the sewage being treated, depending on the process, the treatment desired, and a number of other factors.

Ultimately disposal of the excess activated sludge from the final settling tank as well as from the primary sedimentation basin must be carried out. This can be achieved by transferring the sludge to an anaerobic digester for further bulk reduction. If anaerobic digestion is employed and yielding ultimately, "digested" sludge. Final sludge disposal generally consists of incinerating thickened, dried sludge, or burial. In some cases the thickened sludge is barged for ultimate ocean disposal or dumping. The flow diagram for a conventional activated sludge plant is shown as figure 90 A, as well as for two variations, figures 90 B & C. In a conventional activated sludge plant

**Figure 91.** Aeration tank inlet.

however all of the incoming settled sewage is mixed with the return activated sludge at the head of the aeration tank as shown. With average domestic sewage, the volume of returned sludge is 20 to 30 percent of the volume of the sewage being treated. Depending on the type of aeration being used, aeration basins are designed to provide an aeration time of from six to eight hours using diffused aeration and for longer periods using mechanical aeration, figure 92.

Waste sludge in the conventional activated sludge process varies from 10 to 30 percent of the total aeration solids. Average removal of suspended solids and BOD will vary between 80 to 95 percent, and normally over 90 percent.

Step aeration is a variation employed in the activated sludge process as shown in figure 90 B, the sewage may enter the aeration tank at a number of different points. In step aeration, however, all of the return sludge is introduced with, or without a portion of the sewage, into the aeration basin at the first point of entry. As a result, the sludge solids concentration in the aeration tank is greatest at the first point of entry and decreases further down the basin as more sewage is introduced at subsequent points. Aeration detention time is of the order two to four hours. Return sludge rates are generally maintained at 25-35 percent of the average sewage flow. Since the process provides convenient means of regulating the total solids held under aeration, step aeration has the advantage of being able to reduce the aeration time required for treatment to about half that required in the conventional process while giving comparable treatment efficiency.

**Figure 92.** Mechanical aerator in activated sludge treatment process. Photo courtesy of Haverstraw Joint Regional Sewage Treatment Plant, Haverstraw, N.Y.

While there are a number of variations of the activated sludge process besides step aeration which are used such as, modified aeration, activated aeration, contact stabilization, biosorption, etc., only one "high rate" modified will be noted. The "high rate," figure 90 C, is essentially an accelerated version of the conventional activated sludge process. It is normally used where effluent quality requirements are not so stringent. As shown in figure 90 C either raw or settled sewage mixed with about 10 percent return sludge is aerated for a period of only one or two hours. Normally the suspended solids in the aeration tank are held at lower levels resulting in a reduction of air requirements. BOD removal efficiencies of 50 to 70 percent are typical of the process.

## Trickling Filtration

To the uninitiated, the term trickling filtration is a misnomer, since no obvious filtering action is involved. More appropriately, the term biological oxidation beds or bacteria (or percolating) beds would be more apt and is, in fact, the more descriptive term frequently used in Great Britain.

The trickling filter process, however, may be defined as a fixed bed system consisting of stones, or other media over which sewage or wastewaters are allowed to "trickle." The sewage flows by gravity over a growing, attached microbial population and serves essentially as nutrient. The media to which the microorganisms are attached as slime are loosely packed so that voids or air spaces exist for the passage of atmospheric oxygen. In some cases, forced draft ventilation is

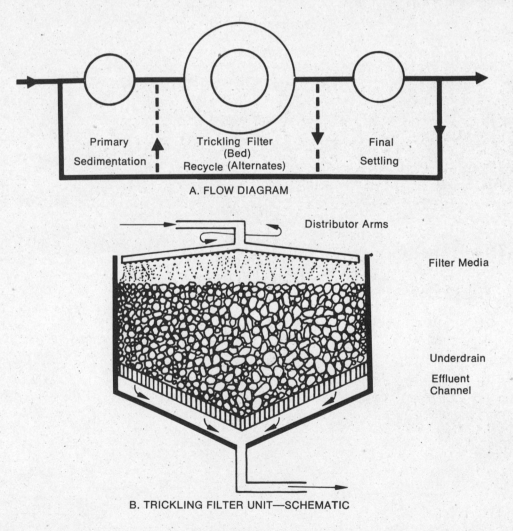

Primary Sedimentation

Trickling Filter (Bed) Recycle (Alternates)

Final Settling

A. FLOW DIAGRAM

Distributor Arms

Filter Media

Underdrain

Effluent Channel

B. TRICKLING FILTER UNIT—SCHEMATIC

**Figure 93.** Trickling Filters

employed. As the sewage passes over the microbial bed, adsorption as well as some mechanical removal occurs. Oxidation takes place long after the sewage particle has passed.

A flow diagram showing the significant features appears as figure 93 A & B. As in the activated sludge process, the "filter" unit is preceded by primary sedimentation and is followed by a final settling prior to discharge. Primary sedimentation is necessary to protect the filter units by removing the coarse and rapidly settling materials. Since the attached slimes periodically slough off of the media and pass into the flowing stream, final sedimentation is necessary to assure a clear effluent. Although trickling filtration involves a number of modifications, the changes are minor and involve variation in recycle, use of "staging" (e.g., use of a primary and second stage filter), forced draft ventilation, or employment of a media other than stone. Other than the dashed lines shown on figure 93 A, no effort will be made to describe these modifications, since the differences are not major as in the activated sludge process.

Figure 93 B is a schematic cross section of a typical trickling filter unit. The significant features include the distributors (or dosing apparatus), the filter media (stones, plastic, etc.) on which the microbial population is growing, an underdrain system, and an effluent channel in which the treated wastewater is collected. The effluent channel will not be described.

The distributors, which serve to distribute the sewage over the filter surface, can consist of fixed spray nozzles or rotary distributor, although alternate methods, such as siphon feed, are sometimes used. The source of filter media is frequently field stones, gravel, broken rock, or any inert material that can serve as a microbial holdfast for the slime material. When stones are used, they will measure from 2 to 4.5 inches in diameter and are selected for their irregular shapes so that adequate voids will be left throughout the filter bed when packed. In general, the filter bed is circular in shape and will range from 3 to 10 or even 14 feet deep, while the average is 6 ft. The underdrain system serves to carry the sewage away from the filter and also to provide ventilation to the slimes adequate to maintain aerobic conditions. The difference in air flow up or down depends on the season, since a temperature differential is involved.

While the use of trickling filters yields BOD and suspended solid removals up to 85 percent, this process has certain other advantages over the activated sludge process, which exceed their removal deficiencies. Notable among these advantages are a capability to withstand increased shock loading and, hence, give a greater stability to the treatment process as a result of organic overload or toxic waste "slugs." Trickling filters, in general, give a more consistent, reliable performance with minimum operational controls.

**Anaerobic Digestion**

Anaerobic digestion is frequently employed as part of secondary sewage treatment in the further reduction of sewage sludges. The basic unit consists quite simply of a covered tank into which the excess sludge can be pumped and held in contact with an actively growing facultative and obligate anaerobic microbial population for a given period of time. An anaerobic digestion unit is shown as figure 94, but with modification for recirculation. Anaerobic decomposition metabolically proceeds as a two phase process during which time the carbonaceous material is first converted into acids or other organic intermediates and then methane gas ($CH_4$) and carbon dioxide ($CO_2$).

Although anaerobic digestion would, at first glance, appear to be simpler than the activated sludge systems previously described, the process, in fact, requires much closer environmental control in terms of pH, temperature, and loading.

Figure 94, although for a high rate unit, contains the essential features of an anaerobic digester. The thickened sludge (primary or from final sedimentation), is heated, normally to 95°F, prior to introduction into the digester. To maintain the temperature at this level for incoming as well as for the tank contents, the sludge is circulated through a heat exchanger. When the digester is mixed, gas can be recirculated and used to keep the tank contents homogeneous. Since the gas produced from anaerobic digestion is approximately 65 percent methane, the resulting digester gas can be used as a source of fuel for the heat exchangers. As indicated, the digester effluent is withdrawn and can be recirculated back into the aerobic process. Thickened or heavier sludge will settle into the bottom of the tank and is withdrawn for disposal. Digested sludge will show a further reduction in BOD, specifically, $CH_4$ production and, in general, is more "stable" and dewaters or thickens more easily.

Depending on mixing, temperature, hydraulic detention and organic loading, anaerobic digesters are classed as conventional, high rate, and anaerobic contact. The anaerobic contact process

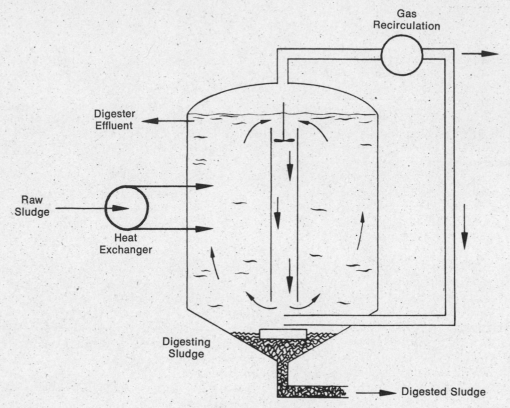

**Figure 94.** Anaerobic Digester

utilizes a separate compartment for thickening, effluent withdrawal and sludge or cell return recycle. Many installations are now equipped with vacuum filters for drying sewage sludge, figure 95.

### The Effects of Wastewater Discharges

The discharge of a pollutant into a receiving water such as a stream, lake, bay, etc., will set off a progressive series of physical, chemical and biological events which eventually result in the overall deterioration of the receiving water ecosystem (i.e., the aquatic environment with its resident creatures as part of a system). The effects of these discharges on water quality can be described by a number of well understood basic responses in the receiving water environment. Generally the observed changes can be classed as physical, chemical, biochemical, and biological. More specifically, pollution can affect the environment in a number of ways:

1. Dissolved oxygen (DO) decrease
2. Ecological changes
3. pH aeration
4. Temperature effect
5. Specific ion toxicity
6. Nutrient enrichment

Quite often, several of these factors will operate together or are interrelated. This will be shown in the discussion that follows. This deterioration can be characterized by a number of changes which occur in the receiving waters, and includes such noticeable changes as alterations in color, turbidity or odor or less obvious chemical changes.

### Dissolved Oxygen

The most important and fundamental effect that a discharged waste will have on a receiving water is probably upon the dissolved oxygen stores of the system. Oxygen is an element that is vital to all life as we know it. Yet oxygen is only slightly soluble in water and is, in fact, found only in small amounts in water. Usually the concentration of dissolved oxygen is measured as parts per million. Physically, oxygen solubility is affected by two fundamental factors in natural waters— temperature and salt concentration. Generally, the solubility of DO changes with temperature, for example, cold waters are capable of "holding" more DO than are warm waters. Another important factor affecting DO in natural waters is the presence of dissolved salts. Dissolved salts lower DO values in water. Although any dissolved salt

**Figure 95.** Vacuum filter for drying sewage sludge, Photo courtesy of Haverstraw Joint Regional Sewage Treatment Plant, Haverstraw, N.Y.

can affect DO solubility, the most common effect can be seen in marine and estuarine waters (a mixture of marine and fresh waters).

*Dissolved oxygen* (DO) concentrations in polluted waters are usually lowered by three general mechanisms; temperature, biological utilization (mostly microbial), and as a result of strictly chemial oxygen consuming reactions (chemical oxidation). While the biological response is probably the most important of all of these mechanisms accounting for DO disappearance, all three are important.

The importance of dissolved oxygen to the character of aquatic life is of paramount importance in an aquatic system. In fact, of all the parameters for measuring the relative pollution of a receiving water dissolved oxygen (DO) content is probably our single most important

criteria. A shortage of available dissolved oxygen (DO) resulting from the introduction of a pollutant will severely affect all of the living systems in the receiving waters and cause a major ecological upset. If the depletion of the DO is sufficient to cause anaerobic conditions, severe nuisance conditions will follow characterized by the presence of odors and the appearance of darkly colored, turbid waters and sediments.

When a wastewater containing available food for microorganisms is discharged into a receiving water, a DO response similar to that shown as figure 96 follows. As shown, the horizontal axis can be viewed either as distance downstream from the point where the pollutant was introduced, or as time required for the biological degradation of the pollutant. Generally, these two parameters, distance and time, can be considered to be the

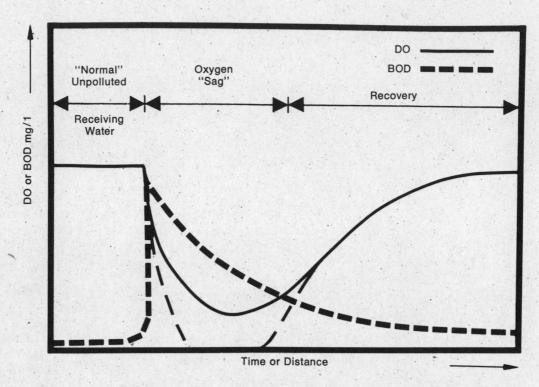

**Figure 96.** Dissolved Oxygen Response in a Receiving Water

same. Under "normal" circumstances, as shown on figure 96, the DO value of a clean, unpolluted water is normally at saturation. Almost immediately after the addition of a waste to a receiving water, however, dissolved oxygen levels begin to drop (thus, this is sometimes called a "sag" curve). This drop in DO or deficit as it sometimes called, is the result of two factors working in opposition to one another: the capacity of the receiving waters to replenish its oxygen stores, and oxygen utilization by the aerobic microorganisms—mostly bacteria. While the mechanisms of the former will not be pursued in any greater detail, the latter reaction of the bacteria resulting in oxygen consumption should be noted.

The biological disappearance of DO in water can be written in general terms by the following equation:

$$\text{Organic Pollutant} + \text{Microorganisms} + \text{DO} \rightarrow \text{More Microorganisms} + H_2O + CO_2 + \text{Oxidized End Products}$$

The biological response shown by this equation is a very natural reaction of the rapidly growing microorganisms (mostly the bacteria) to the abundance of available food (i.e., the pollutant). Essentially this consists of a breakdown of the pollutant and its utilization by microorganisms in their life processes to generate more microorganisms. The microorganisms formed usually make the water turbid. In addition, it is these microbes coming together and dying that contribute to sludges and bottom deposits. In certain situations the capacity of a water to replenish its DO stores may actually be exceeded by the demands of the reproducing bacteria for more oxygen to continue their breakdown of the pollutant. In such cases the waters would actually became devoid of all DO, as shown on figure 96 by the dashed line, and become anaerobic. If this occurs, a major metabolic change follows which will affect the fundamental character of the water and spells severe environmental deterioration to the system. When all oxygen is consumed, the microorganisms again, mostly the bacteria adapt to a different type of metabolism and the following reaction sequence is now carried out:

Organic Pollutant + Microorganisms → Microorganism +

$$CO_2 + H_2O + End\ Products$$

($H_2S$, $CH_4$ $NH_3$ and other reduced end products only partially broken down such as acids,, aldehydes, ketones, etc.)

While the end products of aerobic metabolism are mostly oxidized, stable compounds, insignificant in terms of nuisance or odor, this is not true of anaerobic end products. Anaerobic metabolism produces reduced, unstable end products which can be further broken down or which are quite reactive. Of particular significance is the presence of $H_2S$ (hydrogen sulfide) a very active and odorous gas that goes into solution and combines with many of the partially degraded organic compounds to form still more odorous organic compounds. Moreover, $H_2S$ can combine with many of the minerals present to form insoluble sulfides resulting in the formation of dark brown or black sludge deposits. Thus the general description of gross pollution—odor and color and sludge banks can be understood in terms of an overstressed biotic system that has become anaerobic.

### Ecological Changes

While changes in the DO of the receiving water are generally ascribed to the bacteria, in reality all of the biological community undergoes change. The biological response can be understood by considering two general parameters: species variety and total population or biological mass. Species variety is actually an estimation of how many different species of fish, invertebrates, snails, algae, etc., can be identified in the system. Biological population (or biomass) on the other hand reflects the total population (or weight) of all of these different creatures. A plot of these two parameters before and following the introduction of a pollutant into a receiving water is shown as figure 97. As with DO, the response is idealized and time or distance on the horizontal axis are equatable.

A clean, unpolluted receiving water can generally be characterized by having a high species variety (i.e., many different kinds of individuals) whereas

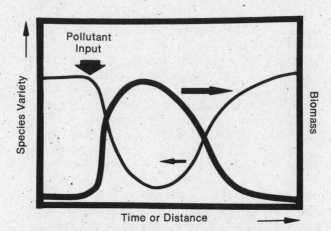

**Figure 97.** Species Variety and Biomass

the sum total of these different creatures will constitute a relatively low total population or biomass. After a pollutant is introduced, however, species variety will drop. The decrease of species variety can be understood by noting that fewer and fewer of the species originally present in the system can cope with the adverse conditions resulting from the pollutant discharge. Almost any factor that can affect biological survival, can result in reduced species variety, for example, low DO, as pH alteration, the presence of toxic materials, hot water, etc. In most instances it is a combination of factors which result in a lowered species variety. In any case, as the pollutant is decomposed and its effects become less severe and conditions improve in the receiving water, creatures (i.e., more species) slowly return and species variety will slowly be regained.

Biomass, or total population as shown on figure 97, shows the opposite response of species variety. Generally, most forms of biological life cannot cope with the adverse conditions created by the pollutant and either die or migrate out of the environment. The species remaining in the environment, therefore have two advantages, first they can either tolerate or utilize the pollutant as food, and secondly their competitors are gone. Thus rapid growth occurs which is the result of only a very few of species types. Typically, the biomass reaches large levels by comparison with levels prior to the entry of the pollutant into the system. As the impact of the pollutant is diminished biomass drops off.

### Temperature

Temperature will affect aquatic systems by affecting two very important processes, dissolved oxygen (DO) solubility and chemical reaction rates.

While the effect of temperature on DO solubility has already been noted, the effects of high temperature on chemical reactions, since thermal discharges constitutes a major industrial pollutant, should be given a brief consideration.

Elevated temperatures to a limit will accelerate all chemical and biochemical reactions. A generalization, stemming from physical chemistry called $Q_{10}$ that applies to all biological systems states that *for each 10°C rise in temperature, the reaction rate doubles*. This generalization has both upper and lower temperature limitations, since at high temperatures living tissues are killed whereas at lower temperatures water freezes and biological activity ceases. However, even before killing occurs as a result of temperature elevations, as the temperature is raised, many biochemical reactions are altered. Generally, high temperatures, those over 75°C, will alter protein configurations. Since enzymes are made of protein and it is the enzymes which mediate all of biochemical reactions in living systems, many reactions cease. The usual nonsymetric response of an organism to temperature elevation is an initial stimulation which may include growth, reproduction, formation of end products, etc., followed by a rapid cessation of activity as the temperature increases.

Since temperature is such a fundamental environmental factor which affects all reactions, the response of a system to a stress such as temperature elevation is not always a simple one and deterioration as a result of DO depletion (e.g., anaerobic conditions) may be accelerated. For example, elevated temperatures, as already noted, will also influence DO solubility. Thus, two general responses follow an elevation in water temperature: decreased DO solubility and an increased activity by the microbes. Since increased microbial growth is also accompanied by an increased DO requirement, all of these factors operating together can rapidly diminish the available DO supplies in an aquatic system.

## pH

Hydrogen ion concentration or pH, like temperature, is a very fundamental environmental parameter which affects all of the biological components of the system. The basis of this effect is the fact that the activity of the enzymes which catalyze biologic change are greatly influenced by pH changes. Most biochemical reactions, in fact, will not take place unless the pH is within specific values. For most biological reactions this is around neutrality (pH 7). Generally any particular biochemical reaction will have a rather narrow pH range for its optimal activity. Usually any departure from a range below pH 6.5 or above pH 8 will affect biological activity.

The term pH can be a deceptive one. This is because pH really represents a logarithmic relationship. For example, a decrease in 2 pH units actually means that the hydrogen ion concentration is now 100 times its former strength. At still another level pH can affect the solubility of minerals, the oxidation-reduction potential of a system, and effectively serve to limit the ecosystem. For example, a lower pH may permit certain forms of life to exist and not others.

## Toxicity

The effect of a specific toxicant or poison in a receiving water upon the biological components sometimes occurs. Most often such toxicity originates from an industrial operation such as metal plating or chemical manufacturing. While toxic effect may be caused by the presence of an organic compound that is very specific in its action such as cyanide, it may arise from the effects of a general poison such as phenol. Among the more common toxic agents are the heavy metals, copper, zinc, lead, chromium, etc. and their dissolved salts such as sulfides or ammonia. Quite often the toxicity of a poison is augmented or intensified by other factors such as temperature, lowered DO, pH (acidity or alkalinity) and the presence of other dissolved salts especially calcium.

Usually all of the biological community in a receiving water will be affected by a toxic agent. However, fish are probably most sensitive to the effects of toxic agents. Usually fish kills are the first and most obvious signs of intoxication. At the more subtle level, however, secondary populations of some of the intermediate organisms such as invertebrates, snails, as well as the smaller microorganisms the algae and protozoa—all of which are usually involved in the ecological web in the system—can be affected either altered or killed.

## Nutrient Enrichment and Eutrophication

Wastewater discharges to a receiving water may not always result in immediate signs of gross pollution. Yet a gradual deterioration of the receiving water may occur. For example, a receiving water may show no obvious DO depletion, scums or odors, or the like and in effect may not appear to be severely affected. However, if the water is closely examined over a period of time, it will be found that the water quality is slowly changing as a result of a gradual and progressive biological development. Most often this occurs in enclosed bodies of water. Usually such receiving waters need not contain large amounts of oxygen consuming organic compounds but will contain nutrients such as nitrogen, phosphorus and other minerals, which are used by microorganisms in their life processes.

This enrichment process is called *eutrophication* and is, in fact, a slow response to the gradual addition of nutrients from natural or pollution sources including industrial pollution in an enclosed or semienclosed body of water. Most often, compounds containing nitrogen such as ammonia or nitrate and phosphate are involved. The process of eutrophication may take place over a period of years or it may occur over a very long period of time, like many decades or centuries. Usually a distinction is made between natural eutrophication which will ultimately happen to all lakes resulting in plant growth, drying and forestation and "accelerated" eutrophication which occurs when man intercedes and introduces masses of nutrients into a system.

What has happened, in effect, is that a relatively unenriched body of water called *oligotrophic* with clean water and sparse biological growth, as a result of the addition of nutrients, such as ammonia, nitrate or other nitrogen containing compounds and phosphates, slowly begins to stimulate a biological population in response to the availability of the nutrients. Photosynthetic algae will usually grow first, using carbon dioxide and perhaps nitrogen from the atmosphere or from bacterial reactions. In the beginning there are not many organisms and development is slow. A biological cycle is initiated however and when they die, their cellular materials settle to the bottom and serve as food for other microorganisms. This forms the basis for other biological subcycles in the enrichment process. Gradually a diverse population is built as other higher forms of life such as invertebrates, fish, etc. arise in the system. Ultimately the lake or basin becomes enriched or eutrophic and the process will continue until the body of water is completely filled and disappears leaving a permanent forest.

Do rivers and streams also become eutrophic? The term cannot be correctly applied to such bodies of water. However, a biological response does occur following enrichment and algae and bottom growth can ensue.

Recently phosphates have been accused of contributing to the process of eutrophication in our receiving waters. No doubt, the obvious increase in phosphates in our domestic wastewaters which followed the growth of the detergent industry after World War II largely stems from their presence in detergent preparations. It is estimated, for example, that approximately two-thirds of the phosphates found in sewage originate from household wash materials.

## REFERENCES

1. American Chemical Society. A Report by the Subcommittee on Environmental Improvement, Committee on Chemistry and Public Affairs, 1969. *Cleaning Our Environment, The Chemical Basis For Action.* Special Issues Sales, 1155 Sixteenth St., N.W. Washington, D.C. 20036.
2. New York State Department of Health, *Manual of Instruction for Sewage Treatment Plant Operators,* Health Education Service, Albany, N.Y.
3. U.S. Department of Interior, Federal Water Quality Control Administration, *Industrial Guide on Thermal Pollution,* (1968).
4. U.S. Department of Interior, Federal Water Quality Control Administration, *Biology of Water Pollution,* (1967).
5. Department of the Army, U.S. Corps of Engineers, *Permits for Work and Structures in, and For Discharges or Deposits into Navigable Waters,* (1971).

Prof. Raul Cardenas, Ph.D.
Department of Civil Engineering,
New York University,
Bronx, New York

# Chapter 21

# Pollution of Streams and Smaller Lakes

Appropriate attention has been devoted to the pollution of the Ocean and the Great Lakes by the scientific community and public information media, but it is the neighboring small lake or stream which represents the meaningful environment for most citizens and officials. Pollution problems of these waters most immediately and directly affect people. It is in this social sense that small lakes and streams may be considered a logical unit of approaching the subject of their pollution. Few small lakes have produced pollution issues which have made their names well known. Indeed, Lake Washington (Washington State) and Lake Mendota (Wisconsin) are recognized names only within a close circle of lake scientists, environmental officials, and staff members of a particularly concerned industry. River pollution, too, seems to have generated only few cases of wide interest. The Cuyahoga of Ohio became known when it caught fire, and the Hudson has been immortalized in a folk song by Pete Seeger. However the fact of the matter is that many small lakes and streams are nothing more than bubbling masses of foul smelling sludge.

Arbitrarily, all lakes in the United States with the exception of the Great Lakes, Lake Okeechobee, and the Great Salt Lake can be considered as small. They can be subdivided infinitely, on the basis of size, depth, basin, biology, etc. There is only limited professional agreement on any taxonomy of lakes, and this does not necessarily correspond to common language titles. The terms "lake", "pond", and "reservoir" all might be applied to a single artificial body of water. Furthermore, the "lakes" created by the impoundment of rivers such as the Tennessee and the Colorado remain essentially as deep rivers in the classification of many investigators. Even the titles of scientific specialties concerned with these waters are inadequate to distinguish the various aquatic environments in any meaningful manner: "oceanographers" work in fresh waters as well, and "limnologists" do not restrict their interest to lakes. Many are also "potomologists" when they are studying running waters. "aquatic ecology" may be the most descriptive general term for the scientific work done in these areas. Within this broad subject, individuals select certain restricted habitats and aspects to examine and interpret. Thus the understanding of the pollution of these varied waters requires a careful selection and synthesis of specific information.

A rough classification of temperate waters will suffice for the subsequent discussion of pollution.

*Deep Lakes.* These are characterized by stratification into two layers of markedly differing temperature, with a transitional zone between. Top and bottom waters may mix regularly once or twice a year or not at all. The layering has many effects, including reduction of oxygen levels, restriction of nutrient movement, and the maintenance of additional biological habitats. *Shallow Lakes*, in contrast, mix at shorter, irregular intervals. *Running Waters* (streams, rivers) ideally lack any time retention. Water should pass any given point in the basin instantly, and never return. In application, there are few ideal lakes or running waters. Even the Great Lakes exhibit a current flow which significantly affects the pollutants put into them. On the other hand both large and small streams develop pools, embayments, and elbows which are much more like lakes than running waters. Consequently over-generalization of types should be avoided in considering pollution problems in particular waters. An appreciation of the gradation and heterogeneity among types is of practical value in determination of pollution abatement policies: the point will be illustrated in discussion of cases later in this writing.

In the history of limnology, an enormous emphasis has been placed on classifying lakes in respect to their biological productivity. Although the preponderance of data gathered has been physical and chemical, the goal of the information has been a biological one: to measure capacity for the growth of organisms, or, more recently, the input, retention, transfer, and output of energy and materials by organisms acting in concert. Three basic classes of lakes were named on the basis of limited experience, and the system has been extended and extrapolated until it constitutes a virtual holy trinity in limnology. These classes were: *oligotrophic* (low productivity), *eutrophic* high productivity, originally applied only to certain highly organic and acid waters. In practice the term *"mesotrophic"* was inserted between oligotrophic and eutrophic. The system may have been overextended far beyond usefulness, and its continued application has had surprising consequences in actual issues of water pollution. Particularly, the term *"eutrophication"*, which appeared in professional literature conspicuously only after 1961, has become a common language word. It is the subject of educational discourse and massive publicity and heated debate, but specialists still gather from time to time and try to define the meaning of the word.

A successional classification is also implicit in an understanding of changes which occur in natural or polluted waters. The skeleton idea of the classification is that the waters begin, acquire a series of biological components which replace one another in a regular order, and finally the basin is filled with soil or dries up. In geological time, the fate sometimes has been innundation by the sea as well. Man's activities have indeed warped geological time and geometry so that salt or brine must be listed among the pollutants. It is critical to distinguish between a *successional scale* and a *time scale.* It is possible in specific cases to determine both for both aquatic and terrestrial situations. Unfortunately there are no general rules which can be applied accurately to many situations without considerable specific study. It is possible for a skilled ecologist to interpret the order and to estimate the speed of successional changes. The layman who must face pollution can acquire the minimum insight for an intelligent approach. The key idea is that succession is occurring, and that it must be taken into account in interpreting recorded changes of lakes and streams.

There are several special classifications of waters which are derived from intended use rather than from comprehensive understanding. These special classifications have been frequently involved in specific issues of pollution as well. The classification of waters as hard and soft has been of some economic importance (syrup manufacture, boiler operation, sale of water softeners, amount of soap used). Water hardness became a weapon to be used in water pollution issues involving detergents, soap, and salt. Color and turbidity have become pollution criteria in the minds of many people instead of simply aesthetic properties of drinking water. While equating

pollution with murkiness is naive, the properties do constitute scientific indices to pollution and to natural condition. A special use category in some states is the "salmon river", or "trout lake". This, of course reflects the high social value of a particular kind of fish. Since a particular environment is required by these fish, the ecological requirements are translated into legal definitions of pollution. Less waste is permitted and a higher level of dissolved oxygen must be maintained in these streams. Other widely accepted classifications have little or no ecological basis, but are important in determining input and use, thus are decidly involved in pollution and its control. Drinking waters, irrespective of chemical or biological quality, are frequently classified as AA (New York) or a similar category to indicate strictest pollution rules. In addition to certain specific prohibition (raw sewage), the restrictions against other inputs to drinking water tend to be vague. In essence, they attempt a high margin of safety to human health by prohibiting almost any input which might be injurious, a nuisance, or aesthetically displeasing. Such is the intent rather than the practice. Finally, the classification of waters as navigable or non-navigable has become of the utmost importance in the regulation of input and usage of the waters. "Enough water to float a log", has been quoted, half in jest, as the Federal definition of navigable which opened many waterways to pollution control. In a surreptitious manner, the classification has determined that national values rather than local will prevail in many small lakes and streams which lie entirely within the geographic boundaries of smaller political units.

A brief definition and classification of "pollution" is appropriate to the understanding of the details which will follow on the effects of pollution and the mechanisms for environmental management of lakes and streams. In the comprehensive view, *"Pollution"* is understood as a change in condition (energy or material level) which disturbs the self-regulatory capacity of an environmental system toward or beyond its operating limits. In very plain language, pollution taxes the cleaning ability of the environment. A convenient taxonomy of pollutant types includes: microbes (particularly human pathogens), organic matter, poisons, fertilizers, and primarily physical factors ranging from silt which blocks light and covers plants, to heat, to radioactive isotopes. Moreover, the concept of pollution is extended to include landfilling and shore alteration, which may have much more devastating effects than the others, up to complete destruction of the habitat—the ultimate pollution.

Both the study of pollution and the practice of environmental management by regulation will be found to conform rather closely to concepts of pollution held by the people involved, and by the kinds of pollution which they may recognize. This relationship will be illustrated in most cases but one example deserves to be cited here. In 1969 the Nassau County (N.Y.) Planning Board would not permit construction of an Industrial Park until convinced that runoff water (storm drainage) from the site would not pollute Hempsted Bay. The developer hired an academic engineer of 20 years' experience to champion his cause. In the consultant engineer's report he quoted a 1917 definition of water pollution which said essentially, "If it stinks and looks dirty, it is polluted," and then declared, "rivers and lakes are nature's sewage disposal plants." He recognized only the problem of how much coliform bacteria and organic matter could be diluted. From this viewpoint of sanitary engineering an era in thinking was leaped within a few years. By 1971 Federal legislation (e.g. "the so-called Muskie Bill") had been considered with the intent of prohibiting the discharge of *any* polluting matter into natural waters. The logic of this legislation is that environmental health is as important as drinking water. The ideological vision is clearly shifted from a limitless environment which serves man, to a closed system in which man plays a critical role.

It is possible to utilize simple models of "a river" and "a lake" for understanding and controlling pollution. The intelligent use of these models presumes awareness of the arbitrary judgments which have been made. The goal of the first part of this writing has been to impart an appreciation of the structure and operation of the real world. In proceeding, the admonition to consider the variations of the real world is implicit, but it will not be repeated in writing. Concepts or models will be used in real world contexts.

Legal concepts of streams strongly favor the property owners on the shore. In navigable waters, the water itself is considered public property, but the right can be exercised only by access to the shore, or from the sea or air. The doctrine of riparian rights is a basic guide: all adjacent property owners are entitled to the "thread" of the stream. In practice, property owners could do almost anything they wished as long as an acceptable share was left to downstream owners. Quantity rather than quality has been the chief concern.

In a typical modern case, the level of a lake may be maintained by a weir, but enough water must be released to turn the ornamental water wheel of a property owner on the outlet stream. A share of quality water is implicity in riparian rights, but the share has hardly been demanded in the past.

The public interest of longest standing in streams is navigation, which has proceeded through rather foul waters at times. Even use as public water supply hardly provided impetus for restricting use or maintaining quality. Almost no restriction whatsoever was placed on non-navigable water, which have been regarded rather strictly as part of private property. Nevertheless, public agencies (a Town administrator, a public health official) have had powers to bring action in cases of pollution where a public nuisance was indicated. Such nuisance actions have been infrequent, and limited to flagrant cases such as strong smell. In addition other property owners could seek damages for nuisance caused by a polluter, but this is an example of equity between individuals, no public interest.

Law designed primarily for settling disputes between property owners has been inadequate for maintaining water quality. One difficulty is the sharing of pollution guilt by a number of parties. Of whom can equity be demanded if many contribute a share of pollution? It may be possible, but it is impractical to sue all contributors. Indeed, the question of shared guilt has hardly been pursued sufficiently for legal precedents to have been established.

The evolution of effective legal regulation began with the embryo idea that the quality of water was indeed a right of the used. With the realization that the resource was not unlimited, restrictions were accepted within the concept of, "best use". This concept still rules most of the legal processes. Essentially "best use" is an economic, not an ecological standard of regulation since use is considered only in terms of a short span of time and space. As ecological content has entered the theory and practice of regulation, a stream is visualized more as part of a water cycle, and as a series of biological communities. This stage of evolution is reflected in attempts to replace the best use concept with the concept of "non-degradation." Non-degradation of the environment is a logical policy under the concept of ecological property. The quality of ecological property can already be defended to some extent by the "class suit." Such actions permit a segment of the population to seek equity—or a degree of quality—for a resource which they do not "own" except in the ecological sense of being in the same ecosystem. In its ultimate ideal, the highest principle of law then becomes the protection of the common property of a population extending indefinitely in time and space.

The development of law has corresponded to ecological knowledge of stream pollution to a degree. Traditionally, it is assumed that stream pollution can be corrected by the natural system which is big enough and has enough time. The model presented by A. F. Bartsch and W. M. Ingram in 1959 has become a highly accepted classic both because of its clear written detail and its graphic illustrations. The chief attention is given to oxygen and organic matter, reflecting the focus of sanitary engineering. Dissolved oxygen and the quantity of organic matter from sewage (expressed as *biochemical oxygen demand* or BOD) are plotted in time and space, above and below a discharge point. The uniform stream model was adjusted to variations in flow, light, and natural or artificial aerations. Similar plots were superimposed on the oxygen-organic matter model, and correlated with it: decomposition products of aerobic and anaerobic decomposition, bacteria, (total and coliform, algae, ciliates, rotifers and crustaceans, and a communal index of species variety, including fish. They also distinguish clean, degradation, decomposition, and recovery zones through selected indicator organisms: Sludge worms, red midges, sow bugs, and aquatic insects.

This model has stood as important in different respects. It has provided objective criteria to judge the seriousness of pollution in many rivers. It provided a yardstick which showed, objectively, that there were decided limits to the number and size of polluting sources which could be placed along the thread of a stream, if the natural system were to clean itself at all. It was no longer possible to preserve the illusion of infinite space where waste simply goes away. The model illustrated that waters remain polluted for some time and distance after pollutants entered. In practice, regulations have given a "free permit to pollute" with certain restrictions on amount. At worst, the permit to pollute is ecologically more enlightened than the doctrine of riparian rights.

Various devices have been used to quantify pollution rights or allowances. All are directed at the ecological goal of minimizing the disturbance to the regulatory capacity of the environment.

Stream standards. Selected ecological criteria must be maintained: Clarity, oxygen, nutrients, acidity. The only restriction on wastes is that they do not alter the water beyond the prescribed limits. The limits are adjusted to correspond with a designated use for the water; a lower quality of water being permitted for farm and industrial use than for recreation. In using this device, the "permit" is granted in common, but without any means of apportioning the share among individuals. Thus the two plants may meet a stream standard, but a third plant would either be denied any right of discharge or else all would need to reduce their contribution. The limitations of standards are many, but it has been of some practical utility in abatement efforts. A serious flaw in reliance on stream standards alone is that they do not protect the quality of downstream areas. Streams may meet their assigned standards and still carry damaging pollution to a shallow lake.

Treatment standards. These prescribe the amount of pollutant which must be removed before discharge into a receiving water, amount generally being stated as a percentage of untreated material. This device is closely associated with engineering design and practice. Treatment facilities are intended to perform a measured change between input and output, and treatment standards require little knowledge of ecology. Obviously a large facility can pollute a given river more than a small one if both remove the same partial percentage. Percentage figures can be falsely reassuring. The removal of 90% BOD by the City of St. Louis sounds commendable unless one reaches the perspective that the discharge is equivalent to putting the raw sewage of 200,000 people into the Mississippi River. Nevertheless, treatment standards have been of practical value, since they provide concrete goals in design and operation. Also several kinds of input waste are known to be reasonably uniform, so that absolute amounts of pollutants can be calculated easily if the information is desired.

Effluent standards. The amount of a pollutant is limited at the point of input either in terms of concentration (percentage of the total input of liquid) or of actual weight or volume, instantaneously or averaged over a period of time. Radioactive materials, for example are regulated by actual amount in given time periods.

A combination of all three basic devices was used in the New York regulations aimed at limiting the phosphate in waters of the Ontario Lake Basin. This is one of the most systematic attempts to apply the permit to pollute in both a fair and effective manner. The highest total standard can be reached by applying the appropriate device, yet minor contributors need not meet some standards if they constitute no threat to the system as a whole. It may be added, that regulations must be written with respect to potential evasion: Effluent standards must specify that permissible concentrations must be achieved by removal—not simply by dilution before discharge.

There is a growing component of ecological consideration in all permit-to-pollute kinds of regulation. The receiving water *is* polluted to some extent, and the environment is, indeed, being used as a treatment facility. The question which has arisen in profuse variation, is, "How far and how long is the pollution to be permitted"? Thermal pollution has drawn an enormous amount of attention to the question. The concept of a "mixing zone" has been prominent in the discussions of design and environmental standards. In this case too, a combination of regulatory devices has been necessary: Effluent standards specify the temperature (and perhaps amount) of water at discharge, and environmental standards limit the mixing zone, inside which there is a reduction of temperature to the stream level.

From the ecologist's viewpoint, the analyses of thermal discharges have revealed the importance of the structure and dynamics of the aquatic community. Two cases will illustrate the necessity of ecological knowledge in judging the consequences of pollution and in devising successful regulatory devices. In rivers, thermal discharges are restricted not only by uniform rules of temperature and distance, but a portion of the particular receiving water must remain unaffected. The same discharge could affect the whole width of the river, or one side of the river to a greater length downstream. In most cases, the biological community suffers less if a continuous thread of the stream is maintained—their natural riparian rights?

The limits of mixing are probably much more difficult to judge in lakes than in rivers. The fundamental difference is that the lake is far more a closed system where actions may have cyclic and cumulative consequences. The ramifications of thermal discharge in a small lake (in the definition of this writing) have received the most thorough-going study, consideration, and debate in the case of the proposed Bell Nuclear Power Plant on Cayuga Lake, New York. The very criteria for measuring heat, temperature and amount of hot water, were brought into question: The fundamental question was asked as to whether the portion of the total annual heat budget was not a

better viewpoint to measure effect. It was apparent too, that the heat question could not be considered in isolation. A major pumping operation would be involved, so that the circulation of fertilizing nutrients might prove to be of greater consequence than heat. The possibility produced design changes. Finally, it was agreed by both proponents and opponents, that the layering period in the summer would be extended for about two weeks. These ramifications illustrate the long range and indirect effects which could be discerned by a large number of highly qualified scientists working under considerable pressure to attack or defend a decision. Unknown ramifications may be much more numerous.

The particular pollution problems of lakes derive from the ecological nature of those heterogeneous environments. Recognition and regulation require some uniformity, however, and the search for ecological models must continue while abatement efforts are already in progress. There is no ready answer to the very first question, "What are the boundaries of a lake?" As the following examples will illustrate, this is a practical question, not one for speculative ecologists. Lakes, psychologically and legally, are perceived as common property to a greater extent than rivers. If the shoreline is held by several owners, their property lines can be projected to a geometric center of the lake, each considered to own a slice, of the bottom at least. Difficulties are apparent both for geometry and the meaningful exercise of ownership. Thus, the practice of public ownership of "land under water" has been widely accepted. "Ownership" may remain nominally with owners of shore property, but the title is almost empty of special priviledge except for dockage: Control is in fact a public function. Pollution questions are often involved in the compromises between public and private ownership. Shore owners have strongly opposed the installation of public boat launching sites with the argument that added motor boat oil, debris, noise, or crowding constitutes pollution. Paradoxically, they may not oppose commercial docks or limit their own horsepower. In essence, the shore owner's view remains, "This is our lake." However, "our" is not private, but a collective for all of the owners.

When the state has established control rights over land under water, it moves into new areas of pollution control. Particularly, it can regulate alterations of the shallow water areas, preventing the filling of marshes, and the conversion of public open waters to private land property. These police powers must be justified in terms of the total ecology of the lake or of the particular area affected. It is in this situation that the boundary question has demanded attention. The question is decidedly an ecological one, although it may be answered by lawyers or engineers. From the tradition of maritime law has come the concept of "littoral zone," now applied to inland waters. In its original ecological context "littoral" means "intertidal," and was meaningful, but there are no tides inland, and water levels not only fluctuate irregularly, but they have been altered temporarily and permanently by man. Frequent use in discussing inland waters has not clarified the meaning of "littoral zone". It generally refers to the ground sometimes covered by water, or else, "shallow water" or "zone of rooted vegetation." This confusion should be seen as an unsuccessful attempt to define the boundaries of inland waters.

Engineers have naturally taken recourse to calculations, and have created arbitrary formulae for determining the boundary. "low mean water level," and "high mean water level," are determined by gathering measurements over a given period of time and treating the data with selected mathematical procedures. The mathematical derivation bears little relationship to ecological realities. Averages in no way describe real constraints. The ecological boundary may be gradual, broad, and changing. Plants and animals are not separated at the low or high mean waterline, but transcend them. Furthermore, the water of the lake is continuous with the ground water and exchanges with it freely near the shore. Thus, the State of Wisconsin recognized that the "lake" extended beyond the water, and applied a margin of error from the engineering averages. Specifically, it regulated usage up to three feet vertically from an arbitrary water level. The greatly increased number of septic tanks created the pressing need to extend the lake boundary. Shores have been economically exploited by developing many cottages on small lots, close to the water table. Drainage of their waste water underground is in effect little different from direct discharge to the open water. There has been no uniform resolution of the boundary problem. Recognition of an ecological boundary has indeed been considered; e.g. the ecological boundary of the lake is the limit of land vegetation (as judged by a competent ecologist). Unfortunately neither the vegetation nor the water have remained natural or constant around many lakes. Ultimately, the boundary must be set by law arbitrarily, but an intelligent setting

is determined from environmental management. This step has already been taken occasionally in the name of public health. For at least a quarter of a century watershed sanitary inspectors have regulated operation of septic tanks in some lake basins. The meaning is clear: The lake and the watershed constitute a single ecological property.

There has been a notable shift in attention during recent years, accompanying increased use of waters for recreation. From microbial organic pollution, attention has moved to nutrient pollution and to poisons. The increased resident population around small lakes in Michigan and Wisconsin were aware by 1950 that their waste nutrients were producing nuisance crops of plants. The increased population of fishermen gained political power enough to achieve prohibitions against pollutants which killed sports fish. Such laws have been remarkably weak instruments. Typically they can be applied only after the fact of a kill and the full burden of proof of cause and effect is upon the State. It has been only the threat to human health which brought drastic preventive action in cases of mercury poisoning. Pollution and the concentrating effect of environment dynamics were simply bearers of the threat. Both mercury and persistent pesticides, notably DDT, have drawn attention to larger waters, not to streams and small lakes. While no imminent threat to human health has been demonstrated for DDT, its universal penetration into the larger environment and the demonstrated devastation of some species has taught splendidly the wider, more abstract ecological lesson that persistence presages pollution. In a rare but exemplary case, residents around a small lake eschewed the use of spray pesticides and returned to screens for protection, when fisheries' biologists informed them that the fish population in the lake had declined because of increased doses of pesticides which pyramided through the environmental system. Restriction of persistent pesticides has now begun, although bitterly opposed.

The enormous increase in motor boats has brought two special pollution problems. The first is simply increased raw sewage, which has been met by requiring either on-board treatment or holding tanks. The second pollutant from boats is petroleum. The effect in streams and small lakes is usually not as obvious as a major oil spill. Nevertheless, outboard motors are inefficient, discharging about one third of their fuel into the water. The effects include some addition to the organic matter to be decomposed by bacteria, poisoning or suffo-

of information and debate preceeded the referendum to approve the bond issue for local funds to match state and federal support. Virtually a full spectrum of issues on small lake pollution was manifest. Although inspection of septic tanks had been instituted 25 years before, moderate to heavy fecal content could be demonstrated in feeder streams. Coliform counts in the open waters has remained low, but the conviction has grown that local pollution around the 15 miles of shoreline could not be controlled. Shoreline septic tanks were often located very near the lake and drained into the water table. In addition, effluent from an overloaded village primary treatment plant entered the lake toward its outlet. Heavy summer blooms of blue-green algae were characteristic, but profuse weed growths were of more concern to residents. Weed poisoning was strictly regulated—essentially prohibited. Mechanical cutting was considered and rejected, not simply on economic grounds (1% of the subsequent bond issue), but on ecological grounds as being futile. A three-cornered dialogue of cause began: Residents blamed farmers for nutrient pollution; cottagers blamed villagers, and villagers replied in kind. Scientific advice was sought, and the conclusion was accepted that cottages and village together were the important contributors, farms being mostly absolved.

It is critical to the concept of pollution and ecology to consider what may be accomplished by completing such a project. Promoters tended to overstate both the threat and the promise. Heavy plankton blooms were observed in 1926, and the same species of both blue-green algae and rooted aquatics were present forty-five years later. The clarity of the water remained similar, apparently to the credit of grazing microcrustacea. Surprisingly, this lake had demonstrated the regulatory ability to clean a punishing load of pollution. It was a practical demonstration of regulatory capacity in a highly productive, integrated, and varied ecosystem. Other Finger Lakes, however, have not handled their increased loads so well. These have shown proliferation of a "weed" aquatic (*Myriophyllum* or milfoil), retreat of the rooted zone and other changes. When the experience with the Madison lakes is added, the indication is that no dramatic short-range improvement should be expected. The Madison lakes are still highly productive. One still has an excess of phosphorus in late summer, probably due to septic tanks not connected with the comprehensive system. Nevertheless, changes have already occurred in both the water chemistry and in the species composition of phytoplankton. The quick improvement of Lake Washington (Washington State) was achieved by replacement of water as well as nutrient removal. Although the Washington achievement is a major one in terms of both scientific information and practical results, it cannot be adapted to many other situations.

Instead of perimeter sewers, advanced treatment before discharge into lake waters has been provided in a few cases. The attempt proved futile at Lake Tahoe, California, and the effluent was diverted from the watershed. A much smaller and different kind of lake in New York (Canadarego) will soon receive highly treated effluent, following a comprehensive multi-disciplinary study of its ecology establishing phosphorus as a limiting nutrient.

Inevitably, the consideration of pollution in rivers must lead to dreams of comprehensive planning. Although a perimeter sewer itself may remove phosphates and organic matter from a Finger Lake, pollutants will be added to streams and to Lake Ontario unless taken out of the system in a treatment plant. The limitations of scientific knowledge, political structure and public information are immense barriers in the way of comprehensive planning. The East is far more crowded and politically complicated than the West. Major planning is relatively simple for the Bureau of Reclamation in the West, whereas the Army Corps of Engineers in the East would need to deal with an almost limitless list of interested agencies and citizen's groups. While it is true that comprehensive water resource planning by federal agencies in the past has not been ecologically based, major internal reforms are occurring, and there is substantial hope that the ecological viewpoint will come to prevail. Comprehensive planning has also been attempted in various forms by individual states. Some proposals, such as the Texas Water Plan, have been labeled as blueprints for major ecological disasters, while eastern states fear a cost barrier with their citizens.

Nevertheless, it is a major advance that officials, scientists, and citizens have begun to give serious consideration to comprehensive planning. Ironically, the industry fight against restrictions on detergent phosphate has brought comprehensive planning to the attention of many people. A good portion of nonsense has been broadcast, to be sure. The whole Atlantic seaboard has been written off as no problem, ignoring the thousands of small lakes and critical estuary areas which may benefit

cation of small animals. These effects hardly seem important, but for the realization that zooplankton can effect water quality. High clarity is maintained in a fertile lake by the grazing action of a micro-crustacean, the common water flea or *Daphnia*—unknown to the motor boaters. Lakes used as drinking water may develop an unpleasant taste after some years of motor waste accumulation.

Near Madison, small "over-fertile" lakes came to the University of Wisconsin's attention. These lakes provided a field laboratory for limnological work directed at understanding the nutrition of plants and the food dynamics of the aquatic community, but the findings led to no immediate managerial action. The model natural sequence of lake history had already been conceived: The gradually accelerating increase in biological productivity and final filling up of the basin. Yet, pollution was not related to natural history at this time. Significantly, the term "eutrophication" began to be popular only in the mid-1960s, possibly because of the influence of D. G. Frey's prestigeous studies.

## Limnology in North America

Few terms have been used so much with so little understanding. Yet, the meaning which has evolved is simple enough in bare outline; most new lakes are infertile, they gradually become more productive, choking then filling, with both organic matter and soil from the surrounding basin. The process is a slow one, but can be accelerated enormously by human activities, particularly by the fertilizer elements of domestic, agricultural, and industrial wastes. Such acceleration was termed "cultural eutrophication" by A. D. Hasler (1969), who also documented his assertion that the process was reversible.

In practice, the decision had been made at least twenty years before to reduce fertilization of European lakes (Switzerland, Scandanavia) by the removal of nutrients, particularly phosphates, from sewage, in addition to sanitary treatment and reduction of organic matter. Research in the United States and in Canada (University of Wisconsin, the Federal agencies in Washington State and in the Potomac River, the Canada Centre for Inland Waters, and others) pointed to the same strategy. In 1967 a symposium of the National Academy of Sciences stated as a summary conclusion that the fertilizing role of phosphate was critical. The viewpoint was established by the International Joint Commission for Pollution on the Great Lakes after a monumental study reported in three volumes (1969). Finally, the American

Society of Limnology and Oceanography confirmed the phosphate reduction strategy in 1971 in summarizing its symposium, "The Limiting Nutrient Controversy."

The controversy still rages in North America, primarily because a major industry has sustained it. Twelve percent of all phosphate produced has been used as the preferred water softening agent along with synthetic detergents for at least twenty years. Its replacement has been resisted on grounds including interference with private enterprise, whitening power, and health hazard. The phosphate controversy has overshadowed a second important controversy: The pollution of waters by farm fertilizer. This problem was brought to public view by Barry Commoner in respect to Illinois rivers in particular. Farm nitrate was defended quite as automatically as detergent phosphate. In addition, there have been some attempts to play one pollution source off against the other (a case will be cited subsequently). These controversies illustrate that pollution is not simply a scientific question but that it is strongly structured by personal and social values. The resistance to control of fertilizer can be contrasted with the ready acceptance of control measures for the relatively recent problem of feedlot waste. Animal manure is "dirty" and attracts little commercial interest.

Under the mountain of chaff generated in the name of science consequent to the nutrient controversies, there lie some unresolved scientific questions. Unquestionably both scientific viewpoint and management decisions in pollution will undergo the same sort of shifts as they have in other subjects. A few selected cases can serve to illustrate some of the open questions both in terms of ecology and management instead of a discussion restricted to theory. These cases also constitute the introduction to the final section of this writing: The systematic approach to pollution of rivers and small lakes.

The first sewering of entire lake perimeters was undertaken at Madison, Wisconsin in 1958. (Monona, Waubesa, and Kegonsa). The scientific leadership of the University of Wisconsin unquestionably produced this management decision. The original project is still in the process of extension to include additional villages and the individual waste of septic systems. Comparatively few communities have followed. New York's first perimeter sewer (Conesus, a small Finger Lake) which is scheduled for completion in 1973. The Conesus project will include an advanced treatment plant giving 80% phosphate reduction. A lively program

by phosphorus limitation. The whole Mississippi River system is treated as a simple system, ignoring the lake-like characteristics of many portions. Behind the face of the controversy is the eloquent concept that control of pollution involves regulation of consumer products. Not only behavior, but the very spacing of humans has become directly involved in analysis of water pollution. There is already a small legal history of cases of contention when lot size is justified by the requirement for septic tank leach fields. Since questions of nutrient as well as microbial pollution are about to be raised (Vermont and probably elsewhere), the case file will grow rapidly.

In the comprehensive abatement programs of the eastern states, it has been discovered that some communities cannot afford, by the current norm, to construct central collection and treatment facilities. Either a subsidy must be supplied by other communities, or else regional zoning must logically be envoked to keep population, agricultural, and industrial inputs strictly within the limits of private disposal systems. There have already been a few temporary suspensions of building permits by health authorities for communities where treatment facilities were scandalously overloaded. It is no longer a long step to consider stabilization of population and production as necessary means of recovering a self-cleaning environment.

Prof. Herman S. Forest, Ph.D.
Department of Biology
State University of New York,
Genesco, New York

# Chapter 22

# The Effect of De-icing Salt on Vegetation

An increasing number of eyesores have cropped up over the last decade or two in the heavily populated areas of North America which experience a period of ice and snow each year. Where once beautiful rows of maple trees lined the highways, now either dying trees or no trees at all grace the landscape. Millions of dollars are spent annually on planting trees and shrubs along major arteries to break the monotony of bulldozed roadsides. But evergreens are turning brown and deciduous trees and shrubs are dying back or not breaking into manes of flowers in spring or early summer. As well, some species of evergreen trees in woodlots adjoining major highways are also turning brown and gradually dying out.

The first possible cause, that usually comes to people's minds, is the exhaust fumes from the cars and trucks crowding these highways. Closer observation of the damage, however, will reveal that there is another cause or possibly other causes. The pattern of injury shows that the side of the tree facing the road is damaged much more than the other side, figure 98, suggesting that the damage is probably not caused by gases. On many young evergreen trees, with branches to the ground, the lower branches appear normal while the top may be almost dead, figure 99. Generally, the trees on the south or east sides of highways are more affected than those on west and north sides. Besides this distinctive characteristic, the injury becomes apparent in late winter or early spring, after a period, when in most areas the flow of traffic has been the lightest.

Analysis of the damaged tissue has shown that high concentrations of sodium and chloride are present. If salt is the causal factor, this would

**Figure 98.** Plantation of red pine on the east side of a major highway. The side facing the road is virtually dead while the other side still shows green needles.

**Figure 99.** A plantation of young white pine on which the lower branches were covered by snow and protected from salt spray.

explain the timing of the appearance of the injury. It has been assumed that most of the salt runs off the highway into the soil and thus affects vegetation through the root system. This is attested to, by areas of soil, adjacent to busy highways or intersections, that are almost completely devoid of vegetation. But this does not explain the pattern of injury at greater distances from the pavement. The undamaged lower branches is one indication that snow cover in the winter yields protection. If one observes the fate of the large quantities of salt used for de-icing purposes, one finds that after a snow fall when the road surface is wet, some of the salty water and slush is whipped up into a spray by the fast moving traffic. This spray may be carried considerable distances by the wind, which after most storms blows from a northwesterly direction.

If salt spray is the cause of the injury then the pattern of damage is explained. In the spring of 1970, the author and a colleague sampled evergreen trees on several sites adjacent to a major highway in Ontario. White pine and white cedar showed the most injury. Needles on white pine showed a definite brown tip and green base. The closer the tree was to the highway, the greater the proportion of brown on the needle. By measuring the green and brown portions, an accurate

estimate of the severity of the damage could be determined. It was observed that in areas where wind flow was not impeded, measurable damage occurred as far away as 500 feet from the pavement. Measurement of the salt content of the tissue revealed that the salt level decreased with increasing distance from the highway as did the injury, the amount of injury being directly proportional to the salt content of the needles, figure 100.

The damage to evergreens is the most obvious because of the browning that occurs in the early spring. Some trees are more susceptible to this form of injury than others. White pine, white cedar and Scotch pine rank as highly susceptible, whereas, Mugho pine, Austrian pine and blue spruce are far more resistant to injury from salt because the salt is not as readily absorbed by the foliage of these trees.

Injury to deciduous trees and shrubs is widespread, even though it is less conspicuous. Only the twigs and buds are exposed to the salt spray and the damage is done before the buds open in the spring with the result that the buds and ends of twigs die. If the tree is not too severely damaged, the branches may leaf out further down the stem, figure 101. This gives affected trees a tufted or "witches broom" appearance. Again certain tree

**Figure 100.** A mature plantation of white pine exhibiting varying degrees of injury depending on the distance from the roadway. The tree in the foreground is about 130 feet from the edge of the pavement.

**Figure 101.** Branches of manitoba maple showing dead branches and new growth budding out further down the stem giving the tree a "witches broom" appearance.

**Figure 102.** An old beech tree about 120 feet from the pavement exhibiting severe dieback particularly on the side facing the road.

species such as beech, manitoba maple and weeping willow are very sensitive whereas trees with thicker twigs, smaller buds or with buds covered with resinous scale tend to be more resistant to injury from salt spray, figure 102.

The use of salt as a de-icing agent has increased exponentially since 1940 with the amount used in the United States for de-icing doubling every 5 years. Applications of 30 to 40 tons of pure salt per linear mile of two-lane highway per winter season are common. This is a tremendous load for the environment to absorb, and absorb it, it must, for there are no other means of dispersal.

A large proportion of the salt is carried by the spring run-off into ponds, streams, rivers and lakes where it creates a new set of problems. Seepage into wells near roadways is also common.

At present there is no satisfactory substitute for salt as a de-icing agent. It remains the most effective and economical. Attempts are being made to find substitutes and derive methods of application, such as the increasing use of sand mixed with salt.

The environmental hazards from the salt application will persist. Research is being carried out to find ways of protecting roadside plantations by such means as spraying trees with a plaster coating which is water repellant. Of course, individual trees and shrubs can be protected in the yards of home owners by wrapping or shielding them. As well, the more resistant trees and shrubs are selected for planting in exposed areas. Considerable effort is also being expended on the selection of salt tolerant grasses and other tolerant ground cover species. However, these are only temporary solutions. A permanent one has yet to be found.

Prof. Gerald Hofstra, Ph.D.
Ontario Agricultural College
Department of Environmental Biology
University of Guelph
Ontario, Canada

# Chapter 23

# Noise Pollution

The word "pollution" has a long and dark history. Most dictionaries list derivatives for it from the Greek and Latin that describe the unwashed, the filthy, and the impure. Word pairs such as "air pollution" and "water pollution" are therefore appropriate for designating the befouling of naturally pure resources.

Even without the added word "pollution," popular denotations of the word "noise" are mostly negative. The noun may be traced etymologically to the Latin variation of the Greek word "nausea." True, "noise" is occasionally used with neutral intent as a synonym for the word "sound," but more commonly it refers to sounds that are unpleasingly or painfully loud, discordant, harsh, or confusing. The negative quality the word has accumulated over the centuries probably accounts for its recent appropriation by the field of communication. Noise is the general term for any disturbance (originally it was specific to electrical disturbances) that interferes with the reception of a signal or of information.

Clearly, yoking the word "noise" to the word "pollution" is redundant. How can one pollute what is already polluted? Yet it serves a practical purpose. In the past few years all two word combinations ending in "pollution" have been given quantitative connotations. It was not until the different kinds of pollution reached alarming levels that anyone stood up and took notice. When, for example, substances like phosphorus and carbon monoxide are released into water and air in sufficient amounts to make air dangerous to breathe and water unfit to drink, then and only then are these substances considered to be pollutants.

Controversy generally centers first on whether a particular substance is, indeed, a pollutant. When it is proved to be so the argument focuses on how much of this type of pollution can be tolerated. In essence this has been the history of society's response to noise. What make noise as a pollutant is that, unlike polluted water and air, it pollutes via one or both of two possible pathways: the soma or the body and the psyche or the mind. Whatever effect polluted air and water may eventually have on the mental and emotional functioning of the organism, the damage circuit is always initiated at the biophysiological level. By affecting the integrity of the peripheral hearing mechanism, noise may also inflict injury through a strictly physical route. This fact could explain why noise pollution is often assumed to be completely analogous to air and water pollution. But because noise is perceived (and interpreted) as well as received, the original damage may be psychological rather than somatic, with any number of possible physiological consequences. We are therefore forced to take into account attitudes toward noise when estimating its evil as a psychological pollutant. Furthermore, hearing loss can itself lead to maladjustment, making noise a double psychological threat.

Data collected in physiological experiments are frequently less suspect than data collected in psychological studies of the "what is your attitude toward low flying airplanes" variety. The former data tend to be more stable and are reputed by some to be more valid. Spelling out the percentage of mercury accumulated in the tissues of a living organism or the concentration of DDT in a plot of land and, more pertinently, counting the number of hair cells destroyed in the inner ear of a sacrificed chinchilla who had been exposed to high levels of sound are tasks that call for relatively straight-forward measuring procedures. It is harder to specify the exact effects of mercury, DDT and loud noise on the behavior of the organism, even in animals.

Occasionally we come by behavioral evidence in tragic ways. The nineteenth century tale of the boiler-makers tells of many who served as unsuspecting victims for unplanned experiments. At that time men hammered away on steel plates, riveting them together to form boilers and eventually developed a type of high frequency hearing impairment which was nicknamed boiler-maker's ear. Examining current audiometric data from this and many other heavy industries in light of case histories we are now able to assert definitively that if the human ear is bombarded for too long by noise that is too loud it will sustain permanent damage. There is much less agreement, on the other hand, about the extent and importance of psychological harm caused by noise—if such harm be even granted to occur—that is not loud enough or sufficiently persistent to impair hearing. Compounding the issue is the fact that people differ in their susceptibility to noise pollution. Authorities thus disagree about what constitutes a damaging level or type of noise.

### Measurement of Noise

In order to assess dangerous levels of a potentially toxic stimulus it is crucial to have available a reliable measuring instrument and a commonly accepted unit of measure. How is noise measured and what are the "polluting" dosages? Sound is produced when energy is applied to a potential vibrator with enough force to initiate vibration, assuming the vibrator is vibrating in some elastic medium such as air and not in a vacuum. As the molecules of air move to and fro like a pendulum around a central point, they impinge upon neighboring air molecules and set them into similar vibratory patterns. Successive groups of molecules become involved in this chain reaction which is called a sound wave.

Speed of molecular movement back and forth around a central point is the frequency of vibration. It is measured in cycles per second (cps), sometimes called Hertz (Hz). One cycle is a complete movement to the right of center, back through center, to the left of center, and back to center again. The faster the movement or the more the cps, the higher the pitch of the sound will appear to the listener. The human hearing receptor is responsive to frequencies as low as 16 cps and as high as 20,000 cps. Vibrations slower than 16 cps are infrasound and those above 20,000 cps are ultrasound. A sound may have only one frequency. It is then called a pure tone. If a sound has more than one frequency—that is, if parts of the vibrator are moving faster than other parts and so more than one frequency pattern is being generated—then the sound is described as a complex tone. Most sounds are complex.

The degree of resistance to vibration in the vibrator combines with the strength of the force applied to the vibrator to determine the extent to which the vibrator and subsequently the molecules

of air move left and right of midline. The greater the excursion the more intense the sound. The more intense the sound, the louder it seems. Because the human ear is a pressure sensitive mechano-receptor, intensity is measured in terms of sound pressure level (SPL), the force applied over an area 1 $cm^2$. The ear is so sensitive it can respond to faint sounds that are measurable in fractions of one-millionth of the atmospheric pressure. The range of magnitude between these barely audible sounds and painfully loud sounds is approximately one million fold. The pressures involved are so small and the range of pressures so enormous that sound cannot be conveniently measured in linear fashion as length is in inches. Instead the observed pressure of a sound is compared to a reference sound pressure level and reported in logarithmic ratios. The unit of measure employed is a ratio called the decibel (dB), which is one-tenth of a bel, named after the pioneer inventor, Alexander Graham Bell. The most popular reference level for the decibel is .0002 dynes per centimeter squared ($dyn/cm^2$). This reference level approximates the sound pressure level of the softest sounds that the pressure sensitive human ear can just about hear in its most sensitive mid-frequency range, the range from about 1000Hz to 3000Hz.

A logarithmic measure like the decibel does not describe equal changes of pressure between sequential numbers in an ascending scale. A sound measured to be 20 dB, for example, does not have twice the amount of sound pressure as a sound measured to be 10 dB. If 10 dB, the bel, is 10 times the pressure of 1 dB, 20 dB is 100 times the pressure of 1 dB. Nor can we add decibels as we do inches or pounds. When two different sound sources strike your ear at the same time and each is 65 db, such as might be the case if two of your friends were talking at once in tones of average conversational loudness, then the total sound pressure impinging on your ear would be 68 dB and not 130 dB. Special formulae and charts are available for determining the overall SPL when sounds of different frequencies and intensities are combined.

The dynamic range of hearing is about 120 dB. We do not start to feel pain in the eardrum—which is the only place we do feel it—until we are subjected to noise levels above 140 dB, which is 55 dB above the 85 dB level judged by most experts to be the uppermost limit for near-total safety from hearing loss.

Few of us are ever exposed to the 150 dB sound pressure level prevailing on the landing deck of an aircraft carrier or the 195 dB blast-off level of a Saturn rocket. But we are closing the gap.

At the lower end of the loudness scale we have your sweetheart's gentle whisper which wafts over to you at about 30dB. That tranquil meadow you always dream about with its chirping birds and soft breezes logs in at around 40–45 dB. Your soft-spoken grandfather hovers around 50 db when he talks, your boss maintains closer to 65 or 70 dB when he is in a good mood, and your spouse when angry roars in epithets of 80 to 90 dB—even louder at times, or so it seems.

Your office typewriter and kitchen blender hit your ears at about 78 and 83 dB respectively. The new garbage disposal unit and lawn mower you just bought grind away at from 90 to 100 dB. Your cousin's unmuffled motorcycle and your uncle's large farm tractor chug-a-lug from 90 to 115 dB, depending on the brands.

Your parents' Friday night cocktail party may register 90, perhaps 100 dB, particularly if too many imbibe too much. The band in which your son plays on Saturday nights is pounding him with 105–120 dB of hard rock at the band stand and when you go to pick up your daughter you will find yourself in the middle of a dance floor that fluctuates between 100 and 115 dB. And that gun you took with you when you went hunting Sunday morning shoots impulsive noise of at least 130 dB every time you fire. With the right equipment you can actually measure the noise around you which, in turn, may stimulate you to abate it.

The instrument used to measure sound is the sound level meter, figure 103. The portable meter is calibrated prior to each measuring procedure to determine that is is working accurately. Depending upon the purpose, we measure the intensity and (or) the frequency spectrum of the sound. Intensity may be measured at different directional points and at varying distances from the sound source, again according to the goal of the investigation. Sound level meters are generally constructed so that it is possible to take into account certain peculiarities of human hearing. Even though most sound level meters have the capability of measuring the intensity of the sound in question in true SPL re .0002 dyn/cm$^2$ on the C scale, they also have adjustment switches which allow for weighted ratings. Today the majority of noises is measured in terms of a special weighted scale called the A scale and sound pressures so measured are reported in dBA. By the nature of the

**Figure 103**

Photo Courtesy of
B&K Instruments, Inc.
Bruel & Kjaer Precision Instruments
Cleveland, Ohio

weighting, this scale most closely resembles the response characteristics of the human ear, which is much more sensitive, as we have already observed, to sounds in the mild range of audible frequencies than it is to sounds of low and high frequency. It is interesting to note that speech sounds have their most important identifying frequencies in the mid range. One may postulate that our capacity for hearing conforms in some ways to the scope and limitations placed upon the production of speech sounds by the human vocal tract. Perhaps we were meant to heed our fellow men.

Many sound level meters incorporate frequency analyzers. That is, they can "listen" to a complex sound and identify which frequencies are participating in that sound and to what extent. Since a meter tooled to identify each separate frequency involved would provide too much information, most of the equipment currently used in noise surveys consolidates abutting frequencies into bands of varying widths. The feat is accomplished by setting up within the instrument a series of filters, each of which allows only a circumscribed group of frequencies to pass through it. The bands are usually an octave wide, although smaller band pass filters are sometimes employed for special purposes. An octave is the distance between one frequency, 250 cps, for example, and a second frequency which is twice the cps—in this case 500 cps. The filter for this band would thus allow to pass through only those frequencies that lie between 250 and 500 cps and it would read out the total intensity of the frequencies in that particular band. Frequency analyses are especially important when the offending noise is suspected of having strong high frequency components. The ear is more susceptible to damage from loud high frequency tones than from loud low frequency tones. Frequency analyses are also of significance in designing programs to abate and control noise. In summary, the sophisticated sound level meter measures and analyzes the noisiness of noises.

### Sources of Noise Pollution

Nowadays one can fruitfully take his portable sound level meter with him everywhere because noise pollution is everywhere. As with other forms of pollution noise pollution is closely linked to the growth of industrial urban centers. A concentration of population in a relatively small area leads in and of itself to more noise, if only from the racket caused by people in close quarters outshouting each other in frustration—not to mention the barking dogs and blaring televisions that accompany people into urban environments.

As urban centers expand, the number of nearby industrial plants multiply. Most industries are notoriously noisy and have been so since the birth of the industrial revolution. The average citizen who is not concerned about the noise-exposed worker may complain if industry sets up shop close enough to the town center to become a pollution problem, including a noise pollution problem.

Trucks transport adds significantly to the maddening traffic noise of cars, buses, taxies, and motorcycles. You can see why in several surveys traffic noise was found by itself to be as annoying as practically all other noises combined—unless you live in the suburbs near an airport. The noise contour patterns for jet take-offs plotted at one airport by the U. S. Department of Housing and Urban Development revealed a very high 130 PNdB (perceived noise decibels) peak level near the plane with a bleeding off in intensity to 90 PNdB at a distance of 7,500 feet. The perceived noise decibel scale is generally accepted to be more effective for evaluating the noise of air traffic than is the dBA scale, since it is weighted even more heavily than is the dBA scale for high frequencies. In any case, as the number of overhead flights increase, on-the-ground annoyance grows geometrically. When the Office of Noise Abatement and Control of the Federal Environmental Protection Agency held hearings across the country in 1971, citizens living in aircraft-impacted towns were among the most vociferous complainants.

Suburbia has noise problems other than those produced by planes and helicopters. Electric lawn mowers roar away at levels that have been measured to be just under 100 dB and hot rod cars and motorcycles without mufflers rev up storms of sound not conducive to peace and tranquility. Inside the home, when operating simultaneously, electric appliances like the blender, the dishwashing machine and the garbage disposal unit produce overall sound pressure levels that match those found in some industrial environments. A mechanized kitchen is not the only source of noise in the countryside. The U.S. Public Health Service reported in 1968 that farmers show a high incidence of hearing loss. It is no wonder since farms have been automated for many years. Farm tractors can produce noise levels ranging from 90 to 115 dB. Many people do not realize that tractors, milking machines and snowmobiles can be as deafening as iron presses and weaving looms.

Alexander Cohen of the U.S. Public Health Service has coined the term "sociocusis" to apply to the totality of physiologic reactions engendered by the unending variety of noise stimuli encountered by the average person from the moment of awakening to the moment of retirement at night.

**Effects of Noise Pollution**

Noise damages our hearing, upsets our vascular and digestive systems, causes us to abuse our vocal folds, by making us shout, makes us nervous, disturbs our sleep, hinders optimal task performance, and interferes with spoken communication by masking out speech.

## AUDITORY EFFECTS OF NOISE

The best documentation for the negative effects of noise pollution is found in the literature dealing with its relationship to hearing disorders. In order to appreciate how noise may affect hearing we must first survey the anatomy and physiology of the ear and then examine briefly how the deviant ear is measured to differ in acuity from the normal ear.

The peripheral hearing mechanism can conveniently be divided into the outer ear, the middle ear, and the inner ear, figure 104. Sound is transmitted through the medium of air along the outer ear canal or external auditory meatus, which is part of the outer ear, to the eardrum or tympanic membrane. The eardrum is set into vibration by the aerially transported sound waves and the amplitude of vibration is proportional to the intensity of the incoming sound. With each vibratory cycle, the more intense the sound is the greater will be the displacement of the eardrum to either side of its original position. There is a relationship between the intensity of sound as measured by a machine and the loudness of a sound as perceived

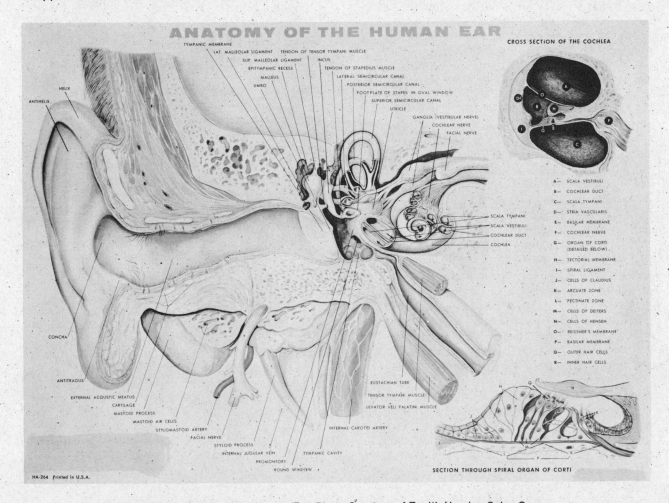

**Figure 104.** Anatomy of the Human Ear, Photo Courtesy of Zenith Hearing Sales Corp.

by the listener. The more intense the sound the louder it should appear to the listener.

On the middle ear side of the eardrum are the three smallest bones in the human body, the *malleus* or hammer, the *incus* or anvil, and the *stapes* or stirrup, in that order, connected to one another in such a manner as to form a synchronized chain. The head of the malleus is attached to the eardrum at one end of this ossicular chain and the footplate of the stapes is flexibly sealed into an aperture called the oval window at the other end of the chain. The vibrations of the eardrum are translated into a concerted back and forth movement of the ossicular chain which is in turn transmitted via the oval window to the fluids of the inner ear wherein is located the end organ of hearing. The entire hearing sensitive mechanism of the inner ear is protectively encased in the hard mastoid bone of the skull lying just behind the visible part of the outer ear called the *pinna*. From the inner ear, sound is transformed into electrical impulses and transmitted along the fibers of the auditory nerve out of the inner ear via the internal auditory meatus to the brain where it is perceived and interpreted.

The area of the stapes footplate in the oval window is at least 20 times smaller than the area of the tympanic membrane. Yet the vibrational displacement of and force upon each of these vibrating areas is equivalent. Thus the middle ear acts as an efficient transformer, stepping up the intensity of the vibratory amplitude of the eardrum and thereby maximizing the pressure per unit area on the fluids of the inner ear.

Fluids of two different composition are contained in three channels within the cochlea, the shell-shaped bony housing for the organ of hearing. One channel, the *scala vestibuli*, runs from the oval window along the two and three-quarter turns of the cochlea to the *helicotrema* or narrow tip of the conch-like structure. There it doubles over the inner channel, the *scala media*, to connect with a parallel channel, the *scala tympani*. The scala tympani returns along the same bony route (at a lower level) to end in a membrane covered opening, the round window, at the lower end of the middle ear.

The *organ of Corti* consists of a double set of hair cells, about 20,000 outer hair cells and 3,500 inner hair cells, which are specialized to respond to sound. The organ is embedded in the basilar membrane which extends along the entire length of the inner insulated column, the scala media, filled with a special fluid called *endolymph*. The

fluid of the two outer passageways, the *perilymph*, is set into motion by the vibration of the stapes in the oval window as are he walls of the inner partition containing the organ of Corti. The mechanical vibrations of the middle ear are thus transformed into to and fro movements of the fluids of the inner ear. The activity of the fluids causes the basilar membrane to be displaced, which, in turn, results in the bending of the hair-like projections of the cells in the membrane, much as stalks of wheat are bent by the wind. Since fluid cannot be compressed the pressure of the liquid waves is released through excursions of the round window.

Most authorities agree that the first sections of the cochlea to receive sound, the basal turn, are those that analyze high frequency sounds. Low frequency sounds are analyzed further up toward the helicotrema and the lowest frequency sounds stimulate vibrations throughout the length of the basilar membrane.

One additional aspect of the anatomy and physiology of the human ear should be explored and that is the *aural reflex* of the middle ear. Nature did not leave us entirely bereft of protection from loud noises. Two muscles in the middle ear, the *tensor tympani* and the *stapedius*, contract when stimulated by mid-frequency sounds of about 80 dB or higher, and lower and higher frequency sounds of about 85 dB or higher. Contraction of the tensor tympani stiffens the eardrum by pulling it towards the middle ear. The stapes is in like manner checked in its movements toward the inner ear by the contraction of the stapedius muscle. The reflex acts as an impedance device, reducing the efficiency of the transformation function of the middle ear for noises of low frequency up to 2000 cps. It offers little if any protection, however, against loud high frequency sounds.

Unhappily the response is sometimes too slow or too short-lived to protect the ear adequately. Gun shots, explosions and certain other kinds of impulse-type noise inflict damage too rapidly for the reflex to serve its purpose. Furthermore, these muscles fatigue if they are overstimulated by extensive exposure to noise. Since we know that damage to the hearing mechanism does occur despite the existence of the reflex, and even when the assaulting noise has its strongest components in frequencies below 2000 cps we might conclude that the reflex affords only limited protection from noxious stimuli that are unnaturally long-lived or sudden in onset.

The hearing threshold for sound is that level of intensity, measured in SPL, to which the ear responds about 50 percent of the time. That is, the sound is just barely audible at this level. Norms for the threshold of hearing are obtained from testing many people who have, upon examination, healthy-appearing ears and who report no significant history of ear infection, trauma or noise exposure.

Norms for the threshold of hearing vary with age. The younger the individual the better the hearing. For many years it was believed that the progressive deterioration of the auditory sense sustained with age, called presbycusis, was only a function of getting older. Now there is some evidence to show that the type of civilization into which one is born may account for the progressive loss. The loss may be due in part to atherosclerotic changes that are, in turn, related to stressful living, and in part to having to endure a lifetime of loud noises at work, at home, and at play. Samuel Rosen, a New York physician specializing in otology, studied a primitive African culture where the tribesmen lived to be very old, retaining into old age healthy vascular systems and acute sensory abilities, including sharp hearing skills. But it is peaceful and quiet there and the pace is easygoing.

Hearing levels may be measured with tuning forks or, more accurately, with a machine called and audiometer, figure 105. The human ear is not equally sensitive to all sounds, as was previously noted. Frequency is related to the psychological phenomenon of pitch much as intensity is related to the psychological sensation of loudness. Pitch, which is usually classified as high, medium, and low, interacts significantly with loudness. On the physical level frequency interacts with intensity or sound pressure levels. For a threshold response to occur we must have delivered to our ears more sound pressure for tones of lower and higher frequency to which our ears are less sensitive, than for tones of mid-range frequency. This inequality of sensitivity over the total range of our hearing—which runs from about 16 to 20,000 cycles per second—is compensated for in the audiometer so that for all frequencies 0 dB re .0002 dyne/cm$^2$ (ISO) on the audiometer is equivalent to the midpoint of the range of normal thresholds obtained from healthy young adults. Degree of hearing loss in any individual is thus expressed in terms of the extent to which his threshold for hearing is raised above the normal zero threshold.

The threshold for hearing is generally tested at discrete frequencies, the most common being 125, 250, 500, 2000, 3000, 4000, 6000, and 8000 cps. Special machines called high frequency audiometers test frequencies as high as 20,000 cps.

Pure tones, that is, tones of only the specified frequency, are most often used as stimuli in hearing tests. From practical experience it has been demonstrated that an average loss of 20 to 40 dB constitutes a *mild* hearing impairment, causing difficulty in understanding faint speech from 40–60 dB, a *moderate* impairment, making normally loud speech hard to comprehend from 60–80 dB, a *severe* impairment, leading to considerable difficulty and over 80 dB a *profound* hearing impairment, resulting in an inability to understand most speech even when shouted or amplified.

We can also find out at what intensity level the individual just barely hears speech by reading to him spondees (words of two syllables of equal stress like hot dog and railroad), and lowering the intensity of the sound until he responds correctly about 50 percent of the time. This threshold level is called his speech reception threshold (SRT). It should be roughly equivalent to the arithmetic average of the thresholds recorded for 500, 1000 and 2000 cps, the frequencies in the so-called midrange of hearing.

The most important frequencies for hearing are in the mid-range because that is the discrimination range for speech sounds. Speech is carried on

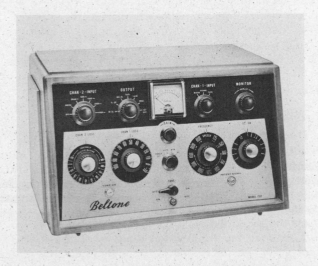

**Figure 105.** Audiometer, Photo Courtesy of Beltone Hearing Aid Company, Chicago, Ill.

frequencies that range from about 200 to 8000 cps. The boundaries for the critical listening range are 300 to 3000 cps. This is the range of hearing that must be preserved if at all possible. Less conservative estimates compress the minimal range for adequate discrimination to include only the frequencies between 500 and 2000 cps. In any case the more intact our hearing is in these middle frequencies, the better are we able to discriminate one speech sound from another. We can actually test the ability of the individual to discriminate speech sounds. We do so by reading to him lists of monosyllabic words (referred to technically as phonetically balanced or PB words) at a most comfortable loudness level (MCL). The patient usually hears a list of 50 words read. The number he accurately discriminates is multipied by two to arrive at the "articulation score" which is reported in percentages.

Poor discrimination is often more significant than average pure tone losses or elevated speech reception thresholds. If the individual cannot adequately discriminate some sounds from others, the speech that he is able to hear whether with or without amplification will be much less intelligible. Discrimination is further reduced if the person also suffers from *recruitment*, which is an abnormal increase in the sensation of loudness despite the fact that his threshold is raised making it hard for him to hear and discriminate speech which is spoken at normal levels of intensity, the individual afflicted with recruitment hears very loud sounds as well as or better than the person with normal hearing. But whether or not he is wearing a hearing-aid he is bothered by the intense signals produced by well-meaning people who shout at him. Their speech becomes less intelligible at significantly higher loudness levels. If you can recall how garbled a message sounds when someone shouts into a microphone you will sense something of the enigmatic phenomenon of recruitment. The individual with a substantial noise-induced hearing impairment frequently has both a discrimination loss and recruitment.

There are two major types of hearing loss suffered as the result of noise. One is called *acoustic trauma* and the other is called as we have mentioned *noise-induced* hearing loss. Acoustic trauma is the sudden loss of hearing following a dramatic burst or period of noise. Soldiers on the warfront after a siege of heavy fire have been known to suffer acoustic trauma. Someone with extremely sensitive ears may suffer a sudden hearing loss after being exposed to only one firecracker exploding nearby. Noise induced hearing loss is usually more insidious. It depresses the threshold for hearing bit by bit over time.

Both acoustic trauma and noise induced hearing loss follow a typical frequency pattern of loss, figure 106 A, B, C, D. Some frequency between 3000 and 6000 cps. seems to be affected first (which one depends upon the individual and the composition of the noise to which he has been exposed) and appears as a dip in the otherwise normally straight line on the audiogram. For noise-induced hearing losses the bottom of the trough spreads out over time and the angles of the sides of the trough become more obtuse until finally enough frequencies are affected to make a difference in the individual's capacity to understand the spoken word.

After sufficiently long exposure speech might not be heard unless amplified. Amplification however would only be a partial answer to the problem since the listener would still have to struggle to discriminate between certain sounds.

When the ear is exposed to loud noise for too long a time some of the sensitive hair-like endings of the cells in the basilar membrane become detached from the gelatinous roof, the tectorial membrane, in which they are embedded. Other mechanical and chemical changes probably occur as well. One result of the disruption of normal mechanical and metabolic functioning is that the threshold for hearing becomes poorer when tested soon after the exposure. This phenomenon is referred to as auditory fatigue or, more recently, as temporary threshold shift (TTS). The effect is called temporary because if the exposed individual is removed from the noisy environment for an adequately long rest period his hearing will return to pre-exposure levels.

Permanent threshold shift (PTS) or noise induced permanent threshold shift (NIPTS), as it is sometimes referred to, is another matter. Irreversible damage to the organ of Corti is sustained, resulting in permanent hearing loss of the sensory-neural type. To date no one has satisfactorily demonstrated that this type of loss can be remediated medically or surgically. The Chinese claim that acupuncture can cure or reduce hearing deficit in some people with this type of loss but such claims remain to be validated by the scientific community in the West.

Just how much noise can we take before we begin to lose our hearing permanently? One may attempt to reply to the question from either a statistical or individual vantage point. It is nearly impossible to make a *priori* predictions regarding which levels of noise are absolutely safe for the hearing mechanism of any particular person. It has been postulated that sounds softer than 80 dB cannot be harmful to anyone's peripheral hearing mechanism regardless of how long one is exposed to them. A moderately quiet suburban street, loud talking, and an electric typewriter have sound levels just under 80 dB. Above this figure there is considerable variation in sensitivity to sound among individuals. Some people are so sensitive they can become permanently deaf after exposure to one round of rifle fire. Others have "tougher" ears and can withstand longer periods in noise without any apparent effect. On the other hand, five hours per day of exposure to jackhammer noises in the 115 dB range for a few years will produce PTS no matter how tough are one's ears. Estimating the risk for damage prior to exposure demands a statistical approach modeled for the "average" person. One asks what percentage of the people exposed to a particular type, frequency and intensity of noise for a specified amount of time is likely to show a significant degree of damage

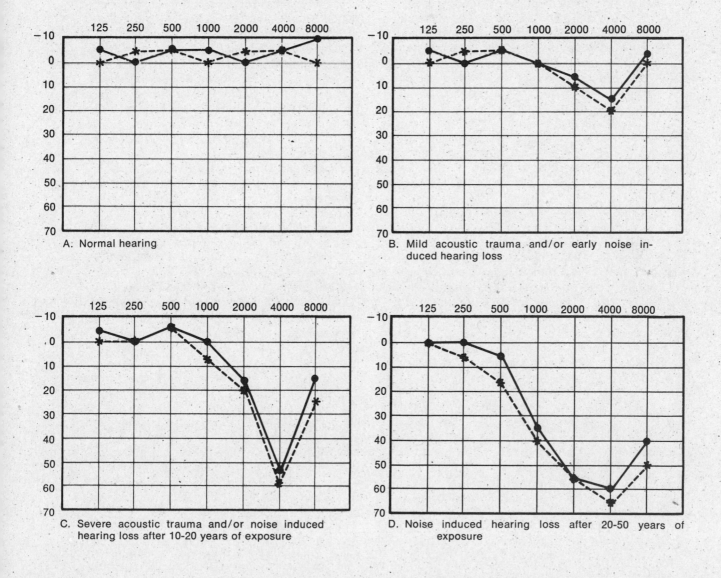

A: Normal hearing

B. Mild acoustic trauma and/or early noise induced hearing loss

C. Severe acoustic trauma and/or noise induced hearing loss after 10-20 years of exposure

D. Noise induced hearing loss after 20-50 years of exposure

**Figure 106.** A comparison of different degrees of hearing loss.

after a lifetime of exposure. The answers to these questions come primarily from three sources.

We can assess the hearing level of people whom we know have been exposed to potentially dangerous levels of noise for considerable periods of time. The measured SPL of the noises in their environment is combined with information obtained from the victim or other knowledgeable source about the length and pattern of exposure periods per day per week per month per year per lifetime to date of assessment. When you have many subjects who were exposed to similar conditions over a lifetime, intrasubject results can be pooled and the pooled data compared with results obtained from control groups of subjects who have not been so exposed. These retrospective studies are only suggestive since it is not possible to know after the fact whether the "subjects" had normal hearing prior to their entrance into the chronically noisy environment nor can we assume to be completely accurate the reported history of the times and patterns of exposure over the years. Despite these drawbacks, much of our information about occupational hearing loss has been drawn from such studies.

A second approach could be to test the hearing of large numbers of individuals before they enter a noisy working (or other) environment where all aspects of the noise-man equation are known in advance. Subjects could be tested periodically for TTS. The audiologist, a specialist in the testing and rehabilitation of hearing impaired individuals, will eventually uncover PTS which will increase progressively as time goes on. Valid correlations and conclusions could then be drawn relative to the effects of noise on man.

There is, of course, one major problem. Such studies are quite unethical. It is inconceivable that anyone would allow people who show significant TTS in the circumstances described above to continue functioning in that environment without either reducing the noise level at the source or protecting the workers' ears. It is even possible that for the extremely sensitive worker, it might be better to advise him to change his task or even his job so that he is altogether removed from the hazardous noise. Indeed what was described as a potentially valuable type of study turns out to be the basis for instituting a hearing conservation program wherein periodic audiometric testing leads to one or more of the three alternative solutions suggested.

Laboratory studies are the best source for reliable and valid data. Animal subjects may be used to test for potentially damaging levels of exposure. Prior to sacrificing the experimental animals for the histological data which will be derived from the microscopic analysis of damaged tissues and cells, behavioral indices of hearing loss may be measured by means of specialized test equipment such as that used for evoked response audiometry (ERA). In ERA the brain wave patterns elicited by sound are amplified with special equipment and averaged by a computer for subsequent inspection. The subject need not cooperate by signaling when he hears sounds as does the human subject in conventional audiometry. ERA has not yet proven to be as valid or reliable as conventional audiometry but its value in animal studies cannot be underestimated.

Until recently, another type of laboratory study was very popular—that of inducing TTS in human subjects and discovering what combinations of sound intensity and time lead to varying shapes and lengths of recovery curves. It was primarily from data collected in TTS studies, correlated at times with data derived from retrospective studies, that damage-risk criteria were set up by concerned groups of scientists and eventually by the United States Government. Damage-risk criteria (DRC) are estimations of the probable damage that is likely to be incurred by a certain percentage of the population exposed to sounds of a specified intensity for a stated duration. We are learning, however, that TTS is not always a good predictor of PTS and that there are subtle but important variations among predication formulae that depend upon the interaction of more than just total time of exposure and intensity level. The frequency level and character of the offending noise, the duration of individual noise exposures and the length of the recovery period allowed after each individual exposure all interact with intensity and total exposure time to affect DRC. Even more alarming is the evidence that we may be permanently damaging our subjects even though they show no PTS and recover in short order from TTS. Noise exposed chinchillas who apparently recovered from TTS (as measured by evoked response audiometry) were demonstrated histologically nevertheless to have suffered permanent damage to significant numbers of both inner and outer hair cells. It is surely a puzzle how so many cells could be destroyed without effecting a permanent threshold shift. Perhaps we have more than a margin of safety in the number of hair cells with which our cochleas are endowed. Even though some hair cells may be permanently damaged after each substantial noise insult, remaining cells (probably neighboring cells) can apparently be enlisted to carry the full load in the identification and analysis of sounds. Eventually, though, we must run out of spare hair cells. When the grace period ends we are left with a permanent hearing loss of the irremediable sensory-neural type.

The histological data forces us to consider a serious experimental question. Should we continue to expose human subjects for even short periods to levels of noise that are potentially hazardous? Furthermore, can we be certain that non-auditory deleterious physiological effects are not also occurring in these experiments formerly thought to be entirely safe?

### Non-Auditory Effects of Noise

Most of the literature on the non-auditory effects of noise on man, particularly with regard to physiological effects, reports results from laboratory experiments with animals. An adequate number of similar studies with human subjects have been performed, however, to suggest that most of the findings in animal research are applicable to man. As with the auditory effects of noise, loud sound has been measured to have the most unhealthy consequences.

Among the psychophysiological effects documented are the following: reduction in the caliber of blood vessels with resulting ischemia or anemia which are conditions of abnormal oxygen transport; elevation of circulating triglycerides and cholesterol in the blood leading to the unsalutary build-up of fatty deposits along the vessel walls; increased cortical adrenal activity with injection of large amounts of adrenalin into the bloodstream; constriction of the iris and dilation of the pupil of the eye; excessive tension in portions of both the skeletal and visceral muscle systems; alternate increase and decrease in gastric secretions; increase in blood pressure (diastolic); subjective feelings of anxiety, nervousness, and irritability. In sum, noise seems to function as a general stressor which overactivates the sympathetic nervous system.

Interestingly, engineers label as noise any disturbance in an integrated equilibrium. The human body is in an exquisitely fine integrated equilibrium called homeostasis. When a stressor upsets the equilibrium, the body instinctively strives to reestablish the steady state. Living organisms function best if subjected to no more than reasonable amounts of stress. Strong and frequently recurring disturbances may result in an adjusted steady state that is a caricature of the normal. To cite two well-known examples, both the neurotic and the high blood pressure patient must operate within distorted homeostatic patterns. Chronic overstimulation, whether due to noise or some other stressful event, results first in temporary habituation and then in permanent maladaptation. Temporary threshold shift, hypertonicity, excessive manufacture of adrenalin and vasoconstriction are some of the short-term effects. High frequency hearing loss, atherosclerosis and early cardiac arrest may follow as permanent residua. In sum, one may be paying too high a biological price in the ecological equation.

Unconscious physiologic responses to stimulation of the auditory nerve by sound have been observed to occur not only in the waking state but also during sleep, under anesthesia, and even after removal of the cerebral hemispheres. Sensory impulses are conducted from the hair cells in the organ of Corti within the internal ear along the auditory nerve on to the cortex by way of the midbrain. Lower level subcortical neuronal systems controlling the autonomic nervous system are stimulated and account for most of the effects already noted. But it is at the cortical level that sound is consciously perceived and interpreted to be pleasant or unpleasant, soothing or nerve-wracking, reassuring or frightening. In some cases the psychological attitude may be secondary to the biochemical changes already induced by the noise, as when a child experiences fear upon hearing a clap of thunder. In other instances certain sounds of only average loudness may irritate some listeners. Of course louder sounds have a greater evaluative communality among listeners. Most people living near airports agree that the repeated whir overhead is unpleasant if not unbearable. Negative attitudes toward noise-producing phenomena, if persistent or frequently reactivated, function as indirect stressors upon the person's own body and hence act to overstimulate the sympathetic network much as do outside stressors.

It should be made clear that research on the non-auditory effects of noise is much more scant than is that dealing with the auditory effects of noise. Particularly controversial is the contention that noise which is not loud enough to induce a hearing loss may nevertheless so disturb the rest, tranquility or mental adjustment of the individual as to cause perceptible damage elsewhere in the body. It must also be stressed that many people today appear not only to be unbothered by loud noises but to seek them out. Witness the levels at which teenagers listen to rock and roll (100–120 dB) or the pride associated with possessing an unmuffled car or motorcycle (100–110 dB). Many of our young are blithely incurring early hearing losses and are probably damaging other parts of their bodies as well. Perhaps they do not take care because they do not really understand what noise is doing to them. Older men and women even into the middle years are sometimes as difficult to convince as is the teenager. Many a worker refuses to wear ear protectors on the job. We must begin to educate our youth about noise pollution when they are very young so that they may learn how to preserve their own natural resources, much as we must teach them that external resources like water and air should be preserved in near-pristine states.

## What Can Be Done About Noise Pollution

There are several approaches to solving the problem of noise pollution. The first was touched upon above and involves education. We must start by persuading many people to stop, to listen, to learn, and then to act upon what they have learned. The subject of ecology and how it is related to all types of pollution should be taught to the preschooler at home and become part of the curriculum from kindergarten through graduate school, much as are reading and writing. It is, after all, a matter of life and death. People should understand what noise is and how it can affect them. They then can see to it personally and through their governmental representatives that something is done about reducing or removing noise from the environment wherever and whenever possible.

The second attack should be directly upon the noise front. Noises everywhere should be measured and evaluated both with sound level meters and through attitude scales. Whenever the noise is found to be dangerously loud or psychologically distressing to many people, ways must be devised to control it or minimize its noxious effects.

Noise control may be directed to the source of the sound or the recipient of the sound. Ideally the source should be looked to first. In industry that means constructing more silent machinery, or baffling machinery that cannot be made significantly quieter. When the results are not adequate for the safety and sanity of the worker, additional measures must be taken such as protecting the ears of the worker with ear muffs, earplugs or other protective devices. Sometimes it may be necessary to rotate the worker's task so that he is not constantly working in the same noisy environment all the time. None of the above suggestions is untried. Several conscientious manufacturers of equipment

and machinery have built quieter models that have no loss of efficiency. Ingersoll-Rand, for example, has an air compressor on the market that is considerably quieter than all air compressors produced in the past. Several companies make effective ear protectors of many different types.

Ideally every person should have a yearly audiometric screening. It is most important to systematically monitor the hearing of anyone who is or might be exposed to noise. Threshold shift could signal the need for one or more of the controls outlined above. As our knowledge of the non-auditory effects of noise becomes more sophisticated we may want to institute additional biological and psychological monitoring programs.

In the community, noise control calls for concerted engineering attention to the racket produced by the combustible diesel engine and the whine of tires in trucks, buses, and cars. Vehicular transportation noises contribute heavily to the overall community noise level. Limiting the number and types of moving vehicles allowed into areas of high density may be part of the solution. Alternate modes of transportation that can carry many people and operate with relative quiet must be designed. Modern high-speed above-ground trains could be one answer.

One of the most difficult problems to solve is the air traffic problem. Certainly some planes being built today are quieter than others. All planes should be required to meet higher standards with regard to noise control. Airports must be built where people do not live and rapid access to such airports from city centers must constitute part of the plan.

Housing developments should employ the latest available information about architectural acoustics so that individuals living in them can retreat to apartments that are havens of quiet. Not only should the residents be protected from street noises but principles of noise isolation should be applied to the construction materials and building plans with an eye toward avoiding the easy transition of sound from the neighbors' apartments through the walls and down the pipes. Finally, repair and construction companies working in residential neighborhoods must respect the rights of residents by using quieter equipment and by not operating while people are trying to sleep.

Recognizing the practical difficulties accompanying attempts at noise abatement and control, we are turning more and more to the legislature for the setting up and enforcement of reasonable standards. Local citizen groups are springing up everywhere, putting pressure on their congressmen to draft and vote for better noise control laws. Several states and major cities across the nation have already passed or are considering important pieces of noise legislation that deal specifically with the noise problems in their own areas.

In 1970 the Administration in Washington set up the Bureau of Noise Abatement and Control as an arm of the Federal Environmental Protection Agency. Congress is continually considering bills drawn up in part as a result of the findings reported by that bureau after it had completed a year-long national investigation of noise pollution.

One major advance which already signaled a new attitude toward noise pollution was the passage of the Walsh-Healey Act in May, 1969, wherein noise standards were adopted for companies signing with the federal government contracts which involved sums of more than $10,000.00. In light of recent research, the standards set forth in that Act have been criticized as being too conservative for workers in certain noise conditions and potentially too risky for workers in other environments. Nevertheless, the 1970 Occupational Safety and Health Act incorporated the Walsh-Healey standards, broadening their application to cover any industry engaged in interstate commerce.

The high level purpose of the Congress as expressed in the Occupational Safety and Health Act of 1970 reflects the attitudes of a nation ripe for quiet. That purpose is ". . . to assure so far as possible every working man and woman in the Nation safe and healthful working conditions and to preserve our human resources." Extend the sentiment to include every man and woman at home and at play, as well as at work, and we will have cut out for ourselves a difficult but commendable task.

## REFERENCES

Bekesy, G. von, *Experiments in Hearing*. McGraw-Hill: New York, (1960.)

Beranek, L. L., *Noise Reduction*. McGraw-Hill: New York, (1960.)

Broadbent, D. E., The effects of noise on an intellectual task. *J. Acoust. Soc. Am. 30,* 824, (1958).

Broadbent, D. E. & E. A. J. Little, Effects of noise reduction in a work situation. *Occup. Psychol. 34,* 133, (1960).

Coles, R. R. A., G. R. Garinther, D. C. Hodge, & C. G. Rice, *Criteria for Assessing Hearing Damage Risk from Impulse-Noise Exposure. U. S. Army Tech. Memo. 13-67.* AMCMS Code 5011.11.84100.

Committee on the Problem of Noise. *Noise, Final Report* H.M.S.O.: London, (1963).

Davis, H. & S. R. Silverman, *Hearing and Deafness*. Holt, Rinehart & Winston: New York, (1966).

Gildston, H. & P. Gildston, Personality changes associated with surgically corrected hpacusis, *Audiology* (In press).

Hirsh, I. J., *The Measurement of Hearing*. McGraw-Hill: New York, (1952).

ISO, R389. *Standard Reference Zero for the Calibration of Pure Tone Audiometers*. I.S.O., (1964).

ISO, R362 *Measurement of Noise Emitted by Vehicles,* I.S.O,. (1964).

ISO, R507 *Procedure for Describing Aircraft Noise Round an Airport,* I.S.O., (1966).

Kryter, K. D. A field experiment on human response to aircraft noise. IEE Conference publication No. 26, Instn. Elec. Engrs. London, (1967).

Kryter, K. D., W. D. Ward, J. D. Miller, & D. H. Eldredge, Hazardous exposure to intermittent and steady-state noise. *J. Acoust. Soc. Am., 39,* 451, (1966).

Lebo, C. P. D. & D. S. Oliphant, Music as a source of acoustic trauma. *Laryngoscope, 78,* (1968).

Lim, D. J. & W. Melnick, Acoustic damage of the cochlea. *Arch. otolaryng., 94,* 294, (1971).

Mills J. H., R. W. Gengel, C. S. Watson, & J. D. Miller, Temporary changes of the auditory system due to exposure to noise for one or two days. *J. Acoust. Soc. Am., 48,* 524, (1970).

Occupational Safety and Health Act of 1970. Public Law 91-596, 91st Congress, S. 2193, December 29, 1970. Part II: *Federal Register:* May 29, (1971).

Rosen, S., Presbycusis study of a relatively noise-free population in the Sudan. *Ann. Otol., 71,* 727, (1962).

Rosen, S. & P. Olin, Hearing loss and coronary heart disease. *Arch. otolaryng., 82,* 236, (1965).

Walsh-Healey Public Contracts Act. Ch. 1, 41 Par 35; *Federal Register:* Par 50-204.10, May 20, (1969).

Ward, W. D., Temporary threshold shift and damage-risk criteria for intermittent noise exposures. *J. Acoust. Soc. Am., 48,* 561, (1970).

Ward, W. D. & J. E. Fricke (Eds.), Noise as a public health hazard: proceedings of the conference, Washington, D.C., June 13-14, 1968. *ASHA Reports 4,* American Speech and Hearing Assoc.: Washington, D.C., (1969).

Welch, F. L. & A. S. Welch (Eds.), *Physiological Effects of Noise*. Plenum Press: New York-London, (1970).

Wilkinson, R. T., Interaction of noise with knowledge of results and sleep deprivation. *Exp. Psychol., 66,* 332, (1963).

Prof. Phyllis Gildston, Ph.D.
Brooklyn College of the City University of New York.
Prof. Harold Gildston, Ed.D.
C. W. Post College of Long Island University
Greenvale, New York.

# Chapter 24

# Radiation Pollution

Exposure to ionizing radiation has been a part of nature throughout evolutionary time. Exposure to ionizing radiation is also rapidly increasing as a result of man's activities. This chapter will outline the sources of exposure to ionizing radiation, review some of the risks associated with exposure to radiation, and comment on regulations governing radiation exposures to the general public and the two largest sources of man-made radiation—medicine and the "peaceful atom." Essentially nothing will be said about the physical processes involved when radiation interacts with matter or the mechanisms through which radiation causes its damage. The first point, that is, the nature of the interactions between radiation and matter, is treated in readily available reference books and texts: for example, the excellent introductory book by Jack Schubert and Ralph Lapp.[1] The matter of mechanisms through which radiation does its damage to biological materials is a highly technical issue and one that is unresolved. Scientists have labored over these mechanisms for many years and continue to do so. The reader who is interested in the various theories that have been put forth and the evidence in support or in opposition to each of the theories must turn to the technical literature—the scientific journals, reports and proceedings which consider these matters.[2]

## Sources of Ionizing Radiation

Radiation exposure is of two general sources, natural and man-made. Natural radiation exposures are those which result from exposures to cosmic rays and to the radioactive isotopes which are present in the earth's crust and are thereby present in greater or lesser extent in physical and biological materials.

Man-made radiation can conveniently be considered in four broad categories: associated with medical x-rays, associated with electronic and other apparatus, associated with military uses of atomic energy, and associated with peaceful uses of atomic energy. In addition, it is necessary to differentiate between environmental exposures which are received by members of the general public and those which are received by individuals who, by the nature of their work, receive occupational exposures to radiation.

A summary of the levels of radiation exposures from various sources, table 11 indicates that, at present, exposures from natural sources approximately equal those from medical and dental uses of radiation, and that the average exposures from military and peaceful uses of atomic energy are considerably lower. It must be appreciated, however, that natural exposures are not changing significantly with time while exposures from medical uses and from peaceful uses of atomic energy are increasing rapidly—or at least have the potential for rapid increases unless vigorous measures are taken to keep them at current levels or to reduce them significantly.

**Table 11.** Some Average Exposures of Ionizing Radiation to Persons Living in the United States.[4]

| Exposure | Millirems per year[a] |
|---|---|
| Natural background (whole body) | 100-150 |
| Diagnostic medical and dental | |
| Bone marrow | 125 |
| Gonads | 55 |
| Thyroid (mostly dental) | 1,000 |
| Weapons fallout (in 1968) | about 2[b] |
| Nuclear industry | less than 1[c] |
| Other man-made sources | less than 1[d] |

[a] Radiation exposures are expressed in units of rems, rads, or roentgens. Although these units are not identical, no significant error is introduced by considering them as being numerically equal for x-rays and many environmental exposures. When considering exposures from $\alpha$-emitters such as plutonium and other heavy particles they are **not** numerically equal.

[b] The dose from fallout is much lower now than it has been in the recent past. In addition, the dose to individual organs from specific radio-nuclides must be considered in order to obtain a representative picture of these exposures. For example, Morgan[4] indicates that the average

dose to the gonads and the total body dose from cesium-137 was 0.54 millirem in 1968, and the average dose to bone marrow from strontium-90 was 2.3 millirems in 1968. Reference 5 contains a good discussion of fallout.

(c) Exposures to members of the general public who live a considerable distance from a nuclear facility are presently low. K. Z. Morgan and E. G. Struxness (in a paper presented at a United Nations Symposium on the Environmental Aspects of Nuclear Power Stations, August 10-14, 1970) have estimated that the average dose to the U.S. population is 0.85 millirem per year. Those individuals who live near a reactor, a fuel reprocessing plant, or whose home is built on uranium tailings may have exposures considerably higher than the average.

(d) Morgan has estimated this dose to be 0.1 millirem in 1966. Some color televisions, however, may give considerably higher doses.[6]

## Medical Exposures

The major exposure to radiation from medical and dental uses is from the diagnostic x-ray. In addition, uses of radioactive isotopes are rapidly increasing both for the treatment of various diseases and for diagnostic purposes. There is a growing debate[3] as to whether current medical practices with respect to the use of radiation reflect the recognition that every exposure to radiation carries with it some risk of damage to the individual. The nature of this risk is discussed later.

Some quotes from recent work of those professionally interested in reducing the hazards from radiation are given here as an indication of the concern being voiced:

• "Regardless of what estimates are used or the arbitrariness of the number selected, it is evident that medical diagnosis is responsible for the saving of many thousands of lives each year. On the other hand, unnecessary diagnostic x-ray exposure to the population very probably accounts for the loss of thousands of lives each year. Just because the medical x-ray facilities in a city are responsible, for example, for saving ten or twenty lives each year, this in no way justifies the loss of one to five lives in the same period because of the use of poor equipment and improper techniques. The average diagnostic dose to the population could be reduced to less than 10 percent of its present value without any curtailment in the beneficial uses of diagnostic x-rays, while at the same time enhancing and immensely improving diagnostic x-rays."[7]

• "It was estimated that restriction of the x-ray beam to an area no larger than the film size would result in a reduction of the genetically

significant dose from 55 to 19 millirads per person per year."[8]

• "... skull films [taken in evaluating head injuries in children] have almost no diagnostic value, are taken primarily to protect the clinician from legal action, and may actually delay vital therapy in emergency trauma cases."[9]

• "[From 1958 to 1963] the amount of x-ray film shipped in America—a fair indication of the amount used—rose from 218 million square feet to 313 million. This is an increase of not quite 44%."[10]

These authors indicate the extent to which exposures are increasing, the degree to which exposures could be reduced without loss of diagnostic information, and at least suggest some of the reasons for the exposure rates now being observed.

There are only three ways in which the hazards associated with medical use of radiation might be reduced:

• The number of examinations might be reduced
• The exposure per examination might be reduced
• A patient population which would be more resistant to radiation might evolve.

The last of these alternatives is rather unlikely and the first two assume an informed medical and dental community. There is considerable question as to the extent that physicians and dentists appreciate the hazards associated with radiation. In spite of a vigorous debate on these matters, and ample discussion of the hazards of low-level radiation, it is still possible to find statements by respected members of the medical community, totally at variance with the facts regarding radiation hazards, stating or implying that the levels of radiation associated with diagnostic tests are harmless.[11]

To combat this ignorance on the part of physicians, the U.S. Public Health Service recently found it necessary to distribute to practicing physicians a booklet outlining the most elementary aspects of radiation and "good [medical] practice."[12]

## Exposures From Peaceful Uses of Atomic Energy

The peaceful atom has found three broad categories of application: the Plowshare program, the use of radioactive isotopes, and atomic reactors for the generation of electric power. The Plowshare program is the use of atomic bombs for civilian activities such as digging canals, mining, stimulating the release of natural gas, and digging holes. Radioactive isotopes are finding increasing use in industry, educational and research institutions, and in medical practice. A recent address by Dr.

Glenn T. Seaborg, chairman of the Atomic Energy Commission, gives some indication of the size of the isotope industry:

"Fifty percent of the 500 largest manufacturing concerns in the United States use radioisotopes. About 4,500 other firms also are licensed to use radioisotopes. Innumerable other companies use exempt quantities of radioisotopes under AEC general licenses. These thousands of firms benefiting from radioisotopes come from virtually every type of industry, including the metals, electrical and transportation industries, chemicals and plastics, pharmaceuticals, petroleum refining, paper, rubber, stone, clay and glass products, food, tobacco, textiles, crude petroleum and natural gas, mining, and the utilities."[13]

Both Plowshare activities[14] and the use of isotopes are now sources of radioisotopes being released into the environment. They are relatively small sources of exposure today, but there are powerful economic pressures acting to encourage the expansion of these industries and expansion will carry with it the threat of large environmental releases unless aggressive regulatory programs are developed.

The atomic energy activity which will probably lead to the largest increase in environmental radiation is not, however, the Plowshare program or the use of radioisotopes, but rather the use of atomic energy for the generation of electric power. The industry is only in its infancy and already it is embroiled in several safety and radiation-related controversies.[15]

Nuclear reactors have been under development for over 25 years, and within the past few years have become a commercial reality. From a handful of experimental reactors for the generation of electricity in the early 1960's the industry has grown such that today over 100 large reactors are either operating, under construction, or awaiting construction permits. The Atomic Energy Commission estimates that by the year 2000 approximately half of all electrical power generation in the United States will be nuclear. In 1970 only 1.3 percent of the electricity used was generated in nuclear power plants.

To understand the environmental impact of any industry or human activity, it is necessary to examine the industry itself and also activities which exist because of that industry or serve as support facilities for that industry. Thus, to evaluate the environmental impact of, for example, a coal-burning electric power plant it is necessary to start with the coal mine, consider transportation and storage of the coal, consider the pollutants and other environmental effects of burning coal at the power plant, and finally consider the disposal of the various residues from the power plant.

In like manner, to understand the environmental radiation arising from the nuclear power industry it is necessary to consider the entire fuel cycle, not just the releases during normal operation of the atomic reaction itself. The nuclear fuel cycle is shown in figure 107, together with notes indicating the nature of the environmental impacts— most of which are radiological in nature.

Details of the radiological implications of each phase of the nuclear fuel cycle are available in several recent books and other publications,[16] and only some of the major sources of environmental radiation will be mentioned here. Uranium miners are exposed to high levels of radiation which has led to lung cancer being an occupational hazard.[17] Large, or potentially large, occupational exposures are part of the working environment in several other segments of the nuclear fuel cycle, particularly in the reactor itself, the chemical fuel reprocessing plant, and the fuel fabrication facilities. Associated with uranium mining is the concentration of the ore—a process which produces large quantities of fine, sand-like, particles called tailings that contain radioactive materials resulting from the decay of the uranium. At several locations in Colorado, these tailings have been used for fill around houses, public buildings and other construction sites and are causing radiation exposures in excess of the standards.[18]

The environmental radiation resulting from mining and milling are primarily of importance in the vicinity of the uranium mines and mills. The radiation released into the environment from other portions of the nuclear fuel cycle are of more general distribution.

There are two categories of environmental radiation associated with the nuclear reactor itself. First, there are the so-called routine releases of low-level radioactive wastes. There has been a nation-wide controversy raging since 1968 surrounding these releases[19] and the situation is not yet completely resolved. It is now clear, however, that these routine releases can be kept as small as is desired (and can well be at such low levels as to make their measurement very difficult) with very little cost. At this writing Atomic Energy Commission hearings are in progress which are expected to lead to a further tightening in the regulations which apply to these routing releases. Of more serious concern is the second category of releases from a nuclear reactor—those resulting from an accident in the reactor. The probability of such

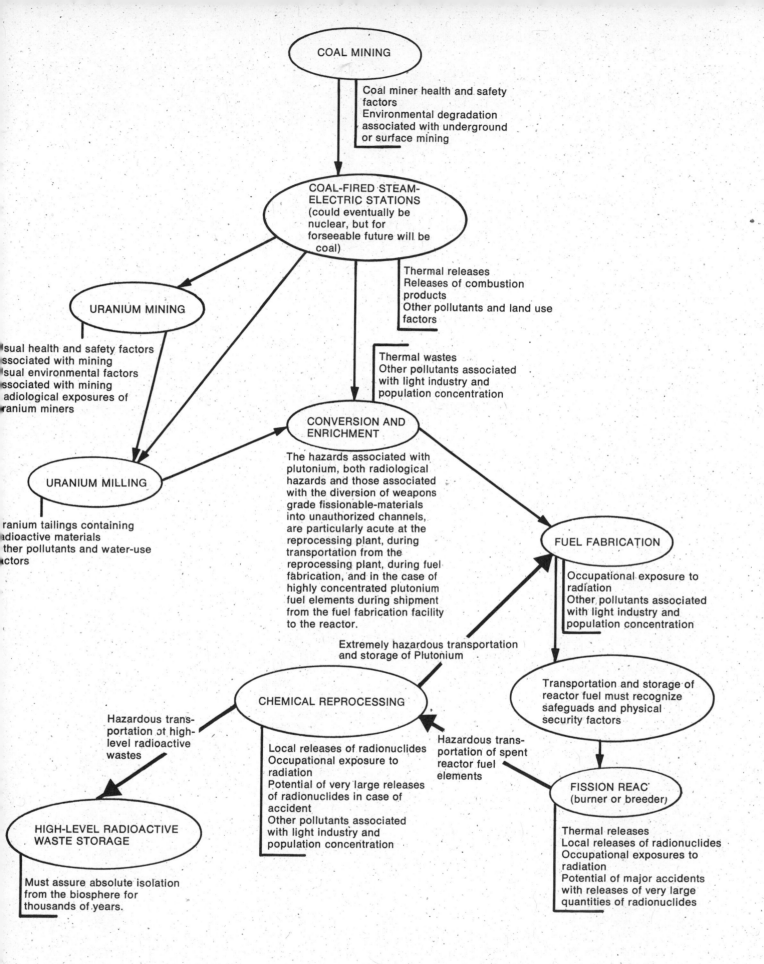

**Figure 107.** Nuclear Fuel Cycle.

The diagram shows the following elements:

**COAL MINING**
- Coal miner health and safety factors
- Environmental degradation associated with underground or surface mining

**COAL-FIRED STEAM-ELECTRIC STATIONS** (could eventually be nuclear, but for forseeable future will be coal)
- Thermal releases
- Releases of combustion products
- Other pollutants and land use factors

**URANIUM MINING**
- Usual health and safety factors associated with mining
- Usual environmental factors associated with mining
- Radiological exposures of uranium miners

**URANIUM MILLING**
- Uranium tailings containing radioactive materials
- Other pollutants and water-use factors

**CONVERSION AND ENRICHMENT**
- Thermal wastes
- Other pollutants associated with light industry and population concentration

The hazards associated with plutonium, both radiological hazards and those associated with the diversion of weapons grade fissionable-materials into unauthorized channels, are particularly acute at the reprocessing plant, during transportation from the reprocessing plant, during fuel fabrication, and in the case of highly concentrated plutonium fuel elements during shipment from the fuel fabrication facility to the reactor.

**FUEL FABRICATION**
- Occupational exposure to radiation
- Other pollutants associated with light industry and population concentration

Transportation and storage of reactor fuel must recognize safeguads and physical security factors

**CHEMICAL REPROCESSING**
- Extremely hazardous transportation and storage of Plutonium
- Local releases of radionuclides
- Occupational exposure to radiation
- Potential of very large releases of radionuclides in case of accident
- Other pollutants associated with light industry and population concentration

Hazardous transportation of spent reactor fuel elements

**FISSION REAC** (burner or breeder)
- Thermal releases
- Local releases of radionuclides
- Occupational exposures to radiation
- Potential of major accidents with releases of very large quantities of radionuclides

Hazardous transportation of high-level radioactive wastes

**HIGH-LEVEL RADIOACTIVE WASTE STORAGE**
- Must assure absolute isolation from the biosphere for thousands of years.

an accident is very small but the consequences should one occur have been described by the Atomic Energy Commission as potentially involving billions of dollars in damage and thousands of persons killed.[20] It is rather unfortunate that the Atomic Energy Commission has seemingly allowed economic pressures, coupled with its own vested interest in seeing rapid commercial development of nuclear power, to lead to a proliferation of the nuclear power industry before the research and development involving the safety features which might prevent large releases of radiation have been completed. The system causing most concern will not undergo full-scale testing until 1974 but the reactors are being built today. Again, at this writing major hearings are being held to determine the manner in which these safety questions will be resolved.

The uranium which fuels a nuclear power plant is not itself particularly hazardous. The radioactive wastes which cause so much concern are created in the reactor by the same reaction which leads to the useful energy which in turn is converted to electric power. These radioactive wastes, created in the reactor, must be removed and somehow stored. From the reactor the spent reactor fuel is transported to a chemical reprocessing plant. The purpose of that reprocessing is to separate the radioactive wastes, the unburned uranium remaining in the fuel, and the plutonium which, like the radioactive wastes, is also created in the reactor. "Routine" releases of radiation from reprocessing plants have been much larger than the releases from a single nuclear reactor. In addition, there is the possibility for accidental releases of very large quantities of radioisotopes.

The radioactive wastes, created in the reactor and removed in the reprocessing plant, must be stored. The nature of these wastes is such that they must be absolutely isolated from the biosphere for thousands of years. This requirement is totally different from any other waste disposal problem which has previously faced mankind. At present, no means for this storage has been implemented and the wastes are being stored in solid or liquid form in tanks at or near the reprocessing plants. This situation, which is regarded by nearly all—friends and critics of nuclear power alike—as being totally inadequate, carries with it the potential for very large environmental contamination. There have been no known major releases of these stored wastes, but the potential is there and increasing.

The last source of environmental radiation from the nuclear industry that will be discussed here is that associated with environmental releases of the plutonium which is created in the nuclear reactor. Plutonium, like uranium, is a fissionable material and, hence, can be used as fuel for an atomic reactor or as the basic material for atomic bombs. In addition to the dangers of diversion of plutonium into unauthorized channels (the "safeguards" and "physical security" problem), plutonium is probably the most hazardous substance from the radiological standpoint with which man has had to deal.[21]

### The Nature of the Hzard Associated with Exposure to Radiation

Ionizing radiation is known to be hazardous and to cause damage to inanimate and animate matter. Although there may be ecological significance associated with environmental releases of radioactive materials and the resulting bio-concentrations that can occur as the radionuclides move through the food chains, only the effects on man will be considered here.

What follows is an assertion of a set of risk estimates, and a statement of the assumptions upon which these estimates are based. The original literature will be cited in only a few critical cases. The estimates of risk presented here are not the lowest that can be found in the literature, but neither are they the highest.

### Specific Radiation Effects

The effects associated with exposures to ionizing radiation include:

- genetic effects: These are associated with mutations, the heritable changes in the germ plasm.
- neoplastic diseases: Leukemias (with the exception of chronic lymphatic leukemia) and cancers are known to be the consequences of exposure to ionizing radiation. Many forms of cancer have been shown to be inducible in humans by radiation.
- growth and development: Children who have been exposed to radiation (for example, the Japanese exposed to the atomic bomb or the Marshallese exposed to fallout) showed retardation both of growth and maturation.
- life span: In addition to the life-shortening associated with premature death from, e.g., radiation-induced neoplasms, a non-specific life-shortening is associated with exposure to ionizing radiation. In animals, exposure early in life may cause greater shortening of life per unit dose than exposure later in life.

- behavioral changes: Although there are papers in the literature which suggest behavioral changes in experimental animals as a result of exposure to ionizing radiation, these effects are usually not considered proven. In my experience, however, behavioral changes are the most readily observable effects of exposure to low levels of ionizing radiation. (Recent evidence seems to indicate that these behavioral changes most probably result from *discussing* the effects of low-level exposure to ionizing radiation rather than exposure to the radiation in question. Further, the changes are readily observed in those who have a vested interest in maintaining the exposures in addition to those actually exposed to the discussion. These behavioral changes might be called the chronic radiation discussion syndrome. An element in the syndrome is the suggestion that decisions affecting the general public be made by "a qualified advisory committee.")

- effects of massive exposures: These effects include the acute radiation syndrome and other manifestations of large exposures. Except as they might relate to accidental exposures these effects play no part in the consideration of exposures relating to diagnostic radiation or other environmental sources.

Of the above, genetic effects, induction of neoplasms and life-shortening are probably the most significant when evaluating the risk from low-levels of ionizing radiation. The other effects associated with massive exposures will not be further considered in this chapter.

## Assumptions, Generalizations and Basic Concepts

Before reviewing the numerical risk estimates associated with the effects associated with low-level exposures to ionizing radiation, it is important to outline the assumptions and generalizations which are made when discussing these risks from a public health standpoint. These assumptions and generalizations may or may not stand the test of time, but considerations of public health require that they be made pending further evidence.

- threshold: It is assumed that there is no exposure below which no damage occurs. This assumption results from an extension of the available experimental and epidemiological data. There is abundant evidence of damage at relatively high exposures to ionizing radiation. This evidence arises from various animal experiments and from human experience resulting from the use of nuclear weapons during World War II, from various accidental exposures, and from experiences involving the medical use of radiation. There is little direct data, however, at exposures below a few rems and to my knowledge no data at exposures below about 0.3 rem. Every commission or agency which has made recommendations regarding public exposure limits has emphasized that it is necessary to assume that no threshold exists. A typical statement from such commissions is that ". . . every use of radiation involves the possibility of some biological risk either to the individual or his descendents."[22]

- latency: There is a period of time, measured in years for man, which elapses between irradiation and the occurrence of an increase in age-specific mortality rate from malignant disease. For example, the latent period for leukemia is about five years, for thyroid carcinoma about thirteen years, and for lung cancer about ten years. The reasons for the latency period are not understood. Following the latency period, the incidence of the effect increases to a plateau, a period which may last for several years, and then for most, but not all malignancies, again decreases,[23] figure 108.

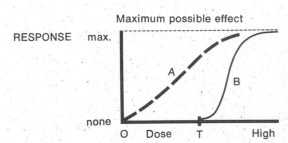

Curve A: Represents a linear dose/response relationship. In the case of exposure to ionizing radiation, there is experimental and other data at high and intermediate doses but very little data at doses at levels usually involved with environmental exposures. That data at higher levels is consistent with the postulate of linear dose/response behaviour at low doses. The same situation pertains with many other environmental contaminants as well.

Curve B. This is characteristic of another large number of materials to which man is exposed. In this case there is no observable response at low doses but at some exposure, T, the response becomes observable in some individuals. T is designated as the threshold dose. The effect then becomes observable in more of the individuals exposed, until at some dose higher than T essentially all of the individuals exposed exhibit the response.

Figure 108. Illustrating linear and non-linear dose/response relationships.

- linearity: While it is obvious that the dose-response curve is not linear at high doses, there is experimental evidence of linearity over a wide range of intermediate exposures. It is entirely consistent with the data to assume linearity at the low levels which we are considering.[24] It is important to point out that this assumption may not prove to be a conservative one, figure 109. Genetic mutation, life-shortening, and the induction of various forms of malignancy are thought to vary linearly with the accumulated dose. When questioning the validity of assuming linearity of the dose-response curve, K. Z. Morgan stated:

Some persons will object to my assumption of a linear relationship between dose and effect and will say that perhaps at very low doses and dose rates, there may be no such effects or the effects may be greatly reduced. To this, I can only reply that the many scientific committees and the national and international agencies which set the radiation protection standards state that since these effects at high dose rates (where the statistics are good and the probable errors small) seem to relate linearly to the dose, it is prudent that we assume this relationship maintains, also, at lower doses and dose rates.[25]

Observed number of cases

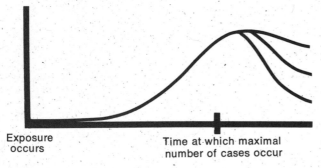

Exposure occurs

Time at which maximal number of cases occur

**Figure 109.** Illustrating the latent period following exposure

In many instances, the induction of neoplasms by ionizing radiation for example, there is a period of time following exposure during which no cases are observed. That is the latent period. The number of cases observed begins to increase reaching a peak several years, in this example, after exposure. The latent period as well as the time when the maximal number of cases occurs varies considerably depending on the specific neoplasm being considered. Likewise, the rate at which the number of observed cases returns to zero following the peak period varies.

- dose-rate: As is implied in the previous paragraphs, it is also assumed that effects are independent of the rate at which the dose is delivered, and that it is the cumulative exposure which is significant.

Keeping these assumptions and generalizations in mind, we can now review briefly the numerical risk estimates associated with genetic effects, induction of neoplasms, and life-shortening.

**Numerical Estimates of Risk**

- Genetic effects: Genetic effects of ionizing radiation, as well as genetic effects of other environmental contaminants, are certainly the most significant from a long range point of view. Various estimates have been made of the extent to which ionizing radiation contributes to our genetic burden. A representative estimate is that of Joshua Lederberg: "This [the background radiation level of approximately 0.1 rem per year] is consensually estimated to account for 10% of our load of mutational damage. . . . We will say then that doubling the background would increase the rate of mutation by about ten percent. There are only minor quarrels about the extent to which mutational damage might be decreased at very low dose rates and there are balancing counter-arguments about the interaction of radiation with other pollutants."[26] Lederberg goes on to say, "The considerable heritable component of many common diseases, taken together with the strong and clear genetic component of many other handicaps, suggests that as much as 50% of our health burden can be attributed to genetic difficulties. This assumption states that, if the entire population enjoyed an *optimal* genotype, including factors that bear on cardiovascular disease, diabetes, cancer, and mental disease, our health bill would be reduced by at least half." Lederberg also assumes a direct linear relationship between mutation rate and the prevalence of genetic defect and between exposure to ionizing radiation and mutation rate.

The literature is replete with discussions of the implications of an increased mutation rate but only one further author will be quoted: "About four million children are born each year in the United States. Of this number, about 80,000 have gross physical or mental defects attributed to mutated genes, and 10% of these, 8,000 per year, are estimated to result from the background radiation of 110 millirem per year. If the American people were as a whole to be subjected to the additional dose of ionizing radiation allowed by the Federal Radiation Council, 170 millirem per year, there would be an estimated additional 12,000 children with gross physical or mental defect born each year. Moreover, the number of embryonic and neonatal deaths attributed to gene mutations is about five times the number of children with gross physical or mental defects from this cause. Accordingly the number of embryonic and neonatal death would, from this calculation, be increased by about 60,000 per year, if the American people were to be subjected to the allowable dose of ionizing radiation."[27] The same author goes on to estimate the uncertainty in this estimate: ". . . it is possible that genetic effects are in fact only one-fifth as great as estimated, or even five times as great."

The estimates quoted above are representative of those in the literature and in the recent reports of the various commissions and councils which consider these matters.

• Neoplasms: The difficulty and complexity of the considerations which go into estimates of the number of neoplasms resulting from exposure to ionizing radiation cannot be overstated. The neoplasms which have been shown to be induced in many by ionizing radiation include: leukemias (except chronic lymphatic leukemia), various bone tumors, various tumors of the central nervous system, thyroid cancer, multiple myeloma, skin carcinoma, lung cancer (bronchiogenic and others), breast cancer, stomach cancer, cancer of the pancreas, malignant lymphoma, cancer of the colon, maxillary and other sinus carcinomas, and cancer of the pharynx. Time does not permit even a cursory discussion of the evidence leading to numerical risk estimates for each of these neoplasms. Several authors have made these estimates and also indicated their estimates of the total number of deaths which would result from radiation-induced neoplasms if the entire population of the United States (taken to be 200 million) were chronically exposed to 0.1 rem per year. These estimates range from 700 to 3000 deaths estimated by K. Z. Morgan,[28] to the 19,000 deaths estimated by John Gofman and Arthur Tamplin,[29] and the 56,000 deaths estimated by Linus Pauling.[30]

Several generalizations can be made regarding the induction of neoplasms. Significant among these is the generalization that the sensitivity to ionizing radiation is inversely related to age at exposure. This is dramatically shown in a recently published study of the relation between childhood

cancers and obstetric x-rays.[31] The results showed that there was approximately a 100% increase in deaths from cancers plus leukemia resulting from approximately four abdominal x-ray films of the mother during pregnancy. When the excess cancer risk was plotted against number of films (each of which was estimated to deliver a fetal dose of between 0.25 and 0.46 rem) to which the mother was exposed, an essentially linear relationship was observed:

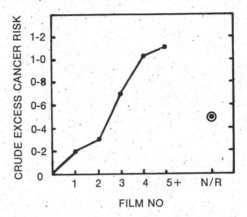

**Figure 110.** Crude dose-response curve. Crude excess cancer risk as proportion of the normal risk (N.C.R.). N/R=no record. [Taken from reference 31.]

A conservative estimate, based on the results shown in figure 110, is that two such films during pregnancy leads to an approximately 50% increase in childhood cancer plus leukemia mortality rates.

This, and other recent evidence indicating a higher than previously recognized radiation sensitivity of the fetus, has prompted the National Council on Radiation Protection and Measurements( NCRPM) to recommend that the allowable, occupational exposure of fertile women be reduced by a factor of ten—from 5 rems to 0.5 rem.[32] It is curious that a similar reduction in allowable exposure of fertile women in the general public was not recommended, although it has been the general policy in the past to recommend that exposure limits for the general public be ten times less than the occupational limits. The maximum allowable dose to fertile women in the general population remains at 0.5 rem.

• Life-shortening: Again, caution must be exercised when discussing life-shortening due to exposure to ionizing radiation. Two types of life-shortening are involved. The first is the shortening of life due to early death from a neoplastic or other

233

radiation-induced disease. The second is the non-specific life-shortening attributed to irreversible injury which rather than producing a specific disease decreases the vitality of those cells and organs in which it occurs. Such loss of viability is presumably somewhat similar to that associated with normal aging process or may accelerate the degenerative changes of aging which determine length of life.[32] Morgan has estimated the life-shortening at thirteen days per rem of exposure at the dose rates associated with diagnostic x-rays.[33]

In applying any of the above estimates, it must be remembered that it is the exposure to the organ being considered which is significant in estimating the risk from damage to that organ. Thus it is the gonadal dose which is significant in estimating the genetic effect, the dose to the red marrow in considering leukemias, the dose to the maxillary sinus in considering cancer of the maxillary sinus, and so forth. Although these considerations should be evident, they are too often seemingly overlooked.

### Standards Applicable to Environmental Radiation

A national debate has been in progress for the past several years regarding not only the appropriate standards which shall apply to the regulation of environmental radiation but also regarding the agency or regulatory body which shall have the responsibility for that regulation. The issue is too complex to consider in detail here, and also the standards themselves and the agencies or governmental units setting the standards may soon undergo revision. At present, the applicable standards are set for members of the general public and for those having occupational exposure to ionizing radiation (Table 12). A very recent Supreme Court decision approved the Atomic Energy Commission as the regulatory agency for standards.

### Summary

Other than from medical and dental sources, the general public is not now receiving exposures to ionizing radiation greatly in excess of background levels. Medical and dental exposures are increasing quite rapidly, however. In addition, the nuclear industry is now leaving its infancy and carries with it the threat of large radiation exposures. Vigorous action on the part of independent scientists and environmental organizations has drawn national attention to the radiation hazards posed by the nuclear power industry in a manner similar to the

**Table 12.** Standards for Exposures to Ionizing Radiation. The units are rem per year.

| Nature of Exposure | Maximum Permissible Exposure Rem per Year |
|---|---|
| Large numbers of the general public | 0.170 |
| Individuals in the general public | 0.500 |
| Occupational exposures | |
| Males | 5.000 |
| Fertile Females | 0.500 |

Note 1: All exposures are in addition to medical and natural background exposures.

Note 2: Medical and dental exposures are not regulated other than by the judgment of the responsible physician or dentist.

attention drawn to the radiation hazards associated with atmospheric testing of nuclear weapons in the late 1950's and early 1960's. A similarly vigorous effort to bring attention to the hazards and benefits of medical and dental uses of radiation is only now beginning. That these efforts are not all carried out in the spirit of scientific objectivity can be deduced from the response of many individuals associated with the nuclear or medical radiation industry. For example, in a recent publication, Dr. Dade Moeller, President of the Health Physics Society, reviewed the potential job market for health physicists in the expanding nuclear industry and then commented:

> Essential to such leadership is that we [health physicists] speak out and make known our positions on such issues as nuclear power safety and radiation protection guides. This includes speaking not only to the public and fellow workers in other fields but also to our Congressional leaders. As in all [Health Physics] Society affairs, the major burden for such actions will fall upon the members of our thirty Chapters. I encourage all of you to be as active as you possibly can. To paraphrase an old adage, *'Let's all put our mouth where our money is* [emphasis added].' "[34]

Environmental radiation is one of the most noxious of all pollutants. It is disturbing that not only must the public be alert to sources of the pollutants, but also to those who seemingly give primary consideration to their purses.

## REFERENCES

1. J. Schubert and R. E. Lapp, *Radiation—What it is and How it Affects You,* Viking Press, New York, (1957) paper edition (1966).

2. A good place to begin is the excellent review article: R. H. Mole, "Radiation Effects in Man: Current Views and Prospects," *Health Physics* (in press).

3. See for example: "X-Ray Debate—Do Physicians Overuse Diagnostic Radiology?" *Wall Street Journal,* (December 22, 1971).

4. This table is based on Table I in: K. Z. Morgan, "Never Do Harm," *Environment, 13*(1) 28-30 (January/February 1971).

5. *Proceedings of a Conference on the Pediatric Significance of Peacetime Radioactive Fallout,* Supplement to *Pediatrics, 41*(I, Part II), January (1968).

6. *Electronic Products Radiation Control,* Hearing report of the Subcommittee on Public Health and Welfare of the Committee on Interstate and Foreign Commerce, U.S. House of Representatives, Government Printing Office, (February 1, 1968).

7. Morgan (cited as reference 4 above).

8. U.S. Public Health Service, *Population Dose from X-Rays, U.S., 1964,* USPHS Publication No. 2001, (October, 1969).

9. F. F. Roberts, "Is This X-Ray Really Necessary?' *Medical World News,* pp. 15-16, (October 23, 1970).

10. American Medical Association, "The Wasted X-Ray" (editorial), *American Medical News,* (September 21, 1970).

11. See, for example, W. C. Alvarez, "X-Raying Teeth Has No Harmful Effect on the Body," column in the *Minneapolis Star,* p. 7C, (March 30, 1971).

12. U.S. Public Health Service, X-Ray Examinations: A Guide to Good Practice, USPHS Document No. BRH/P-6, Government Printing Office, (May, 1971).

13. G. T. Seaborg, "The Atom's Expanding Role in Industry," Atomic Energy Commission News Release S-24-69, Washington, D.C., 20545, (July 31, 1969).

14. A. R. Tamplin, "Plowshare as a Radionuclide Source," Paper given at the Workshop on Radioactivity in the Marine Environment, National Academy of Science and the National Research Council, Woods Hole, Mass., (September 6-15, 1967).

15. C. E. Barfield, "Radiation Standards Dispute Imperils Nuclear Program," *National Journal, 2*(46), 2483-2497, (November 14, 1970).

16. See, for example, *Environmental Effects of Producing Electric Power (Parts I and II),* Hearings before the Joint Committee on Atomic Energy, U.S. Congress, Government Printing Office, (1969) and (1970).

17. *Radiation Standards for Uranium Mining,* Hearings before the Joint Committee on Atomic Energy, Government Printing Office, (March 17-18, 1969.)

18. H. P. Metzger, "Dear Sir: Your House is Built on Radioactive Uranium Waste," *New York Times Magazine,* (October 31, 1971).

19. The best source for following current nuclear reactor activities is *Nucleonics Week,* a weekly newsletter published by McGraw-Hill, New York.

20. U.S. Atomic Energy Commission, *Theoretical Possibilities and Consequences of Major Accidents in Large Nuclear Power Plants,* Report No. WASH, 740, Government Printing Office, (March 1957).

21. D. P. Geesaman, "Plutonium and the Energy Decision," *Bulletin of the Atomic Scientists,* pp. 33-36, (September 1971).

22. Federal Radiation Council, *Background Material for the Development of Radiation Protection Standards,* FRC Report No. L, (May 13, 1960).

23. J. W. Gofman, A. R. Tamplin et al., "Radiation as an Environmental Hazard," presented at the 1971 Symposium on Fundamental Cancer Research, M.D. Anderson Hospital, Houston, Texas, (March 3, 1971) in press.

24. S. Jablon and J. L. Belsky, "Radiation Induced Cancer in Atomic Bomb Survivors," paper presented at the Xth International Cancer Congress in Houston, Texas, (May 1970) proceedings in press.

25. Morgan (cited as reference 4 above).

26. Joshua Lederberg, "Cost-Analysis of Genetic Disease From Radiation," manuscript, (January 5, 1971).

27. Linus Pauling, Affidavit submitted to the State of Vermont, PSC Hearing on the Vermont Yankee, (September 9, 1970).

28. Morgan (references 4 and 16 above).

29. J. W. Gofman and A. R. Tamplin (reference 16 above); and Gofman, et al., (reference 23).

30. Pauling (reference 27).

31. A. Stewart and G. W. Kneale, "Radiation Dose Effect in Relation to Obstetric X-Rays and Childhood Cancers," *The Lancet,* pp. 1185-1187, (June 6, 1970).

32. National Council on Radiation Protection and Measurement, *Basic Radiation Protection Criteria,* NCRPM Report No. 39, (January 1971).

33. Morgan (reference 4 above).

34. Dade W. Moeller, "The President's Message," *Health Physics, 21*(1) (July 1971).

Dean E. Abrahamson, M.D., Ph.D.
Associate Professor and Director
Center for Studies of the Physical Environment
University of Minnesota
Minneapolis, Minnesota

# Chapter 25

# Visual Pollution

*"We are what we see."*

In a solemn ceremony on Tuesday, April 20, 1971, Secretary of Transportation John S. Volpe oversaw the removal of a billboard on Interstate 95 near the town of Freeport, Maine. It was the first billboard to be taken down under the Federal Highway Beautification Act, passed by Congress and signed into law by the President in 1965. Three days earlier six militant ecologists from Ann Arbor, known as the Midnight Skulkers, took the law into their own hands and clandestinely removed their 167th billboard from along the highways of central Michigan. The young billboard bandits had been leading a crusade to rid Michigan's highways of garish billboards which they considered illegal and unsightly. The activist group might also have been aware of the fact that between 1965 when the federal law was enacted and the first symbolic removal by Secretary Volpe more than 250,000 new billboards has been placed along the nation's roads, bringing the total to over one million. During the week that the Secretary of Transportation and the Ann Arbor Six took their separate actions, eskimos in Kotzebue on the Bering Sea were parking their snow-mobiles below the beaming face of Colonel Sanders, taking out buckets of Kentucky Fried Chicken from one of 3,600 similar looking establishments located around the world. Meanwhile Detroit, the motor city, was two months into a four month ban on the construction of new gasoline service stations. While Detroit was fighting a rear-guard action against gas stations, advertising men for Texaco were developing a program to "cut the clutter." Their program would shortly blossom under the teasing headlines, "Can the oil industry be accused of 'visual pollution'" and which would go on to state proudly, "we've started a nationwide program to encourage (dealers) . . . to rid their stations of superfluous signs, pennants and trappings." Not able to stop at that the ad would continue, "However, when Texaco retailers have real news about our products which can be truly helpful to customers they will present it clearly and in good taste." Good taste, however, is a highly subjective judgement and one which does not lend itself easily to consensus, particularly when corporate and public interests collide.

On opposite ends of the country, two communities—Concord, Massachusetts and Carmel-by-the-Sea, California—continued to enforce local zoning ordinances against garish strip development which had become prevalent elsewhere. Opting to preserve their unique historic and residential character, both towns decided that they would be happier without a new MacDonalds (1,850 locations in other places), Dairy King (4,253 outlets, mostly similar) or Burger King (800 stores and coming up fast).

Sensitive to increasing community resistance, Burger King vice-president John Hollingsworth acknowledged that his company was beginning an extensive remodelling program. Hollingsworth's explanation of his firm's action was that it helped Burger King relate "to the external imagery of the 'now' world."

The "now" world is a visually polluted world, filled with increasing numbers of billboards, junkyards (screened and unscreened), abandoned buildings (including ever present gas stations which have outlived their advertising usefulness), transmission line corridors festooned with overhead wires, blatant signs reaching higher and higher for the American public's more and more difficult to catch eye and culminating in sprawling roadside strips which often succeed in making all places equal in vulgarity. As fecal coliform counts rise in urban waters and as nitrous oxides increasingly saturate American air, the eyes of the nation are at the same time being forced to view greater and greater levels of visual pollution.

To pollute means to defile; to *pollute visually* means to defile and degrade what we see in the landscape. This definition immediately suggests two things which require analysis. First, is there anything intrinsic in a landscape which by itself is polluting? Signs, billboards, junkyards—are these objects and features intrinsic visual pollutants? The second consideration has to do with subjective evaluations. Can people agree on what is and what is not visually polluting? Are standards of beauty and ugliness commonly accepted?

Setting aside for the moment an individual scale of values, visual pollution begins when the landscape loses its coherance to a viewer. If a street, roadside or valley possesses visual clarity, if it has a relatedness between its parts and if these parts are seen with clear limits, then it is safe to conclude that the landscape—as a whole—to the single judging viewer is visually unpolluted. However, given clarity, relatedness and a sense of limits, the introduction of any element which diminishes those qualities would be visually polluting in *that* landscape.

A large billboard, placed at the entrance of an Italian hill town would be a visual pollutant, destructive of the scale and texture of a community which had evolved its form over a long period. The same billboard placed along a roadside strip on the way into Phoenix, Arizona, joined with other similar billboards, repeating motifs, patterns and colors in an area where these elements constitute the architecture of the place (since there would be few other elements of comparable size or frequency to compete with them) would contribute to the rhythm and texture of the roadside experience. If these billboards alternate with hugh roadside signs, sprawling gas stations, neon-fringed, pastel-colored eateries, acres of mutli-colored cars parked in front of gigantic shopping plazas and they all in turn alternate with the flow of steel-toothed vehicles, then what would be visually polluting in such a landscape would be a single large living oak or cactus, since such a tree would undermine the esthetic "integrity" of the plastic landscape.

A pollutant, therefore, is not an absolute. Something is not a pollutant simply because it is intrinsically ugly, nor is it a non-pollutant simply because it is intrinsically beautiful. A visual pollutant is a *misplaced* element. Whether or not something is misplaced depends on context, and context has to do with the character of visual field as a whole at any one place. One billboard in a scenic valley may be so galvanizing in its visual insistence that it effectively destroys the entire setting. Power lines moving across a pastoral landscape—one marked by small farms, silos, domestic animals, narrow roads and old villages—can have an instantly urbanizing effect, figure 111.

**Figure 111.** Visual Pollution in rural America. Photo courtesy of Clarence Davis.

The visual field is fragile. Single misplaced elements can destroy whole vistas. Billboards along a sprawling roadside "strip" may almost work, preparing the driver for the arrival of a commercial district, announcing that stores, bars, motels and eateries will be coming up. Such billboards are a preview of coming commercial attractions and an obvious form of advertising. However, a billboard placed on a scenic road, much as a television commercial inserted in a fine film, will wreck the mood, interrupt the continuity and will probably have an effect well beyond the time it is actually seen.

A proper context or fit in the landscape does not, however, require the exclusion of all non-conforming elements. A landscape which posses visual clarity may comfortably absorb forms and objects which in themselves are non-conforming. The risk of sameness is boredom. A landscape can be so filled with strident forms which attack and bombard the eye that a kind of gross conformity is created in which all the elements in the landscape scream so loudly for attention that visual monotony ultimately is produced.

Conformity becomes a visual pollutant when sameness dulls the landscape and our ability to respond to it. Non-conforming elements—the unexpected and surprising juxtapositions of man-made with the natural—do not necessarily constitute visual pollutants. If the non-conforming structure is sensitive in scale and character to the existing landscape, the viewer may be helped to readily sense the consistent quality of the place.

Earlier we touched on the question of individual response to the landscape. Obviously, what one person might regard as scenic and beautiful another might consider repetitive and boring. There are some who find nature and "scenery" boring; they would prefer the jazzy rhythms of an intensely developed roadside strip. A resident of the inner-city, accustomed to the constant clash and apparent discontinuity of man-made forms might well have a higher tolerance for the diverse city environment, finding in it order, pattern and clarity. A country-bred person could easily be confused and disoriented by the same city landscape. What would be home to one would be visual pollution to the other. Any definition of visual pollution, therefore, must allow for cultural and environmental differences.

Standards of beauty—natural and otherwise—resist consensus. The English art historian Eric Newton claims, "except within the vaguest limits, beauty cannot be measured . . . therefore it cannot be defined. It cannot be measured either in quantity or quality; therefore it cannot be made into the basis of a science." The perception of beauty is always elusive, fleeting, personal and subjective. A range of personal associations, one's past history, determines our ability to find beauty. We can surely recognize elements of fashion (enlarging of lips, stretching of necks, carving of skin) which certain cultures regard as beautiful and which others find repulsive. Tastes in popular objects such as clothes, cars, homes and furnishings are inevitably influenced by cultural (and often commercial) values. Some of these values are the product of a long, slow evolution. Others appear to be created overnight by mass media. The question of beauty (and indirectly, the question of what constitutes visual pollution) is more related to the ability of an object or setting to call forth good associations in the receiving party than to any inherent physical form in the object itself.

Given an individual's own set of values—and that *must* be given—it is possible to work toward some sense of beauty for that individual. Robert Kates has wisely pointed out that rather than attempting to "maximize beauty" in the landscape, we should "minimize ugliness." And to come full circle, Kates confirms the view earlier expressed that ugliness is more a question of misfit and context than any other factor.

The struggle against visual pollution is, therefore, essentially an effort to preserve the *integrity* of the visual environment. As has been pointed out, there is often an integrity to the roadside strip such as found in downtown Las Vegas or at grandiose jet airports. Insistence upon banal screening, cosmetic plantings and restrictions on sign heights may actually serve to render these unique American landscapes less potent, less rhythmic and ultimately perhaps even less functional. The danger, and it is a real danger, is that at a certain point strip "integrity" will fall over into vulgarity. Perhaps the greater danger, however, may be that by repetition of forms and franchises, all roadsides will get to look alike. William Shaffer, an IBM executive from Kansas City, once told a New York Times reporter, "With the cars, the roads, the road signs, you can't really tell whether you're driving from the airport downtown in Memphis or in Minneapolis. Hell, I can't tell whether I'm in Houston or Des Moines."

An homogenized American Landscape is a dreary prospect, figure 112. Every possible encouragement should be given to those communities which resist the strip developer, providing that the community in question has a recognizable historical or cultural tradition. Concord and Carmel

**Figure 112.** An homogenized American landscape offers a dreary prospect, a strip in lower California. Photo courtesy of the New York Times.

have every right to making the zoning choices that will prevent strip sameness from coming into their areas. Communities such as these have a visible cultural and architectural history which, if preserved, can enrich all our lives.

Many other communities are newly expanded, if not newly created, possessing a mobile population with no sense of roots. Such places often lack many physical structures from an earlier era. These towns are defined in contemporary terms, fortunately or otherwise, by the intensity of strip development both in residential as well as commercial areas. Visual obsolescence is built into these places. Renewal for such towns would constitute a remodeling or more accurately, a re-packaging of the nationally conceived, nationally promoted franchise developments. Such places then, as is true of automobiles and skirt lengths, are constantly vulnerable to the caprice of fashion. One can only imagine how a community which once took pride in its new MacDonalds feels when a neighboring town gets a newer MacDonalds (one with smaller architectural ribs). No doubt the first community would suddenly feel older, out of fashion, anxious to have its MacDonalds updated. These are trade-in communities. They find their character expressed in a conscious desire to shed what is old and replace it with what is new.

The ecological implications of such attitudes, apart from considerations of visual pollution, are staggering. Nature editor J. B. Jackson once commented on this issue as follows:

> The attitude is a hopelessly superficial one. The affront of the auto junkyard to our sensibilities is much more than esthetic: the junkyard testifies to waste and inefficiency, to technological incompetence. The billboard is an affront not merely because it is ugly but because it is a form of economic parasitism, depending on the presence of the highway with no service in return. The dilapidated town or village is an affront to our social conscience that no amount of beautification can assuage.

How effective will current campaigns for recycling of paper, metals and glass be if whole communities continue to assume a disposable attitude toward the major structures which fill their roadsides? Perhaps the time has come to adopt policies of planned rehabilitation rather than constant updating of roadside facilities. What must also be understood is that the Concords and the Carmels are the exceptions. Most American towns, particularly those in the fast growing suburbs have very little to preserve apart from the increasingly obsolescent" structures along the strip.

If uniqueness is being lost in the structures placed by man upon the land (and we have every reason to believe it is) then it becomes doubly important to preserve visual uniqueness in the natural habitats which have not been altered by man. If the man-made American landscape is in danger of becoming almost everywhere the same, then the struggle against visual pollution must take place where the land forms themselves still possess regional character and individuality. Visual pollution must be prevented from attacking two particularly vulnerable areas: first, the scenic views along roadways, and secondly, natural areas threatened by transmission lines. These areas are vulnerable simply because it doesn't take very much to destroy the visual field. One misplaced billboard destroys a view. One seventy-foot utility pole can take the "naturalness" out of a natural setting.

Roads are public corridors. The earliest roads were personal paths, defined more by individual behavior than by public planning. As social systems elaborated, the concept of spatial roads—land set aside for persons and goods to be moved along from point to point—evolved. Some geographers are inclined to think that new roads are once more becoming a function of behavior, though not in the old ways. They see new roads leading nowhere (as older roads once led to town), starting everywhere and leading on without a clear destination. These new roads are behavioral in the sense that they emphasize a "goingness," a texture of activity lacking a target. This concept suggests that roadside facilities are located like pearls on a necklace, strung out rather than focussed, targets in themselves rather than being associated with a particular community. Such a concept permits, and in some cases encourages, the locating of roadside facilities everywhere and anywhere because the aimless motorist is likely to stop almost anywhere.

This characterization of the new road is less critical, however, than geographers would have us believe because the issue isn't whether there should be advertising and selling establishments *or* scenic vistas along the roadway. The larger issue is whether or not the public has a right to make the decisions controlling what will and what will not be found along the roadside. On this point the courts have repeatedly held that no private person or company has an inherent right to advertise on a road which was created with public funds. Private roads, we might add, are rare.

Billboards take on a commercial value only when they are seen. Placed in the center of a wild, inaccessible forest billboards would be valueless. Billboards are totally parasitic. Without the public road and hence the public eye there would be no

reason for constructing billboards. Since it is the public itself which has already paid, and paid well, it is inappropriate (as the New York Times once editorialized) to make "motorists on many (federally funded) roads (be) forced to pay for them a second time as members of a captive audience who have to look at advertising signs while they drive, whether they want to or not."

The Federal Highway Beautification Act of 1965 began the national effort to bring billboards under moderate regulation. Among other features, the act established buffer zones of 660 feet in which no billboards could be erected. Needless to say the intent of the legislation is regularly circumvented huge signs are often placed just beyond the 660 foot limit imposed by law. The original draft of the law forbid the placement of any billboards "visible from the highway," a feature that many conservationists think should be restored. A questionable feature of the legislation is the provision for mandatory compensation for removal of signs from private property. Estimates are that fulfillment of that requirement will cost the federal government more than 300 million dollars.

The inadequacies of the federal legislation, coupled with the slow pace of enforcement has led at least one state, Vermont, to adopt a model code for the preservation of roadside scenery. Vermont's newly enacted law is a notable success, dealing as it does with the problem of roadside scenery preservation on a comprehensive level. Highlights from a discussion of the Vermont code (written by Neil Williams) plus excerpts from the code itself are included in the box. Other states and perhaps even counties should find useful guidelines and standards in the Vermont approach to scenic road preservation.

The formal language of the Vermont code includes specific standards for placement, size and lighting of signs. Areas of special scenic value are defined and criteria elaborated. Designation of such scenic areas is made by the State Planning Director with the advice of a Scenery Preservation Council. In addition to types of facilities named (see box) the Vermont code prohibits the placement on designated scenic corridors of amusement parks, eating establishments (except those located in residential buildings), gravel pits, open storage of non-agricultural products, motels, quarries, trailer parks, trucking terminals and used car sales lots. The law also carefully defines the state's powers to acquire rights and interests in land in order to accomplish the purposes of the act.

Charles Morrison Jr., Director of Community Assistance Programs for the New York State

Department of Environmental Conservation, once wrote, "Roads, by themselves are not scenic. But the identification, preservation and enhancement of scenic assets and elimination of deficits in the corridor is the real concern." The State of Vermont has perhaps moved furthest in the direction of successfully reckoning with that concern.

The second major visual pollution confrontation is taking place over the location of transmission line routes, figure 113. Dr. P. H. Rose in *Science* (October 16, 1970) indicated that in 1967 about seven million acres of land were in use for the rights of way of transmission lines and that twenty million acres will be needed in the near future as a result of increasing power consumption. The dedication of such large numbers of acres to transmission line corridors will certainly effect growth patterns of many communities and perhaps more to the point here, will certainly effect the visible landscape.

**Figure 113.** Power lines in a rural landscape. Photo courtesy of Clarence Davis.

The gravity of this problem was underscored in a staff report of the Hudson River Valley Commission which commented, ". . . electric power lines are one of the more significant despoilers of natural beauty . . . and utility companies are under constant pressure to avoid stringing wires indiscriminately over the landscape. Legislators often threaten laws to force power lines into costly undergrounding, pressuring the power companies even more. Whether the esthetics of latticework 'high-tension' towers are better or worse," the report continued, "than many other affronts to the urban and rural landscape is not the question. Billboards to some people's eyes are much worse. But to many, paths cut through the countryside to accommodate transmission-line towers and poles are just as unacceptable."

Acceptability seems hardly to be the issue as communities all across the nation are being required to accommodate the often pernicious intrusion of transmission-line corridors. Citizen action groups spring up wherever utilities propose to take new lands for the construction of "power towers" and numerous, often protracted, ugly confrontations take place.

The usual height for 345-KV towers is 60 to 70 feet except on either side of a river, highway or railroad crossing where the height might be increased. Structural steel is normally used for these large towers. Steel pipe or wood poles, ranging in height from 45 to 70 feet are used for 115-KV and 230-KV lines. The major concern, of course, is visibility. There is no easy way to hide poles.

## SCENIC ROAD PRESERVATION IN VERMONT

Vermont's scenery is among the state's most precious possessions, and an asset of the regional and national significance. For Vermont is the pleasantest part of that broad green strip which lies between the rapidly growing metropolitan areas along the Atlantic coast and in the St. Lawrence valley: we are the flavor in the sandwich, as it were . . . All over the state there are long drives extending through unspoiled countryside for an hour or two, in a continuous experience of ever-varying beauty.

The economic value of these scenic assets is of major importance to the state, particularly in two ways. First it is the quality of our environment which persuades many young Vermonters to stay here to live and work, thus cutting down on the outward drain of talented youth. Second, these same assets play a leading role in attracting into the state new permanent and part-time residents, tourists, new industries and new cultural facilities . . .

More intangible values are equally important, and perhaps more so; for now that the basic problem of economic want is beginning to approach solution, more attention can be properly focused upon qualitative aspects of life. A day or month in the Vermont countryside . . . not only adds to the pleasure of life, but promotes physical health, mental health and working efficiency.

### Protection of Roadside Scenery

. . . Our proposals are based upon the (following) classification of scenic values . . .

a) Distant Views—Glimpses of mountains or mountain ranges, usually through a gap in nearby ridges and broad views of valleys, lakes and mountain passes or from ridgetops.

b) Particularly striking features in the near and middle distance (gorges and gulfs, notches and gaps, cliffs and rock outcrops, waterfalls, rivers, lakes and island marshes.)

c) "The typical Vermont scene", normally a valley extending from ridgetop to ridgetop, and including such scenic features as the following: farms and barns, fields and forests, fine trees and groups of trees, wild flowers and ferns, small cemeteries, covered bridges and villages and churches. Other features may be more spectacular, but these typical Vermont scenes represent something almost unique, one of our prime scenic values—a landscape adapted to human use, with infinitely varying patterns . . .

d) Settlements (including) approaches to towns —steeples, fine old building or groups of buildings, cultural centers and historic spots and relics. A comprehensive program for protecting and enhancing scenery in Vermont will require further development and continuous rethinking of these categories, together with a full esthetic inventory of the state . . .

### Scenic Corridors

There are two kinds of scenic corridors along major highways through the state where immediate protection is both necessary and feasible . . . (along) the emerging federal interstate highway system (and along) intermediate traffic highways (which) extend for substantial distances through scenery of great splendor . . .

The use of land along such routes may appropriately be regulated by public action since the three primary requirements for action under the police power are satisfied. First, the scenery involved is unique, not common all over the state. Second, the regulations will be uniformly applicable to all land similarly situated . . . Third, the routes involved are small in number and of uniform character . . .

In both types of corridors, listed obnoxious uses are forbidden, ie., billboards, junk yards and dumps, storage tanks, filling stations and garages and drive-in theatres . . . for a half-mile on either side of the road or up to the ridge top, whichever is less . . .

Industrial designer Henry Dreyfuss was commissioned by the utilities in a project coordinated by the Edison Electric Institute to develop more esthetic designs for transmission towers. Dreyfuss came up with more than 100 designs, using a wide variety of materials, including laminated wood, aluminum, concrete, reinforced plastics, steel pipe and others, some of which might be considered an improvement. None, needless to add, achieved the ultimate goal of invisibility. In terms of pure esthetics, the Dreyfuss designs had a certain elegance. Some single element poles achieved a notable simplicity.

The difficulty with the Dreyfuss designs and the inherent problem of any design for transmission line poles is that the height and frequency of the poles effectively destroy the natural beauty of any undeveloped area, regardless of the esthetics of the poles themselves. Undergrounding is often proposed as the ultimate solution, but here too the landscape will suffer, though perhaps less grievously. Questions of relative cost and reliability aside, transmission line undergrounding requires the maintenance of an open swath, readily accessible to utility vehicles in the event that service or repairs are necessary. Undergrounding, therefore, means constructing and maintaining an open roadway for the entire length of the transmission corridor, a solution which many informed conservationists (including leaders of the Sierra Club) regard as no real solution. This is not to suggest that in certain specific, highly unique cultural and historic areas, which might be relatively open in character, undergrounding might not offer the most acceptable solution.

Whether electric power is transmitted above or below the ground is perhaps less significant than the fact that twenty million acres will be taken for the location of these corridors. As an aside, the extent of land taking might also provide one more reason for a cut back in power consumption. The magnitude of the projected land use has forced utilities, government agencies, town boards and individual citizens to reckon with a problem that possesses no simple solutions. Open cuts in the land resist easy landscaping. Power towers have a permanently urbanizing effect. Urban character, for many, is closely related to numbers of people exclusively. This is a highly restricted viewpoint. Urbanization is as much a factor of scale and structure as it is density of people. A series of 70 foot high towers, placed into a rural, pastoral landscape transforms that landscape as much as a series of apartment houses. While some local

support services, gas stations and highways to name two, will not follow on the placement of transmission line towers, the intrusion of high, man-made elements towering over natural foliage, swinging over recreational areas and straddling lakes and streams, will inevitably diminish the "naturalness" of natural areas. Because vision is such a fragile sense, easily intruded upon, such towers become constant reminders of the mechanical, technological element. Some utilities claim that corridors open new areas for hunting and snow-mobile use but conservationists note that such recreational boons are of questionable value.

A recent phenomena has been the development of efforts to preserve landscape types—prairies, pastoral lands and areas of special ecological significance. Struggles against visual blight caused by power tower corridors, therefore, have taken on more than a local esthetic importance. What good is it to preserve a grass land or a prairie, conservationists ask, if a transmission line corridor can effectively upset the ecology of an area and perhaps, more importantly, prevent visitors from experiencing a quality of place evoked by the unique environment.

The Hudson River Valley Commission's report, cited earlier, offers eight principles relating to the choice and design of transmission routes, choices which up to now have been largely made exclusively by the utility companies. These principles, which can provide the basis for informed public participation in the process of route selection, are as follows:

1. Avoid sites of great scenic value, including prominent ridge lines, lakes, barren sides of mountains or hills.
2. Keep alignments along the bottoms of lower slopes and valleys between hills.
3. Avoid crossing hill contours at right angles: avoid steep grades which expose the right-of-way to view.
4. When crossing principle roads, jog alignments to avoid extended tangents along the right-of-way on either side of the road.
5. In rough or very hilly country, change the alignment continuously in keeping with the scale of topographic change.
6. Where feasible and in keeping with the prior criteria, run lines along the edges of differing types of land use, with special emphasis on preservation of forest growth.
7. For lower voltage sub-transmission lines, undergrounding is desired when alignments parallel major highways or scenic areas. In these instances, undergrounding should be considered when construction is undertaken.
8. If a proposed route realistically can be shifted from a scenic area to one already industrially developed, the route should be placed through the latter.

Buckminster Fuller has predicted that in the next fifty years we will replace nearly all of the man-made structures around us. The potential for increased visual pollution is high if Fuller's predictions are accurate and if past history is a reliable guide. If we move wisely, however, to establish criteria for the preservation of natural and historic areas, then the prospects of containing the blight of visual pollution are great.

In a world preoccupied with massive problems of population control and technology out of control, a concern for the visual world, what some have disparingly called "visual amenities" might be thought of as well meaning but undeserving of high priority. Many years ago a wise sage suggested that we are what we eat. An equally wise philosopher might today offer the thought that we are what we see. The environment is everything that surrounds us. This environment gives shape to our character. The events of life must all take place somewhere and the "whereness" affects the perception of the event. Our environment is more than a passive backdrop; it is the stage on which we move. The objects and forms on that stage shape our actions, guide our choices, restrict or enhance our freedom and in some mysterious way even predict our future.

In a report on room decor and its ability to calm disturbed children (*Journal of the American Medical Association*, January 31, 1972) Norbert I. Reiger, MD and Director of the Children's Treatment Center at Camarillo State Hospital, California, concluded "the structure and quality of the physical space surrounding man affects his emotional well being. In fact the physical space immediately surrounding a man becomes part of his body image." Our view of ourselves, our sense of potency, our sense of self-worth—these images of self—are all profoundly affected by the physical environment in which we move. The struggle to maintain diversity and quality in both the natural and the man-made environment is more than an esthetic struggle. It is a battle to preserve a body-image of man, unique unto himself yet functioning harmoniously in a coherent environment.

Alan Gussow
Artist
Congers, New York

# Chapter 26

# Solid Waste Pollution

Solid waste pollution, often termed the third pollution (after air and water) may be generally defined as a variety of discarded, useless materials resulting from normal human activities as well as industrial, commercial, mining and agricultural operations. Although industrial, commercial and domestic wastes are of prime importance in urban and industrialized areas and are usually thought of as the "solid waste problem", they account for only about 10% of the total waste load. Each year over 4 billion tons of wastes are generated in the United States. This is equivalent to 100 pounds of solid wastes per person per day. Of this total, only about 3.5 pounds are actually produced by the average citizen while mining and mineral operations account for 31 pounds and agricultural wastes contribute 58 pounds.

### Agricultural and Mining Wastes

Agricultural wastes, the largest contributor to the waste load, is composed of, for example, crop residues, wooden debris, animal fecal matter and pesticide residues and containers. Mineral and mining wastes stem from the mining of minerals and fossil fuels together with their milling and processing. The copper and iron and steel industries alone produced over 700 million tons of wastes in 1965.

These two sources of wastes, which account for 90% of the total, have received only limited attention outside their spheres of generation. Exceptions have arisen when other forms of pollution and health dangers have resulted from inefficient disposal or lack of attention to proper disposal. Particulate air pollution associated with the burning of crop residues, waster contamination and accelerated eutrophication as a result of fertilizer and feed lot runoffs as well as unstable waste heaps are but a few examples of instances when these wastes have been the subject of concern.

### Urban Wastes

The major focus of solid waste management is upon the over one and a half billion pounds of urban wastes which is collected on some regular schedule and disposed of at an annual cost of over $4 billion per year.

Urban or municipal wastes comprising domestic, commercial and industrial wastes as well as demolition and construction debris, street sweep-ings and litter, can all be collectively called 'refuse.' Subdivisions under this general term are:
1. garbage, which is waste from the preparation, cooking and serving of food, market waste and waste from the handling, storage and sale of produce; 2. rubbish, which includes both combustibles such as paper, cartons and wood and non-combustibles such as tin cans, dirt and glass; 3. ashes, which consist of residues from the furnaces used for heating and incineration.

The composition of municipal refuse is a function of the mix of industrial, commercial and domestic units in a given area. However, on the average, about 50% of all these wastes are paper and paperboard with glass, garbage and metallics accounting for about 10% each.

Industrial wastes are generally divided into two categories: first, there are the wastes which are common to all industries, generated in normal operations such as packaging, shipping and maintenance. These consist of metals, fibers, wood, plastic and glass. Second, there are those wastes which are specific to a particular industry. For example, a textile mill's wastes will contain cloth and fiber residues while printing and publishing establishments will involve the disposal of inks, glues and paper.

Commercial and domestic wastes are similar to the first type of industrial wastes but contain a higher percentage of garbage and packaging materials. Commercial wastes are defined as those stemming from service-oriented establishments such as restaurants, hotels and hospitals, as well as general offices, wholesale and retail outlets while household or domestic wastes are generated from any multi- or single-unit residence.

### Collection

Many industries must bear the responsibility of disposing of their wastes, especially when they pose special problems. On the other hand, commercial-type wastes are normally collected with the domestic wastes by the municipal or private collection system, operating in that area. The minimum frequency of collection is dependent on the volume of wastes generated and the putrescibility of the refuse. Certainly household garbage which will rot quickly must be hauled away more quickly than inert industrial maintenance wastes.

The obvious need to collect wastes has resulted in a shortsightedness in the national approach to the treatment of solid wastes. A 1968 survey of U.S. solid waste practices revealed that 79% of community funds were allocated in support of collection services, leaving a scant 21% for disposal. An average municipal collection system must spend about $20 a ton to collect and transport the wastes to the points of treatment and disposal. Some cities, notably New York, spend a great deal more, the exact amount depends on the nature of the traffic and road patterns and the types of residences that are being serviced. In efforts to cut costs, the old compactor trucks are being replaced by experimental, automatic pick-up trucks and large centrally located containers to reduce the number of pick-up points on the route and the number of manhours required for collection.

An often overlooked factor is the occupational hazard associated with the handling of refuse. The National Safety Council and American Public Works Association reported in 1967 that the work injury frequency among solid waste employees is nearly 900% greater than the average for all U.S. industries. The injuries arise because of the hazardous nature of the material the workers must handle and the physical exertions required of them. Also, there is little mandatory training for solid waste management personnel.

Recent advances made in Sweden point the way to more efficient collection systems through the use of pneumatic and slurry pipelines. In urban areas, where the population density is high and there is a corresponding high traffic density, there is a clear need for an automatic system of collection that can transport the discarded materials directly from the point of generation to the point of treatment through underground pipes or sewers.

### Treatment

Once collected, wastes may be disposed of directly or first treated to reduce their mass and volume making their disposal more economical and more compatible with their surroundings.

Presently, the most common means of reducing the wastes before disposal is by burning them in an incinerator. The controlled combustion occurring within the incinerator will dispose of some of the wastes by converting them to water and gasses that are released to the atmosphere leaving a residue of primarily inert glass, ashes and metallics. On average, volume and mass reductions approach 80–90% and 75–80% respectively. Incineration is in limited use due to its expensive capital and operating costs. In 1968, the average incinerator cost in excess of $7,000 per ton of rated capacity to construct and had an average operating budget of $4.50 per ton. Also, of increasing concern are the air emissions from incinerators. These consist chiefly of particulates and the oxides of nitrogen with trace amounts of sulfur dioxide, aldehydes, hydrocarbons and ammonia. Particulates may be controlled using existing collection devices but the oxides of nitrogen may only be somewhat reduced by operational modifications while the trace emissions are proportional to the refuse content. Usually no attempt is made to curtail these emissions.

The advantages of incineration include its continuous operation during all weather conditions, the need for small amounts of land space and its ability to accept nearly all kinds of refuse and produce stable, highly reduced, inert products.

In line with current trends to use all available resources, research has focused on alternative systems of waste treatment. These include methods of generating electric power or steam heat from the energy value of refuse (4000–6000 BTU/pounds) and the development of pyrolysis as a means of making a fuel out of the refuse. Plants either built or planned in Montreal, Chicago and St. Louis burn refuse as in standard incinerators with the addition of a system to use the heat produced to drive turbines or heat water for steam. Not only will the heat value of the refuse be used but operating costs will also be reduced by the sale of the power. Pyrolysis is the thermal decomposition of refuse in the absence of air for the production of gaseous or liquid fuels as well as the production of inert fractions of ash and inorganic residues. Pyrolysis systems are still in the development stages and it is yet to be shown to be the superior method of waste treatment. Operating costs range as high as $6 a ton varying inversely with the size of the facility. In this case also, projections indicate that products and residues will offset some of the costs.

Another waste treatment process is composting, which is the controlled, accelerated decomposition of putrescible wastes by aerobic microorganisms. Composting will produce excellent soil conditioners and will add biochemical stability to a fill. However, as a method of large-scale treatment, composting has failed to be competitive since there are few markets for the product and the costs of production are too high for normal usage. Furthermore, to compost wastes in urban areas where the refuse problem is the most serious would only compound the economic problem due to the high costs to haul the product to rural areas where it may be used.

## Disposal

Untreated wastes and the residues from waste treatment facilities must be disposed of in some environmentally safe manner. The most common method of disposal is the landfill which may range in efficiency from nothing more than an open dump where wastes are brought and dumped without any precautionary methods, to a sanitary landfill in which wastes are systematically compacted into trenches of existing formations with safeguards to protect against air and water degradation. To achieve environmental compatibility, the wastes must be covered each day to guard against the breeding of rats and the feeding of birds and insects; there must be no open burning at the site; and all ground and surface waters must be mapped and protected from contamination by chemicals and bacteria leaching from the site.

Within the landfill, the aerobic and anaerobic microorganisms will decompose the refuse into natural earthy materials with the production of methane, carbon dioxide and carbon monoxide. Hazards associated with these gases as with the settlement of the refuse complicate the procedure but may be overcome with careful planning. Once landfilling operations have been terminated and final settlement has occurred, the land can be developed for low intensity uses such as recreational areas and parks. Sanitary landfilling costs are always less than $5 a ton and for large sites may be as low as $1 a ton. Nevertheless, according to the last survey of U.S. landfill sites, only 6% of all sites were judged sanitary. The prevalence of bad examples have discouraged the acceptance of landfilling as a sound method of solid waste disposal and as a viable means of reclaiming land for public use.

Other methods that have been used for disposal of refuse include hog feeding, ocean dumping and deep well disposal. Environmental and health effects associated with each of these have made them less desirable than sanitary landfilling.

## Recycling

Resource recovery or recycling has become an integral part of present approaches to solid waste problems. Recycling may be defined in this context to include: 1. the reuse of materials in their original form; 2. materials reused in some altered form; and 3. the treatment of all wastes and the reclamation from the residues of fractions, to be used in the secondary materials industries. An example of the first is the returnable bottle which may be reused over and over until breakage occurs. The use of glass as a paving material is an example of the second definition and the third is often called the total system approach to recycling.

A typical total resource recovery system grinds up the refuse, separating out those portions that may be easily reclaimed from the solid wastes. The remainder is either pyrolyzed or incinerated. The heat content of the refuse is used either at the treatment plant itself or is sold. All salable fractions of the residues recovered before and after the refuse is treated, are also marketed. A system that is operating well should be able to reduce the quantity of the wastes to only 1% of the original. A pilot plant in Franklin, Ohio, has been demonstrating the effectiveness of this approach and has met with only partial success. Many engineering difficulties must still be overcome before the system becomes fully operational on a large scale basis and attains the reliability needed to treat municipal wastes.

## Legislation

In 1965, Congress passed the Solid Waste Disposal Act providing local governments with capital grants for demonstrating projects and the upgrading of existing collection, treatment and disposal systems. This Act was amended in 1970 by the Resource Recovery Act which places emphasis on the development of recycling techniques as an accepted part of the government's approach to solid waste management.

Future legislation promises to center on curbing the generation of solid wastes. Use and disposal taxes are being considered both to discourage the wasting of resources and the production of wastes as well as to spur the use of recycled materials. Solid waste pollution is now becoming a major problem in the United States as urban areas are faced with an acute shortage of landfill space for the disposal of their refuse and incineration is viewed as a threat to air quality which in many areas is below the federally mandated standards. Adequate solutions do not presently exist and proven technology which is environmentally sound is not yet available. Greater commitments will have to be made by both the private and public sectors before substantial and effective progress will be made.

Louis Slesin, M.Sc.
Scientists' Committee for
Public Information
New York, New York.

# Chapter 27

# Transportation and Pollution

In ancient times, the man who travelled 50 miles from his home was something of a world traveller. Today, the man who commutes 50 miles daily to his job is a commonplace. Society, in the space of a few centuries, has evolved from one where almost all food and goods were produced by farmers and workers living next to their farms and workshops and then consumed by local people, to one where food and goods are routinely drawn from hundreds and even thousands of miles away. Fast rail, truck and air travel have made it possible for residents of Chicago to eat Mexican tomatoes the day after they are picked while New Yorkers routinely drink orange juice squeezed from Florida fruit. Vacationers think nothing of jetting from the United States to Switzerland for a week of skiing while others drive hundreds of miles each weekend to neighboring ski resorts.

This mobility has strong economic implications. A factory or office located in a metropolitan area may be within commuting distance of a million or more potential workers. With this kind of pool to draw upon, the employer can usually fill even the most demanding position, while the person looking for a job has a large number of companies among which to search. The system also allows concentration of talent and production facilities in a few central locations. A corporation can manage a large number of far-flung plants using just a few top executives, with the executives staying informed by mail, telephone, and—when necessary —by a quick air flight to a trouble spot. Similarly, the output of these factories can be shipped throughout the country or the world, allowing for a concentration of production means. The first legal attempt to desegregate America's public schools has relied upon daily school busing, thus making the public school busing fleet the world's largest transportation system.

This great mobility is not achieved cheaply. Transportation consumes 20% of our total energy, produces much of our air pollution and noise, kills over 50,000 persons a year in accidents, and takes up a minimum of 30% of the land in our cities.

The large and still growing environmental impact of transportation is due largely to a continuing shift from the physically efficient train and bus to the physically inefficient car and jetplane. In general, the car and plane require more land, consume more fuel, produce more air pollution, and generate more noise than the train or bus. In many cases, the differences in physical efficiency are immense: the auto requires ten to one hundred times more land to move the same number of people the same distance than does the train or bus, while the plane creates ten to one hundred times more noise than do ground transportation systems. Planes landing and taking off at New York's Kennedy Jetport are estimated to disturb more than one half million people. Since every metropolitan area has at least one major jetport, noise from air travel disturbs many millions of Americans daily.

But while increasing use of autos and planes has in many cases degraded the quality of life of non-travellers, it has increased the quality of life of travellers. The auto is faster, more private and more comfortable than existing means of mass transportation. The jet plane, by increasing travel speed, has decreased the hardship of travel. So the desire of residents in an urban or suburban community to eliminate air pollution by reducing auto use or to reduce jetplane noise by curtailing or banning flights runs counter to the travellers' desire for increased speed and comfort.

If the costs and benefits of transportation were distributed among the same people, there would be less of a problem. If heavy users of air transportation lived near airports, then the people who benefit from air travel would also suffer from its effects. But things usually do not work out this way. Air transportation is most heavily used by people in the upper income groups and they generally do not live near airports or other sources of noise and pollution. This maldistribution of benefits and environmental costs is most marked in the case of the Interstate Highway System. As originally conceived, the Interstate System was to provide fast, safe highway travel between major U.S. cities. In practice, the system has penetrated into cities and many Interstate roads are used primarily for daily commuting; suburbanites living on the outskirts of cities use the roads to commute to their jobs in the downtown area. It is dangerous to generalize about cities, but many of them consist of a central business district surrounded by an older and decayed inner residential ring, now usually a ghetto for minority groups. Further out is a ring of prosperous suburbs with many of the inhabitants of this ring commuting to central city jobs via Interstate roads.

The construction of these roads through ghettos reduces the available housing in these areas. Since many of the ghetto inhabitants cannot move for economic reasons or because they are discriminated against in the housing market, the reduced housing must absorb the same number of people, increasing crowding. In addition, although the roads cutting through the city may increase the mobility of the suburbanites, it decreases that of the inner city dweller, who is less likely to own a car and who now lives in a community physically split by a six lane limited access road. The road may also increase air and noise pollution in the area although in some situations it will draw traffic from local streets, improving conditions.

As a result of the spreading transportation-related environmental problems caused mainly by increasing use of automobile and jetplane, a demand has developed for more use of mass ground transportation. This is actually an attempt to reverse a trend. The auto over the past few decades has all but eliminated bus, trolley and commuter rail systems in many metropolitan areas while the airplane has done the same to long distance train and ship travel. But it is now argued that the large scale adverse environmental impact of auto and plane require a return to mass transit. To judge the depth of this need and the improvement that can be expected from increased use of mass transit, let us compare the environmental impact of the auto and plane with that of the train and bus.

*Land use.* The auto requires between ten and one hundred times as much land as a train or bus system to move equivalent numbers of people. In other words, a highway lane devoted to passenger cars has 1/10 to 1/100th the passenger capacity of a highway lane devoted to buses or to a train track. Passenger capacity is the number of passengers that move past a fixed point in an hour. It refers to the flow of people and has nothing to do with the length of the road or track or with the number of people or vehicles on that road or track at any one time. To see what determines the passenger capacity of a highway lane devoted to

autos, let us do a simple calculation. If the cars are moving at 60m/hr., a rule of thumb (one car length for each 10m/hr. of speed) tells us there should be about 100 feet between cars. Since the cars are moving at 60m/hr. (88ft/sec) an observer on the side of the road will count a car about every 1.2 seconds. This means that in one hour about 3,000 cars will pass. Because cars move at differing speeds, pass each other, and must enter and leave the highway, peak capacities are closer to 2,000 cars per hour than to 3,000. In rush hour traffic, the average car carries just the driver (non-rush hour traffic has about 2 people per car). This means that the maximum passenger carrying capacity of a highway lane during rush hours is 2,000 persons per hour. If the cars were carrying the over-all average of about 1.5 persons per car, the peak capacity would be 3,000 people per hour.

By contrast, a ten car subway train in rush hours carries 200 persons per car, for a total of 2,000 persons per train. The trains commonly run as close as 30 trains per hour, for a capacity of 60,000 persons per hour per track. This gives a subway track a 20 to 1 advantage over a highway lane before the land needed to park the cars is calculated. However, there are difficulties with subways. For safety reasons, the speeds during rush hours are low since the trains are run very close together. In fact, during rush hours a subway system resembles a conveyor belt more than a railroad. In addition, packing 200 people into a car causes a great deal of discomfort. As a result, most people who can afford it, change to automobiles.

In an attempt to overcome these disadvantages, the San Francisco area has constructed a commuter rail line called BART (Bay Area Rapid Transit), which is meant to provide a fast, comfortable trip in order to attract people from their cars. When operating at full capacity, BART will run trains every 90 seconds, carrying 72 seated passengers (no standees) in each car of the six car train, for a total of 432 passengers per train. Since there will be 40 trains an hour, a BART track will have a capacity of 17,280 passengers per hour, the equivalent of six highway lanes. Since a railroad track is about the same width as a highway lane (12 foot right-of-way), BART needs no more than 1/6th the land needed for auto travel. BART trains will average 50m/hr., making them competitive with the auto in terms of speed. But rail transportation is at a disadvantage in the low density suburbs. No matter where the line is built, few people will be within walking distance and its

249

high volume capacity will be wasted. Transportation experts, therefore, believe that only the bus has the possibility of competing with the auto. The bus can move through low density areas picking up commuters and then travel via highway to the central city. Since each bus can carry 50 seated passengers and since a highway lane can accommodate at least 800 buses per hour, a lane devoted to buses has a theoretical capacity of 40,000 persons an hour, figure 114.

### Energy Use and Air Pollution

The movement of people requires energy. In fact, transportation consumes approximately 20% of the total energy used in this country. The consumption of this energy results in air pollution. In the case of the auto, bus, and plane the pollution is emitted directly by the vehicle. Electric powered commuter railroads and subways emit no pollution themselves, but the electric power plants do pollute.

Energy efficiency is expressed in passenger-miles per gallon of fuel. A passenger-mile is generated when one person travels one mile. If a car carrying two people travels 25 miles, 50 p-m's are generated. Table 13 lists the fuel efficiencies for

**Table 13.** Transportation Fuel Efficiencies

| Transportation Means | Fuel Efficiency —P.M./Gallon |
|---|---|
| Helicopter | 1.5 |
| Supersonic Plane | 13.6 |
| Boeing 707 | 21 |
| Boeing 747 | 22 |
| Automobile | 32 |
| Subway | 75 |
| Commuter Train | 100 |
| Bus | 125 |

(Source: "System Energy as a Factor in Considering Future Transportation," by Richard A. Rice, ASME Winter Meeting, 11/29-12/3/70, N.Y.C.)

several types of transportation vehicles. In the case of electric powered vehicles, the electric energy they consume has been translated into equivalent amounts of gasoline.

**Figure 114.** Cars moving at 60m/hr. need a buffer space between them of 90 feet. As the diagram shows, a passenger car occupies less space—and holds upward of 60 people—than an auto, which holds an average of 1.5 people. Since a highway lane and railroad track each have 12 foot rights-of-way, this gives raid transit a big space advantage over autos.

Clearly, air transportation is the least efficient user of fuel, followed by the auto. Subways, buses and computer railroads are fairly efficient. The more fuel used, of course, the more damage due to such things as strip mining of coal and oil spills and the faster we deplete our fossil fuel reserves. But the largest impact of fuel use is air pollution. It is estimated that in 1968 the auto produced 40% of America's air pollution, emitting 59 million tons of carbon monoxide, 16 million tons of hydrocarbons (unburned gasoline), and 7 million tons of oxides of nitrogen. An auto without any pollution control devices emits about one pound of pollutants for every three miles travelled. The hydrocarbons evaporate from the gas tank and carburetor and are emitted as crankcase blowby or from the exhaust. Carbon monoxide and nitrogen oxides are formed during the combustion process in the cylinders and are emitted in the exhaust.

Table 14 lists the various pollutants emitted by uncontrolled autos and the upper pollutant levels

**Table 14.** Automobile Generated Air Pollution in Grams/Mile.

| Source Model Car | Uncontrolled | 1968 | 1970 | 1972 | 1975 | 1976 |
|---|---|---|---|---|---|---|
| Exhaust hydrocarbons | 13.1 | 5.6 | 3.5 | 3.1 | 0.46 | 0.4 |
| Carbon monoxide | 110 | 50 | 35 | 31 | 4.7 | 4.1 |
| Crankcase blowby— hydrocarbons | 4.1 | 0 | 0 | 0 | 0 | 0 |
| Fuel tank & carburetor hydrocarbons | 3.0 | 3.0 | 3.0 (a) | 0.16 | 0.16 | 0.16 |
| Oxides of nitrogen | 6.4 | 7.1 | 5.5 | 4.4 (b) | 2.0 | 0.5 |

(a) 0.49gms/mile in 1971.
(b) 2.7gms/mile in 1972 and 1973.

allowed under Federal law for different model cars. The tests to determine whether these standards are met have been challenged by some experts, who claim that the pollution control devices degrade quickly in use and that most cars on the road do not meet the standards. It is not yet certain what standards will be set for post 1974 cars nor is it known if auto manufacturers will be able to meet stricter standards than those in effect for 1972 cars with the current internal combustion engine. The possibility exists that they will be forced to go to a different power plant, such as an external combustion engine (a steam engine) or to a Wankel rotary engine, figure 115.

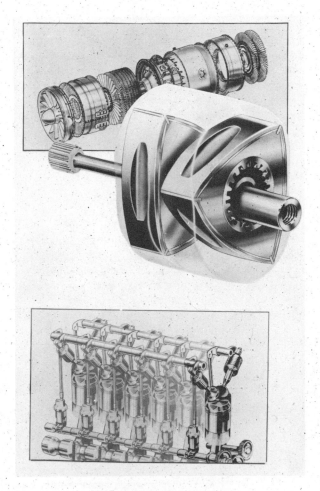

**Figure 115.** Comparison of the Wankel Engine with Turbine and Reciprocating, piston Engine. Photo courtesy Curtis-Wright Corporation.

Steam engines are very clean because they generate their energy by means of a continually burning flame which boils the liquid into vapor. The steady flame of the external combustion engine is much easier to tune than the intermittent, explosive flame of the internal combustion engine. While the rotary Wankel engine is also an internal combustion engine, and therefore dirty, it generates more horsepower per pound than the conventional reciprocating internal combustion engine. It also runs much hotter. This combination makes it possible to attach pollution control devices which can clean up the exhaust. In the case of the conventional internal combustion engine, there is simply no room under the hood for control devices.

Both the Wankel and the steam engine run very well on low octane unleaded fuel, giving them an additional advantage over the conventional internal combustion engine. Lead compounds such as tetraethyl lead are an important component of gasoline for the internal combustion engine as it reduces the combustion phenomena known as "pinging" or "knocking." Lead is both a pollutant in its own right and fouls the catalytic converter, which is one of the devices used to clean up exhaust emissions.

Comparison of auto pollution with bus or train pollution is difficult. Buses are mainly diesel powered and government standards for diesels are in terms of density of smoke emitted. The situation is also complicated for electric power plants. These power plants may generate electricity using water power or gas, both of which are very clean, or coal or oil, which vary in their ash and sulfur content. Autos emit little particulate matter (ash) or sulfur oxides while power plants produce practically no hydrocarbons or carbon monoxide. In addition, cars emit pollutants at ground level while power plants emit their pollutants from stacks which may be several hundred feet high. Despite these differences, it is still of interest to compare the overall amount of pollution given off by autos with that emitted by power plants. This is done in detail in table 15. It shows that the total pollution emitted by an average power plant in producing 100 p-m's of travel by electric powered train is $8.4 \times 10^{-2}$ lbs. A car meeting 197? standards would emit about 6.2 lbs. of pollution, or 73 times as much as the train. In addition to the other differences mentioned, it should be noted that power plants produce a great deal of sulfur oxides, which pound for pound do more harm than carbon monoxide or unburned hydrocarbons.

**Table 15.** Air Pollution Emitted by Automobiles and Electric Powered Trains per 100 P-M's.

| 1972-1974 | HC's lbs | CO lbs | NO$_x$ lbs | SO$_x$ lbs | Flyash lbs | Total lbs |
|---|---|---|---|---|---|---|
| Autos | 0.5 | 5 | 0.7 | 0 | 0 | 6.2 |
| Electric Powered Trains (a) | 0 | 0 | 0.014 | 0.056 | 0.014 | 0.084 |

(HC refers to hydrocarbons, or unburned gasoline; NO$_x$ to oxides of nitrogen; SO$_x$ to oxides of sulfur; and flyash to particulate matter.)

(a) The trains are assumed to be powered by a power plant emitting the average amount of pollutants per kilowatt-hour produced in 1970.

## Noise Pollution

Noise occurs wherever large amounts of energy are released. Energy release occurs not only in the proverbial boiler factory but also on highways, airport runways and in subway tunnels. As our use of energy increases, the noise levels in the community increases, with much of the increase caused by our increasing mobility. For example, from 1954 to 1967 average noise levels in suburban residential areas doubled and in some cases quadrupled. In addition, maximum noise levels occurring in the community increased by a factor of 64. In large part this was due to the introduction of jets and a great expansion in auto and truck traffic.

Table 16 lists some common sources of noise as well as the noise produced by various transportation vehicles. The dBA is a weighted measure of noise which takes into account sounds which humans are most sensitive to. It is a logarithmic scale and a 3dBA rise corresponds to a doubling in sound intensity. Table 16 illustrates why existing and proposed jetports have been fought so strongly by citizen groups across the country. A jet flying over a community at a height of 1,000 feet is 6 times noisier than an air compressor at 20 feet. In addition, the jet will disturb tens and in some cases hundreds of thousands of persons.

The area around Kennedy Jetport in Queens, New York, has perhaps been most intensively studied for noise impact. Instead of the dBA scale used in table 16, a scale has been developed which is directly related to human response. The Noise Exposure Forecast (NEF) scale predicts the human response to a given noise level. For example, exposure to NEF 30 increases nervousness and irritability, makes it more difficult to concentrate or to relax, and will disturb the sleep of half of those exposed to this noise. About 700,000 of Kennedy's 'neighbors' are exposed to at least

**Table 16.** Some Common Sound Levels in dBA's.

| Source | Level (dBA) |
|---|---|
| Rock and Roll Band | 110 |
| Jet at 1,000 feet | 103 |
| Inside 35m/hr Subway Car | 95 |
| 20 feet from Compressor | 94 |
| 25 feet from Motorcycle | 90 |
| Prop Plane at 1,000 feet | 88 |
| 50 feet from 40m/hr Diesel Truck | 84 |
| 100 feet from 45m/hr Diesel Train | 83 |
| Clothes Washer | 78 |
| 25 feet from a 65m/hr Car | 77 |
| Dishwasher | 75 |
| Near Freeway Auto Traffic | 64 |

NEF 30, with 120,000 of the 700,000 exposed to NEF 40. Experts recommend that areas experiencing NEF 40 should be used only by businesses housed in soundproof buildings. But this is not the case in the Kennedy area. There are 35 schools with 40,000 students in the NEF 40 zone. It has been calculated for one of these schools that the students lose a full day of schooling per week due to noise. Teachers in such schools use a 'jet pause' technique: they stop talking as the jet approaches and pick up in mid-sentence once it has passed.

As our transportation systems expand, the noise generated will also increase unless steps are taken to decrease the noise level. Automobiles, trucks, buses and their tires can be designed to be quieter. Auto manufacturers have put a great deal of effort into making the interior of cars quiet and practically none into decreasing the exterior noise generated by cars. Jetplanes can also be quieted and in addition the total number of flights can be reduced by building larger planes (2,000 seat planes are possible today) and by scheduling fewer flights so that average seat occupancy rises from its present 55 to 60%. If planes instead flew 90% full, one out of every three flights could be cancelled.

## Solving the Problem

Even after the environment-transportation problem has been characterized to the nearest million tons of air pollution and db's of noise, disagreement exists about the broad outlines of the solution. A possible approach is the purely technical one, in which—for example—new model cars are built to emit less pollutants and noisy planes are gradually taken out of service to be replaced by quieter ones. In general, this is the path we have so far been taking.

A more radical type of solution would consist of banning autos from cities and replacing them with mass transit. Jetports would either be moved far away from cities or air travel would be replaced with high speed ground transportation.

Still others suggest that we must begin by deciding how we want to build our metropolitan areas and then build transportation systems to fit the agreed upon pattern. They say that at present development follows the highways and so we are letting the road builders decide how urban areas will develop. Because of this lack of over all planning our urban areas have sprawled and we have maximized our transportation needs. This built in need for a large amount of transportation has in turn led to our environmental-transportation problems.

But even among those who agree that planning must come before construction of transportation systems, there is no unanimity. Some see metropolitan areas remaining essentially suburban in character, with the auto and to a lesser extent the bus being the main means of transportation. Others see the need for so-called 'new towns' with homes, jobs, and schools in close proximity to eliminate the need for transportation. Still others say that as communication (picture-phone, cable television, electronic mail, etc.) improves, the need for physical movement will decrease. They suggest, therefore, planning for a communication centered city, which will be different from one organized around physical movement. The type of solution a person favors depends in part on how he or she perceives the present problem. To some, such things as noise and air pollution are comparatively minor problems afflicting what is a basically sound, efficiently functioning transportation system. These people, therefore, favor a narrow technical approach to eliminate the few environmental problems. Others see the same noise and air pollution as symptoms of an inefficient and physically and socially destructive system. They view the car and —to a lesser extent—the plane as dangerous, wasteful machines which are consuming our resources, our land, and our peace and quiet in the name of some meaningless mobility. Obviously, the questions raised are not just technical. They involve values and the solution to the problem will depend on the values the nation brings to bear on it. In this sense, the environmental impact of transportation is similar to a host of other problems we face.

Prof. Edwin H. Marston, Ph.D.
Department of Physics
Ramapo College, Mahwah, New Jersey

# Chapter 28

# Ecological Engineering

Ecosystems come in all sizes, shapes and degrees of both complexity and interaction. Each —whether it is a meandering rill, a mighty river, a farm pond, or even aquarium, an inland sea, an immense ocean, a woodlot, a trackless forest, or even a hamlet or megaloplis—can be considered and is essentially a dynamic, living entity which like other living entities is more or less but never simply equal to the sum of all its parts. Today, we are everywhere immediately accosted by vital problems of pollution and the resultant degradation of ecosystems, especially aquatic.

The survival of our very civilization and, possibly our lives depend upon the solution to these problems. There are three basic causes: The first two are, burgeoning human population and the fact that modern "western" culture is fundamentally based on high power output, with accompanying low efficiency.[1,2] For an example of this power/efficiency problem: Our high yield agriculture which is becoming ever more vitally necessary, basically amounts to converting fossil fuel[3] into food inefficiently in terms of energy. Because energy is extremely cheap at present, we mistakenly delude ourselves that efficiency has to do with time and dollars. Consider the energy budget of mined Chilean fertilizer, insecticides, etc., all put into a corn field in Iowa. Simple addition of all the calories (units of energy) utilized in mining or manufacture, transportation to, and finally application to the field itself brings one to the realization that all these calories, wherever actually expended, can be considered concentrated in that field. This results in perhaps 4 to 6 times increase in yield but (energetically) at very inefficient cost vs. the horse and plow.

This brings in the third basic cause: The short future of "world carrying-capacity," specifically, how many of us this really rather tiny globe can support on any sustained basis. It is true even considering nuclear reserves that unless some dramatic breakthrough is made in the sources or resources of energy. Such a breakthrough is almost completely unforeseeable today except through faith alone, in itself a very slimming diet. Lack of this breakthrough remains a specter even should

carrying capacity (energetically) be maintained or even increased by limiting pollution and stopping or dramatically slowing the degradation of natural systems and recovering those already badly fouled. Except for the possible "lemming" effect and mutual human extermination, the prophets of doom are completely wrong though the death-rate will dramatically increase and Ehrlich's[4] famines will occur. Human overpopulation will not in itself lead to human extinction but overpopulation by consumers does lead to environmental degradation and diminished carrying capacity for those consumers. The study of ecology is replete with numerous field[5,6] and some laboratory examples.[7] Coupled with our fossil-nuclear fueled, short-term, artificially inflated carrying capacity: An eventual population crash is obvious and unavoidable but human population will stabilize at a new, drastically lower level relative to the new, altered and "fuel-less" carrying capacity, having initially overshot that mark by some considerable margin. Assuming even that carrying capacity will be cut by 98% to about 60 million people world-wide and that at the interim low point it approaches even 5% of that figure or 3 million people world-wide: Biologically neither is by any means a small population in danger of extinction. However civilization as we know it would be extinct.

As ecologists and population biologists we can advise, predict and help popularize solutions to the strictly human numbers problem. Desmond Morris[8] in popular fashion nicely summarizes much of our problem in this respect. However, actual solution is obviously the job of governments acting in concert or coercion to public attitudes and practices. Similarly, the new energy source question is out of our sphere and must be left to other experts—excluding those who would cover the earth with solar panels that because sunlight comes in on an areal basis we are always unpleasantly faced with the second law of thermodynamics.

So long as energy has not yet become limiting *we* ecologists and population biologists, engineers, sociologists, economists, urban planners, etc., and private citizens can directly attack environmental degradation, show how it may be reversed and proceed to do so. Retiring American Association for the Advancement of Science (AAAS) president Spilhaus' recent call for development of a new "ecolibrium"[9] is completely correct. So too is his demand for both new "atmoculture" and "hydroculture" analogous in principle to our modern ". . . culture of our renewable plant and fiber crops which we call agriculture . . ." Indeed, if we do not, others may not have time to work out solutions to the other two basic problems and latter day carrying capacity may even be considerably less. In reality it is this possibility which may lead to the extinction of the human species, nature's recovery lag-phase being what it is.

Somewhat belatedly we are beginning massive programs to actually minimize pollution in many areas, rather than merely somewhat de-germ and cosmeticize our environment it as in the past. The mechanism is through greatly enhanced pretreatment, i.e. more, larger, and better sewers, treatment plants and the like. While vitally necessary, these steps alone are not sufficient though they can buy time by ameliorating the people's effect on systems not yet completely fouled. Given time and a static or diminishing population they might even be sufficient for many systems still only marginally polluted. However, populations are neither static nor diminishing and what is to be done for those large systems already badly fouled? Natural recovery lag-time for these must be measured in generations, Lake Erie is but one prime example.

It might be said that our environment is going downhill at an ever increasing rate towards an eco-disaster at the bottom, in one way or another. Further, as in a careening trailer-truck in similar straits, we are now attempting to brake via greatly enhanced pretreatment of our wastes as noted above. However, even at the outset we must remember two things: *First*, that whatever we do we are putting energy into the environment, into one system or another, as in our oversimplified trailer-truck example braking energy might be applied in either tractor or trailer or better yet, both. *Second*, life and ecosystem processes exhibit curves. Often these curves are sigmoid, "S-shaped" in our truck anology. Unfortunately we cannot yet jump clear to another planet, so we had best begin to steer, that is, ecologically engineer. Ecological Engineering was defined some years ago by H. T. Odum et al[11] as ". . . environmental manipulations by man using small amounts of supplementary energy to control systems in which the main energy drives are still coming from natural sources." In analogy

to mechanics its essence may be said to be based on sufficiently strong ecoleverage in conjunction with properly placed ecofulcrums. Also, because ecosystems are dynamic, living entities "moving" toward climax, eutrophication, and such that, in a sense, we must further incorporate the concept of ecomomentum.

Grossly oversimplifying for conceptual purposes, Ecological Engineering then becomes something like jujutsu whose successful application is also a direct function of momentum, leverage and fulcrums, properly applied. In essence, jujutsu amounts to using your opponent's momentum or off-balance potential energy against him. Finally, like jujutsu, no two situations or systems are necessarily alike and though the method and principles generally applicable, entirely different specifics may be called for.

Ideally and ultimately, ecological engineering and environmental engineering are the same in the sense that we must begin to engineer our environment, but in practice, they are quite different. Civil engineering colleagues insist that environmental engineering already exists. It is a flourishing, well-funded, rapidly developing discipline by virtue of officially changing the name of the field previously known as sanitary engineering.[12] As such, this field is rich in technological know-how and is immediately concerned with the immense environmental problems we face. These include urban and industrial pollution, acid mine drainage and many other areas of concern. Unfortunately, know-what and know-where are still lacking in large degree; both being dependent upon more and better data, inter-disciplinary synthesis and basic research. We see around us the effects of past shortsighted and uninformed interference with ecosystems. Concurrently acting and interacting with our now "environmental engineers" is the U.S. Army Corps of Engineers, also with tremendous technical know-how and apparently with some sort of fetish to either dramatically impede or assist gravity in its action upon flowing waters. Environmental engineering today is at worst a fiction, at best embryonic. Its development can only await, or be concurrent with, what we must now call ecological engineering and although imperative to our future: ecological engineering does not yet exist except in very rudimentary form. Ecological engineering demands, at the very least, empirical and statistical predictability in dealing with and utilizing in optimal fashion the world in which we live in all its systems, levels, interactions and complexities. Decisions must be based upon accurate knowledge

and reliable prediction concerning eventual implication and effects of each relevant alternative. It
being obvious that there are several alternative
ways to achieve a given optimum, of which we
would naturally desire the best result. One other
point may not be so obvious. That is: There may
well be, and probably are, several optima—possibly
even as many as there are people seriously considering any given case. However, decisions
concerning *which optimum* are, and must be
political in the broadest and best connotation of
the word. In short, having once decided which
optimum is desired, relative to any given case,
whether by referendum or representation, those
we delegate to make the necessary decisions and
ovrsee the work must have quantitative, reliable,
real-time odds in reference to each alternative or
combination thereof.

At best, ecological engineering also implies
fundamental syntheses and comprehensive understanding. Today we stand somewhere in between,
far short of either. It can be said that environmental
decisions without benefit of a full deck are the
rule today, generously considered largely by how
we must play the game. We have considerable
knowledge of how many cards comprise the deck
and some knowledge of their distribution. Unfortunately even today we are forced to play with most
face down. Bringing more cards face-up into the
hands of those we delegate to play for us is therefore essential to rapidly deducing both actual rules
and successful strategy. Put another way, our data,
of necessity, are generally narrow in scope, non-
continuous and short-term, in a word, spotty; our
syntheses, models and theories limited and unequal
to either task. Just now via erudite committees are
we even beginning to presume definite standards
for water quality, largely on the basis of obvious
need[13,14] and somewhat in the vein of locking the
barn door after the horses were stolen. The result
might be said to be a nothing more than a "camel,"
the definition of which is ". . . a horse designed
by committee."

Concurrently we are at last beginning to apply
current technology to the problem.

Since today almost everyone appears to be or
calls himself an ecologist, some short crystalization
of descriptive modifiers appears in order. Here the
term "applied" denotes those senior, field and
technical workers primarily concerned with specific
individual or general problems—the gatherers of
data and writers of technical reports for specific
agencies or branches of federal, state or local
government. These include fisheries and pollution

biologists, public health people, etc. "Basic" here
refers to "pure" researchers whether in academia
or government and the line separating each from
the other is very hazy and ill defined.

It is regrettably important that "basic" versus
"applied" ecologists, those of greater and lesser
eminence and the supposed experts in dealing with
natural systems, have effectively been last on the
scene in relation to the problems of "pollution."
Perhaps we "true" ecologists have been too concerned with the dynamics of individual tree growth
to note the approach of the approach of the woods-
men, complete with chain-saws and caterpillar
tractors. Why? we must ask.

Even today, many "pure" ecologists still regard
the intrusion of the human element into their
natural systems as being most unnatural, opera-
tionally ignoring the fact that today such "natural"
systems are (or at least soon will be) the unnatural
ones. This truism is reflected in our participation in
the International Biological Program.[15,16,17] By no
means is this statement meant to, nor does it
detract from the fundamental value of such studies.
It does suggest that we had best also initiate
analogous programs in populated areas.

Fortunately, some scientists who may almost be
dubbed "ecoprophets" have been preparing the
way. Slobodkin's,[18] Jordan's,[19] Ehrlich and Holdren's[10] and Deevey's[3] recent contributions are
most welcome and extremely important in just this
respect. They herald the approach of the basic
ecologist in this vital area, contribute significantly
to the development of seeds sown by H. T. Odum
and others and, finally, describe some of the
factors necessary for the continued maturation and
eventual flowering of ecological engineering.
Slobodkin emphasizes automated physical and
chemical environmental monitoring, emperical
(e.g. multivariate) analysis and the development of
more and better trained environmental scientists.
Jordan, on the other hand, stresses the development of environmental scientists and introduces the
extreme relevance of human behavior and ecology.
Ehrlich and Holdren introduce several important
axioms and concepts including: the threshold
effect, non-equality of "bodies" (i.e. differential
environmental effect of different people, population
types and populations/unit area) and that population control while vitally necessary, is not sufficient.
Deevey leavens the loaf with insight, perspective
and humor.

Because we must deal with people both as
individuals and members of groups, this becomes
a fundamental question. The answer can only be

implied as yet. It has not been fully formulated. Perhaps the best alternative is to cite what should *not* be done.

Failure of the University of Pennsylvania's Institute for Environmental Studies, WRA and Regional Science Research Institute's *Plan and Program for the Brandywine* provides almost a classical example. As reported,[20] obligatory easements to the Chester County Water Resources Authority were fundamental to the plan. Two socio-political axioms were ignored. The first assumes that Machiavelli[21] on "patrimony" can be translated into the present frame of reference as, "People (especially Americans) are and will be more concerned with property rights than with 'conservation' and/or 'pollution.' " Currently, perhaps more acceptably, it can be termed "the territorial imperative." The second is dredged from the dark recesses of undergraduate political science lectures. Even partial state or agency ownership is not necessary for control. This second principle can be considered fundamental to capitalism as we know it, perhaps biologically the best yet reconciliation of territoriality with complex social needs.

A good analogy may be made between the activities of ecological engineering and medical science.

To stress not only the importance but, as well, the kinds of elements we need in imperfect analogy, ecological engineering may be compared to medicine if we view it as modern human engineering. Temperature in the ecosphere is analogous to the temperature of the patient as it deviates from normal. Quantity and transport data equals pulse rate and/or electroradiograph. Diurnal variation in oxygen[23,26] and pH ($CO_2$),[7,27,29] the metabolism of the system, approximates gross symptoms exhibited by the patient. Species diversity compares with red and white blood cell count, type and ratio. Average concentrations, and non-diurnal changes in the levels of phosphate, nitrate, oxygen, pH, light, electrical conductivity, turbidity, etc., compare with urinalysis, blood chemistry, electroencephalographic and a myriad of other tests.

Practitioners of space medicine recognize the value of these measurements and continuously monitor diverse variables in the human subjects they manipulate. On the other hand, ecological engineers have only the first two measures; emphasis being on amount of water, water potability and flood control. We have no vast backlog for range of normalcy in others; only short-term or periodic spot checks here and there; for exam-ple, the Pennsylvania Department of Health water monitoring system which by present standards is extremely good, consisting of 175 stations, most sampled quarterly. From what little we have we may deduce general condition but are neither is a position to reliably diagnose nor prescribe, to say nothing of understand. Relative to the progress of modern medicine we stand somewhere in the late 19th century.

Four examples will serve to illustrate what some of the problems are like and how far we have to go.

(1) In cases of acid mine drainage: The question is, which will suffice, chemical neutralization and/or biological control? For example: Is there some organism, perhaps a plant which requires low pH (high acid (alkaline conditions) modifies pH upward and cannot tolerate high pH)? Off the top of his head, Dr. W. E. Manning of Bucknell University suggested cranberries. Should they work for this purpose they might well be ideal, providing new economic gain in the area while at the same time ameliorating acid water conditions in the upper reaches of various streams and tributaries. While not suggesting that cranberries are actually the answer, they do conceptualize the sort of thing being suggested. Chemical methods of neutralization are enormously expensive in the long run and, at least the simplest, add salts to the system— either dissolved or as piles. Are we substituting one type of pollution for another? Biological control of this problem is virtually uninvestigated[30-32] but should be earnestly pursued, either alone or in combination with a chemical solution. After enough microcosm (i.e. "little worlds" in which we can virtually play God) studies have been completed,[7,25,28] fieldwork could begin in pilot areas.

(2) The case of oil spills:[33-36] Recent reports indicate that use of detergents to disperse oil from the "Torrey Canyon" disaster did more damage to marine life than did the oil itself when left undisturbed.

(3) Active de-eutrophication of large polluted systems. At this time, what is still meant is "research toward" this end, again using microcosms. One approach might be air-bubbling, where forced air is pushed through the stagnant, standing water. Unfortunately, though, there have been many bubbling and artificial de-stratification studies pointing toward "water quality improvement" in small and moderate-sized natural systems, virtually none deal with the biology and dynamics of what actually happens under such circumstances. Previous work by Kamann's group[37] aerating large

(6′) diameter vertical fiberglass cylinders in the State University of New York at Brockport's dystrophic McCargo Lake showed several extremely important results of broad and significant implication. These are: (a) Unaerated control cylinders do not operate significantly differently than the lake as a whole, (b) periphyton (the stuff that grows on exposed surfaces) standing crop organic matter biomass in aerated cylinders greatly exceeds that in unaerated ones and open water, generally by 3 to 5 times, (c) both indirect and some direct evidence that primary productivity (Pg) is decreased by aeration in many cases and, (d) strong evidence that the observed biomass increases are in fact due to "pumping" "fossil" organic matter off the bottom in an aerobic environment where appropriate, "desirable" organisms can live to utilize it.

The ultimate implications for a large polluted aquatic macroecosystem like Lake Erie are abundantly clear: Taking "fossil" and current pollution and sending it up through the trophic levels to be removed as usable export, strongly and positively affects economic feasibility of reclaiming such systems. This is especially true when we consider that this method offers the least expensive, most promising mode of energy input throughout large systems on a 10-20 year sustained basis. But, unfortunately for accurate prediction, detailed dynamics and hard data giving all facets are needed, and work thus far has been in open cylinders not completely isolated from the lake. This is because of settling, and other phenomena. Further, how can one measure primary productivity in continuously air-bubbled systems? Standard methods are Carbon-14[38], and diurnal variation of either oxygen $O_2$[23], or pH[27]. [14]C values are open to serious question except when related to each other[39-52]. The answer is extremely simple: We must replicate the closed systems in pairs with the open ones, each closed system having an apical clear plastic "bubble" whose "air" is pumped to the bottom of the cylinder for aeration. Monitoring [14]C productivity and standing crop of organisms in both open and closed cylinders and diurnal variation in $O_2$ and pH in the closed ones will give good data and summarizes the basic concept. Open [14]C rates can then be accurately corrected, meaningful turnover rates calculated, and other basic measurements made.

In practice this will require at least two additional pairs of cylinders, in addition to the pairs already mentioned. Of these eight cylinders three pairs will be vertically circulated, two by air bubbling and one by a submersible pump. One air bubbled pair will have 25 mcm. plankton netting or finer at the sediment-water interface. Energy input from the submersible pumps will be equal to that by air bubbling. In this way contributions via air vs. pump and organic matter vs. nutrient leaching may then be measured and quantified so that the data may be related to the whole. Although this sounds a bit complicated, it is still back in the horse and buggy era when compared to our knowledge of medicine.

(4) The problem of human population density and pollution: the numbers of diverse species which can be found in streams relates inversely to the nearby human population density[53-58] but with considerable variance. What kind of human population, business, industry, etc?

Perhaps we can eventually construct a "population type" pollution index, weighting various types and refining the significance of past empirical relationships into a predictive tool. This scale, used in conjunction with similar analyses on environmental aspects, essentially a complementary tool, could lead to specific recommendations and alternatives regarding "zoning" of population type, density and/or kinds, time of year and degree of water treatment necessary in any given area to preserve species from disappearance. Extending our analogy to include military medicine: It immediately becomes apparent that our emphasis on new sewers, treatment plants and pollution control while vitally necessary, is not nearly sufficient. These compare with splints, bandages, sulfa and penicillin on the battlefield. The best we can call ourselves today is ecocorpmen. We must rapidly set about learning to become ecophysicians.

Since we know what we need for the success of our ecological engineering endeavor, we must look more closely at a number of other major and minor sorts of data which are not now available but which must be gathered. They are "people"-data, breadth and mass-data, and the system to relate them. In effect once we have the first two, the third will be the method by which we draw the information together and arrive at valid measuration.

*People-data,* people pollute and it is now obvious that different people, and population types, affect the environment differently. Further, from what has been said above, it also follows that the need for ecological engineering is directly proportional to how many people there are. In short, it is obvious that people, their makings and leavings have become major variables in the delicate "equations" of natural systems. Further, in natural systems the whole is not necessarily simply equal to the sum of its parts, and it is fundamental to science that discrimination beyond the resolution capability in any system is false. We need some way to tie human variables to ecosystems at all levels in meaningful, quantitative and effective fashion—a hierarchial, national grid system.

Point data, while excellent for retrieval purposes in relation to many environmental and biological variables, just won't do for people. People are moving points and therefore must be considered by area. We need a system to accurately delineate ecosystem areas of all sizes, shapes, levels and combinations—in relation to the "people" occupying and/or affecting those areas. Further, any such system obviously can never completely reflect nor accurately delineate natural boundaries because we will be imposing some sort of artificial categorization on the unbroken continuum of nature. However it is possible to design a system with minimal, known error limits which in itself does not unduly illegitimize or bias the data from real ecosystems and their boundaries, so that we obtain one that can follow natural boundaries with a minimum of error. Hierarchial grid arrangement, squares within squares, and so on, is the key that will facilitate coding, retrieval and accurate cumulation, amortization, comparison and synthetic analysis of all variables for any area or areas regardless of size. This can make it possible for us to find out what we need to know within any given area, regardless of size.

Such a system is impossible today, except for very large systems like whole river basins, or each of the Great Lakes. For example: With exception of some large cities where block (or block-group) data are available, the best we can do at present is often political subdivision, census tract or enumerator district which by no stretch of the imagination can truly reflect either affairs in nature especially in subecosystems. Further, basically it is in these subecosystems where the action is. Documentation for this basic concept appears below in our amplification of a model program.

Integrated with the others this grid concept not only permits formulation of a practical rationale for pioneering in this vital field but could also provide virtual complete anonymity for statistics in comparison to the considerable amount of prying and downright nosey information, now being sought by the census people. Replies would then be addressed from grid squares, not residences.

(PU's vs. PE's): A short aside is necessary here in order to avoid confusion before formally introducing the other two major factors. The grid concept essentially deals with people or population units (PU's) or, in short: How many "people" do what, to a given area? A new factory employing X-number of people around the clock actually equals X PU's *plus* how many more in terms of additional environmental stress?

The concept of population equivalents, (P.E.'s)[59], is already developed and it is now routinely applied to various pollutants. Today, for example, we can say that company or factory Y's effluent contains so many PE's worth of phosphate/day. However, we still do not quantitatively know what such a figure means in terms of ecologic impact upon a specific environment or to the specific system to which the PE's are contributed. Also our PE standards may soon be out of date relative to current, real and active people today or tomorrow as standards and modes of living change in our society. Nevertheless they do provide at least some means of comparison and this is a step forward. Even so, the PU effect of factory or company Y is not solely limited to the sum of its effluent PE's. Even though its effluent be purified to the quality of distilled water, factory or company Y still affects the environment simply by actively being there.

In any event, when applied to the environmental systems of metropolis and megalopolis, PU seems a far more accurate and descriptive term.

Breadth and mass data is the second necessity. This really amounts to the other side of the same coin. Presuming a rational grid system for people data and hence differential PU values for different areal systems (watersheds) both large and small: The need for quantitative measures of their effects is obvious. Basically it must amount to extensive, intensive and continuous (or continuing, as feasible), monitoring of as many environmental, biological and human variables as possible throughout each and every macro- and mini-system.

Slobodkin[18] is quite right in maintaining that as yet we do not know what the best predictors are, and we had better very soon begin to find out. This point is emphatically emphasized by Rosenzweig's[60] recent work, illustrating that even with some of our classic ecological generalizations —the glaringly obvious and up-to-now accepted explanations are simply not correct.

Finally, one other point demands specific mention here. It too is documented below in the model program. That is: Masses of these kinds of data will be needed anyway for legal purposes so long as this nation remains one of law where the accused are innocent 'til proven guilty. Why not then gather such data in some comprehensive, routine, logical and consistent fashion. That is, with a plan, in order that they may amount to something more than specific evidence and dusty archives. There still remains the third requirement for the building of a system for relating and interpreting the data.

It may be said to be axiomatic in science that masses of unanalyzed data are useful only to those analyzing them. Fortunately our technological capability has recently become equal to the task. When analyzed by modern mathematical and computer methods, this mass of data can form not only the basis of detecting change in its early stages (hence applying corrective measures in time) but also for comparative synthesis, simulation, modeling and pragmatic projection[18,56,61]. Unfortunately, the precise details of these modern methods and how to fit them together must still be worked out concurrently with the others.

The three fundamental, major elements outlined above essentially amount to the mass of scut-work still needed. Put another way, they comprise the "applied" research aspect of any such a plan, essentially analogous in nature to the bulk of the actual work done by the United States Weather Bureau. However that agency's mission would be far from complete without the many theoretical insights supplied by basic research meteorologists involved both within the agency and outside of it in academic institutions. Further, without the input from trained personnel input from their research in academia, the Bureau would be hardpressed to adequately sustain itself much less make progress.

There are three other lesser requirements the importance of which becomes immediately apparent. These are: 1. More and better trained environmental scientists; 2. Basic research producing new syntheses and understanding and, finally; 3. A place to start. More and better scholars/researchers, technicians, and better basic research go together. Pragmatically these two are connected because the enlightened age of indentured servitude is not over. It is quite viable and flourishing today. We call it Graduate School and graduate students make the best slaves because they are smart, interested and willing. We are going to need more environmental scientists at all levels and all the chiefs and subchiefs must be more broadly but deeply educated[62] whatever their primary discipline in order that they better understand each other and more effectively interact. This must be so at the technician level also if the chiefs are ever going to effectively lead. We also need more and more basic research consisting of: new hypotheses, models and theories derived from intensive sampling and experimentation at the microsystem

Our need to set a starting place. It should be obvious that we must begin somewhere and that such a comprehensive approach cannot suddenly appear full-blown on the scene. Further, some few natural systems will be nearly ideal, many marginally so and most distinctly unsuitable because they are already densely populated and polluted, essentially amounting to environmental Gordian knots. These we must eventually learn to untie, but in the meantime we must prevent them from becoming even more complex problems if at all possible. They are not generally the best places to start, because they would overtax our technology while we are still trying to construct it. Therefore we must ask on how big a scale we should begin. All other things being equal, the answer to this question is primarily economic and a matter of choice. In essence this need for a start and its requisite alternatives can probably be best expressed using real examples: One large-scale, one small.

A prospective Maxi-Approach: Part of a major river basin would be the logical choice, but few such areas exhibit that unique combination of characteristics necessary. This is especially important if the syntheses reached are to be applicable to the entire system and even larger systems. The middle Susquehanna River Basin in Pennsylvania is one. It consists of major portions of the North Branch, West Branch and Central Susquehanna subbasins. This macro-system is: 1. Relatively undeveloped and under-populated resulting, at least in part, from scarring and pollution; the legacy of past practices. Virtually all major types of pollution are present. In this respect the area is not unique though in parts, some factors reach unparalleled proportions. It is unusual in that all stages are represented—from clean and almost untouched to indescribably foul.

2. Subdivided, especially through interaction of mountain barriers and human exploitation into a

differential mosaic of smaller systems. These are of varying size, complexity, degree and dynamics of interaction; providing numerous interfaces between and among all levels. These interfaces, in a few meters, often mirror similar changes in several miles of stream and provide a new, manipulatable and replicative tool to get at dynamics and causative factors.[63] All macro-ecosystems are actually this way. However, as illustrated by any topographic map of this region, few posses such broad scope or are so nicely differentiated and delimited internally.

3. Compact. Distance between extremely diverse areas is never great. Radius from a central location is never more than about 60 miles. In the near future, it could make efficient and economically feasible both extensive, sophisticated monitoring and intensive sampling of representative levels, subsystems and both pilot and experimental areas. This would insure that proper integration and perspective could be maintained, valid conclusions reached, and eventual optimal, workable solutions achieved.

4. Beginning to undergo rapid development. It is not only the last such area equidistant from the whole east coast megalopolis from New York-Washington but, considering foci of megalopoli to the northeast, north and west, it sits as an underdeveloped inverted triangle equidistant from all. The new Interstate Highway 80 traverses the region and other major highways are proposed or under construction. One immediate result has been skyrocketing real estate values as new companies and personnel move into the area. Population projections[64,65] indicate even more rapid growth in the near future. This is of fundamental importance. Not only will these changes affect the ecosystem in general but, because by nature they must be differential, different subsystems will be affected in different ways. Through time and with base-line studies[62] throughout the macrosystem we could consider various subsystems as "controls," others as "experimentals"; providing another outstanding opportunity to get at both the factors involved and their dynamic relationships throughout the macrosystem with both time and base-line studies.[62] Finally, some institution of sufficient strength, breadth, diversity and interest must exist within or very near the region and must make a strong commitment for such a program to be logistically feasible. Unfortunately, no such institution now eixists within this basin.

A Prospective Mini-Version: It is virtually a truism, especially regarding lakes, that increased population density of whatever sort will accelerate eutrophication and decrease water quality. Beeton[66] effectively, though only qualitatively, has demonstrated this for the Great Lakes. However, attempting an integrated, comprehensive program would constitute a mega-version at this time and difficulties would be compounded by the plethora of projects already being done and the different types of political territories, national, state and academic, involved. Canadaigua Lake in Upper New York State, on the other hand, is a medium sized lake with a well defined watershed and is still relatively clean while some it its neighboring Finger Lakes are already encountering major pollution problems. Further, the lake is currently subject to rapid change. Current construction of a 500-family unit condominium promises significant increase in human stress, even of itself, to say nothing of its opening the door for others.

The plan here is for: Studies to be made of past organic and inorganic sedimentation rates by short cores and radiometric dating using carbon-14; while studying present rates by means of sediment traps placed in 1 to 5 year intervals at coring station locations; diurnal productivity (metabolic) studies via $O_2$ and pH. These should be accompanied by continuing environmental monitoring at the outset, eventually becoming "continuous"; the making of current population census. Past population estimates should be made from USGS maps and other data as well as the gathering of local USDA information regarding amounts, types, and locations used both for fertilizers and pesticides.

Using this data we should eventually be able not only to extrapolate past productivity, etc., but also to predict future productivity; constantly checking prognoses and newer results against newly developing human stress. The factors making this miniversion even more attractive are that the "critical mass" need not be so large in the first place and assimilation by the parent institution is much easier in the second.

A full program should then draw all these requirements together. Given all six elements in a two-fold program combining both basic and applied research toward eventual synthesis, water will emerge as the overall unifying factor. This is dictated by its dynamic relation to man and all other living things. There must be six apparently different aspects combined, interrelated and interacted in unique fashion in a synthetic approach. It is expected that they will do so synergistically. The design is basically fitted to the maximum approach because it is easier to scale a program down rather than up. These are:

1. Extensive, Detailed And Continuous Monitoring Of Environmental Variables which are shown in Table 17. Probably all of them are needed, at least from some proportion of the stations. Initially, however, to keep individual automatic-station price within reason, "continuous" (i.e., ca. 10 minutes/hour) monitoring should probably be restricted to the 20-26 primary channels in Table 17 directly related to water and ecological energetics. The variables considered necessary are those whose importance can easily be documented by reference to a standard ecology text[6] or primary papers[23,24,27].

Present facilities and equipment such as those of the U.S. Geological Survey should not be duplicated although monitoring of *all* variables should be greatly extended, especially to smaller streams. Other "feasible" variables, such as those concerned with atmospheric pollution should not be included in these stations because all will occur at relatively low points in the system. Soon after starting up, several atmospheric monitors should be scattered at high points throughout the system for base-line data and to detect possible changes. Continuing, weekly, monitoring should be utilized for the primary variables inappropriate for continuous analysis and can easily be accomplished incidental to necessary, routine station maintenance. Continuing or periodic monitoring and/or sampling of primary aquatic biological variables should occur in similar fashion or as shown in Table 1, again incidental to station maintenance. Twice-a-year sampling is put forward as a minimum compromise with economic realities and the need for storage. The sorting and analytical recording of all such samples is not feasible. It is necessary for economy that an eventual and continual, thinning of routine samples be carried out while still providing adequate material for long range comparison and detailed analysis. One suggestion is that routine samples not utilized be canned and deposited with the Smithsonian Oceanographic Sorting Center (SOSC). All samples would be maintained for six years. After that, odd-yeared 7 to 21 year-old samples would be discarded. After 21 years, alternate even-yeared samples should be phased out. Applying our Maxi-proposal to the Susquehanna River basin, we find the total eventual number of automatic monitoring stations needed is roughly estimated at 500. However, all 500 would be required sometime soon in the life of the project, certainly about the fifth year. The best game plan would be to initially minimize the number of stations throughout

the Basin and maximize the number of stations in a few key watersheds. After about three years, the number of stations could be lessened to a minimum in some areas while maintaining base-line studies throughout, without significantly impairing information quality and synthesis. Some stations, of course, would remain concentrated in areas of intensive research, (see below) but those freed could be used for installation in new watersheds. The total number in use at any one time should not exceed 600. Station locations should be retained in the event reconcentration ever becomes necessary and long-range comparisons become essential. Applying our Mini-proposal, we find that 8-15 automatic monitoring stations should be sufficient.

The assimilated data, analyzed through multivariate statistical techniques (e.g., human population density in the area, biomass or species diversity in stream vs. degree and/or type of pollution and the like, will yield some significant empirical relationships. Computer modeling and simulation will also be possible. With such relationships and models we may then cautiously begin to predict events refining both the accuracy and delineation of many of the parameters throughout, thus becoming capable of increasingly more accurate predictions.

2. Design, Development And Testing Of Appropriate, Reliable And (over long-term) Expendable Equipment For Such Monitoring. The equipment must be portable and easily installed, unlike the heavy permanent installations in use today. These permanent installations cost at least $15-25,000 per station and, therefore, are generally used only on major streams or rivers. Significant local effects in subsystems may easily be masked. Often the specific cause and effect are generally not objectively discernible. The proper design and construction of the stations should lower the cost by at least a factor of four, perhaps further as the usage of such equipment becomes well recognized and mass production becomes feasible. Of necessity, monitoring equipment like this must come into broad usage in the near future, if only to provide legal evidence for the police function of enforcement agencies.[67] Here, as with ecological engineering, the heavy expense of real-time data handling is superfluous with court backlogs being what they are, so, too, are expensive on-site digital coders installed at each station. On the other hand, electronic sensors are cheap, most average less than thirty-five dollars ($35) per unit. Even those which are now very expensive are so only because they are new and not widely employed. In bulk they

## Table 17. VARIABLES NECESSARY AND OTHERS FEASIBLE TO MONITOR

| Type and Source | Importance | Variables | Scope[1,2] | Method | Ecological Relevance | Input, Storage and Accessibility | Component Availability | Monitor Channels Necessary |
|---|---|---|---|---|---|---|---|---|
| Environmental Aquatic | Necessary | Alkilimity | Continuing | Titration | Established | Manual, Computer | | |
| | | Phosphate | | | | | | |
| | | Calcium | Continuous | Probe | ... | Automatic, Computer | CLEC[3] | 1 |
| | | Chloride | " | " | | " | " | 1 |
| | | Conductivity | " | " | Diss. Solids | " | " | 1 |
| | | Current Vel. | " | " | Established | " | " | 1 |
| | | Depth | " | " | ... | " | " | 1 |
| | | Nitrate | " | " | ... | " | " | 1 |
| | | Oxygen | " | " | ...8,9 | " | " | 1 |
| | | pH | " | " | 6 | " | " | 1 |
| | | Sodium | " | " | ... | " | " | 1 |
| | | Sulfide | " | " | ... | " | " | |
| | | Temperature | " | " | ... | " | " | 1 |
| | | Turbidity | " | " | ... | | CLEC/SEBD[4] | 2-4 |
| Atmospheric | | Radiation | " | " | ... | " | Tanner-Soumi[10] | 2 |
| | | Rainfall | Continuing (-ous?) | Gauge (±Probe) | ... | Manual or Automatic, Computer | CLEC (or SEBD) | (or 1) |
| | | Rel. Humidity | Continuous | Wet Bulb Thermistor | 7 | Automatic, Computer | CLEC or SEBD | 1 |
| | | Temperature | " | " | 7 | " | CLEC | 1 |
| | | Wind Vel. | " | " | 7 | " | " | 1 |
| Biological Aquatic | | Bacteria | (?) Periodic Demand (Inst.) | Std. Tech. | 7 | Manual, Computer | | |
| | | Biol. 0. | 2-4 x/yr. Demand | Std. Tech. | 7,8 | " | " | |
| | | Bottom 0. | Continuing | Odum, 1957 | 7,8 | | CLEC | (1) or more |
| | | Bottom Inverteb. | 2 x/yr. | Sampling | 8 | Manual, Computer and/or Storage and grad. pt. phase out | " | |
| | | Pesticides | 2 x/yr. | " | 7 | " | " | |
| | | Phytoplankton | 2 x/yr. | " | 7,6 | " | " | |
| | | Zooplankton | 2 x/yr. | " | 7,8 | " | " | |
| Terrestrial | | Human | Periodic, | Cycle Survey and | Obvious | Manual, Computer | | |
| Environmental Aquatic | Feasible | Ammonia | Continuing | Std. Chem. Tech. | ? | Manual, Computer | | |
| | | Current Dir. | Continuous | Probe | eddy size? | Automatic, Computer | CLEC/SEBD | 1 |
| | | Bromide | " | " | ? | Automatic, Computer | CLEC | 1 |
| | | Cupric | " | " | ? | " | " | 1 |
| | | Fluoride | " | " | ? | " | " | 1 |
| | | Iodide | " | " | ? | " | " | 1 |
| | | Perchlorate | " | " | ? | " | " | 1 |
| | | Aldehydes | " | " | ? | " | "in use" PBAP[5] | 1 |
| Atmospheric | | Carbon Monoxide | " | " | "pollution" | " | "in use" PBAP[5] | 1 |
| | | Carbon Dioxide | Continuing | Van Slyke | ? | Manual, Computer | CLEC | |
| | | Fine Particulates | Continuous | Probe | ? | Automatic, Computer | "in use" PBAP[5] | 1 |
| | | Hydrocarbons | " | " | ? | " | " | 1 |
| | | Hydrogen Sulfide | " | " | "pollution" | " | " | 1 |
| | | Oxides of Nitrogen | " | " | 7 | " | " | 1 |
| | | Pressure | " | " | ? | " | CLEC | 1 |
| | | Radioactivity | " | " | obvious | " | " | 2-3 |
| | | Sulfur Dioxide | " | " | "pollution" | " | "in use" PBAP[5] | 1 |
| | | Total Oxidants | " | " | ? | " | " | 1 |
| | | Wind Dir. | " | " | ? | " | " | 1 |

1. Continuing here designated or meaning weekly.
2. Continuous here designated or meaning ca. 10 minutes/hour.
3. Commercial Laboratory Equipment Catalogs, e.g., Catalog 68, Arthur H. Thomas Co., Phila., and Catalog 16, Forestry Supplies, Inc., Jackson, Miss.
4. Specific Equipment to Be Developed.
5. Pittsburgh Bureau of Air Pollution; two stations operating, Network of 18 stations to be in operation ca. 1970.
6. Byers and Odum (26).
7. Odum (6).
8. Odum (22).
9. Odum (23).
10. Tanner-Soumi Net Radiometer, Seaman Nuclear, Milwaukee, Wisc. 53208.

would easily become available at a lower price. The necessary electronic circuits are relatively simple and the data can be polled on inexpensive, standard cassette-type tape. The use of such equipment combined with inexpensive standard tape recorders; relatively inexpensive, lightweight, rechargeable batteries, signal phasing and design modularization provide the key for economic use of the system. Necessary and expensive signal sorters and digital translators may be reduced to the mere number needed for efficient processing at the computer site. On-(sampling)-site calibration at the beginning and ending of each tape, roughly a week, and computer correction would keep field aspects simple and manageable. However, some

data channels will be lost or have to be discarded because of individual component failure, perhaps as high as 50% in the early years. Nevertheless we would still be incomparably better off data-wise than we are now.

Nereus Corporation[68] at one time had prototypes for just such equipment under way. The cost was estimated as being roughly equivalent to that of a new American car. InterOcean Systems has an *in situ* monitoring system operational with capacity for up to 7 channels.[69] The manufacturer has made assurances that when ordered in sufficient numbers price/unit will approach and eventually match a reasonable price estimate for a unit with 19–23 channels capacity, figure 116.

**Figure 116.**
Monitoring System, Photo courtesy of InterOcean Systems, San Diego.

3. Concurrent Monitoring Of Population Density, Distribution, Composition And Organization Throughout The Region By Watersheds.

Accurate data concerning these human variables, their dynamics and development are fundamental. The grid system forms the key for their use. An immense amount of data is already available, taken by different agencies and levels of government, business and private organizations in different ways, on different bases and for different purposes. Much is "attic" data in the sense that we may know where it is but cannot get to it or use it without digging through, moving or even completely rearranging in immense number of "boxes, bags and trunks." Further, both it and the rest are in the form of apples and oranges, etc., that's federal census data by tract (equaling political subdivision in most cases), state and municipal tax, license, and similar data by address. Results of uch data are reflected in the analyses and projections made, the only criteria being related to political subdivision or urban center.[64-66,70-72]

Pragmatically, continuing periodic data gathering is most economically feasible here. The summers could be utilized for active sampling and must involve at least one professional sociologist, two to four semi-professional and 30 to 50 non-professional workers. These latter could most effectively be recruited from the ranks of students or other temporarily unemployed individuals. Eventually, by example in this model area, we would hope to induce necessary slight modifications in methodology and format of data gathering by agencies, organizations and institutions involved. Should this happen, later large-scale field efforts in conjunction with the over-all program would be largely unnecessary. We could then interrelate their data. At present, there is no way to do so.

4. Intensive Investigation, Sampling And Experimental In Individual Diverse Component Sub-systems. I hesitate to presume what basic research should be done by anyone. It would seem, however, that investigation into the relationships between ecological energetics, species composition and diversity might be the most enlightening. Microcosm studies, both local and those already done[7,28] could play an important role here. Computers can be used not only to plot diurnal curves but also to empirically derive best-fit equations to such simple sinusoidal types of curves. Comparison of such equations and their parameters from both controlled environment microcosms and nature may yield new insights into the dynamics and relationships of the variables involved. Even

if not, reducing an ecosystem to an equation does seem rather elegant and satisfying. In any event, basic research will not only make use of routine monitoring data but will add new data for comparison and combined analysis. Most important will be new hypotheses and models which can then be tested. At the very least, these will aid us in more clearly delineating the variables involved, hence predicting more accurately. The line between basic and applied research will become even more nebulous as the research proceeds.

5. Computer Integration And Analysis Of Mass Data Accumulated. This will obviously require medium to large computer capability. However, complete and fully integrated software for such a system is as yet unavailable. Furthermore, a valid grid system enabling effective and meaningful integration of all factors is still to be worked out. All the operations of such a system must be fully compatible with the computer programs of STORET I and II.[73] This will require the services of highly sophisticated biometrical-programmer personnel. The hierarchial grid system must be easily expandable to national scale and should probably be metric, based upon squares; the smallest of these should probably be 0.5 km on a side. This would permit comparison and contrast of both the whole and its component parts at, between and among all levels. Only minimal warping in analysis of systems at all levels should be allowed to occur. Models of large systems could truly, and without injustice, reflect cumulative addition and amortization of both the blocks and subsystems which comprise them.

The Spindletop Research Center[74] has worked out some such system (Water Management, Information System—WAMIS) for the State of Kentucky. Perhaps it is significant that the Pennsylvania Department of Forests and Waters has awarded Spindletop a contract for Phase I, a feasibility study, for a similar system in that state. Further, the former Governor Raymond P. Shafer announced a program for computerization of all water sample data gathered by the state. The rationale for this as reported in the press[75] centers about the policing and enforcement functions of such a system. Both are indeed encouraging because the basic concepts of such a system were mentioned to the Pennsylvania Departments of Health and Forests and Waters in early 1968.

Unfortunately, Spindletop's Kentucky system is based on largescale areas approximating larger watersheds of the state. Though point recording and retrieval are included, reduction to an integratable grid system is not yet clear. Even when this is included, grid size in this system would be such that major problems in cumulative retrieval and analysis would occur. These problems will be analogous to those created by the National Oceanographic Data Center's[76] early reliance solely on Marsden Squares,[77] areas so large that data retrieval for local sub-areas promised to become lost in a deluge of print-out. Integration and analysis between all levels and components must be guaranteed. Otherwise, most facets of any such program become emasculated, contributing largely to the existing mass of inaccessible "attic" information not readily available. Resolution and discrimination, even in a microscope, are limited by its component parts. Therefore, if only in fashion of the British "Loyal Opposition," some program must not only interact with institutions like Spindletop to stimulate production of an optimal system but, should theirs prove inadequate to the task, an alternate (hopefully compatible) scheme must be immediately available, lest we become locked into our own data banks.

Fortunately the Cornell Aeronautical people have recently come up with one[78] at least, a grid system which through small modification should do the job nicely. Grid squares thereon can be as small as 0.1 km$^2$.[79]

6. Graduate training Expanded and increased in A) quality and breadth through interaction and cross-fertilization of ideas among and between departments, B) quantity and level of support in order that a sufficient amount of the best talent may effectively be recruited to do major portions of both the basic and especially the applied research involved. This second point is fundamental to success in any large university-affiliated research program and it insures that all resources are utilized most efficiently.

Basic differences in drive, contribution (i.e., responsibility, reliability, and new ideas) and cost exist between mere technicians and graduate research assistants and/or technicians, especially at the same salary level. This is because graduate students are deeply involved in the program, personally and intellectually. Further, with some planning, considerable portions of both data gathered and conclusions reached may relate directly to their own basic research (theses). Because their work is more than a job, it automatically stimulates quality and breadth. Finally, two conclusions become inescapable in connection with any pioneering model program: Not only do applied aspects form a logical vehicle for supporting basic research but *in such a*

*restricted area,* a university can do a better job less expensively than any combination of agencies.

In essence, items, 1, 3, and 5 above form a core of applied research whose results will immediately and ever increasingly be directly applicable to both the immediate and long-term environmental problems we face. Item 5, computer integration and analysis, is placed next-to-last because it also constitutes both key and bridge permitting maximum synthesis between applied and basic research aspects of the program, item 4. Together, these all combine to insure and make possible not only increased graduate training but also increased quality, scope and synthetic perspective in such training.

Program Development: No such program can, nor should intend to promise immediate, short-term breakthroughs or final solutions to practical problems. It does promise new weapons with which to attack them. Scope and complexity of the total problem demands that any such comprehensive program be planned and organized on a continuing, renewable, three- to five-year basis.

However expressed, the initial period must essentially be one of tooling up. At least three years will be necessary to develop, test and produce sufficient environmental monitoring equipment. Mapping, grid system, human variables, computer programming and their effective integration with environmental data must also show similar lag time. The grid system, central and essential as it is and simple as it sounds, must actually be a synthesis of different aspects, techniques and disciplines. Biologists and engineers may adjust station locations somewhat in order to conform to a grid system, but the human ecologist (sociologist) may not. His data are of a different sort and he must design his program concurrently if optimum results are to be achieved. He must ascertain what data are already available and can be rendered usable, what data must be collected in the field through observational or survey techniques, devise and test appropriate techniques and apparatus and insure that observational units are readily relatable to both those in the biometric monitoring system and those already in use, e.g., by the Bureau of the Census.

Environmental monitoring should, therefore, begin on a manual, continuing rather than continuous, pilot basis in conjunction with basic research and testing of equipment in a few selected streams and small watersheds. It can evolve and increase significantly in scope after about three years, in phase with other facets of the program. Five years should see a mature program operating at or near maximum efficiency and cost, beginning to yield practical results other than those directly attributable to basic research. Direct total cost, exclusive of monitoring equipment in the middle Susquehanna Basin would level off somewhat below five hundred thousand (present) dollars per year which is minimal considering the magnitude of the problem.

**Conclusion**

Briefly, we need to know: What "kind" and how many hypothetical "bodies" occupy each natural unit area of whatever size; how they effect and are affected by that and surrounding areas; what they think and want, or think they want; how best to minimize or remove the environmental stress they cause; and finally, how best to sell them on the actions that are needed. As Slobodkin[18] and Ehrlich and Holdren[10] point out, the "best" solution may well be impossible or impractical now or even in the future. I use the word here to mean the best that is pragmatically possible. To judge meaningfully involves realistic estimation relative to each alternative. Accurate modeling and reliable projection are therefore essential.

In terms of planning and implementation, political subdivisions and their component parts are artifacts of civilization and generally do not reflect natural boundaries. Information on them can easily, quite accurately and perhaps more discriminatingly, be generated from data taken in a grid system of appropriate size such that natural boundaries are not significantly distorted. The reverse is patently impossible, discrimination beyond resolution inherent in any system plainly being spurious.

# REFERENCES

1. H. T. Odum and R. C. Pinkerton. *Amer. Sci.* 43: 331-343. (1955).
2. H. T. Odum. *Environment, Power, and Society.* Wiley-Interscience, New York. 331 p. (1971).
3. E. S. Deevey Jr. *Bull. Ecol. Soc. Amer.* 52(4): 3-8. (1971).
4. P. R. Ehrlich. *The Population Bomb.* Ballantine Books, New York. 223 p. (1968).
5. R. L. Smith. *Ecology and Field Biology.* Harper and Row, Publishers, New York, 686 p. (1966).
6. E. P. Odum. *Fundamentals of Ecology.* W. B. Saunders Co., Philadelphia. 574 p. (1971).
7. R. J. Beyers. *Ecol. Monogr.* 33(4): 281-306. (1963).
8. D. Morris. *The Human Zoo.* Dell Publishing Co. ed. McGraw-Hill Book Co., New York. 204 p. (1969).
9. A. Spilhaus. *Science* 175(4023): 711-715. (18 February, 1972).
10. P. R. Ehrlich and J. P. Holdren. *Science* 171(3977): 1212-1217. (26 March, 1971).
11. H. T. Odum, W. L. Siler, R. J. Beyers and N. Armstrong. *Publ. Inst. Mar. Sci. Univ. Tex.* 9: 373-403. (1963).
12. There is at least one good reason: At least in New York City garbagemen are now titled "Sanitary Engineers". Light is beginning to appear at the end of the tunnel however. H. T. Odum has recently joined the *Department of Environmental Engineering* at the University of Florida.
13. Federal Water Pollution Control Administration. *Report of the Committee on Water Quality Criteria of the National Technical Advisory Commission of the Secretary of the Interior.* U.S. Govt. Prtg. Off., Wash., D.C. (1968).
14. J. L. Wilhm, T. C. Dorris. *BioScience* 18(6): 477-481. (1968).
15. National Research Council. *U.S Participation in the International Biological Program.* Report No. 2, U.S. Committee for the International Biological Program. Wash., D.C. 26 p. (1967).
16. S. Friedman. *BioScience* 18(9): 882-883. (1968).
17. A. L. Hammond. *Science* 175(4017): 46-48. (7 January, 1972).
18. L. B. Slobodkin. *BioScience 18(1):* 16-23. (1968).
19. P. A. Jordan. *BioScience* 18(11): 1023-1029. (1968).
20. P. Thompson. *Science 163(3872):* 118-1182. (March 14, 1969).
21. N. Machiavelli. *The Prince.* (1513). The passage cited is found in the section on cruelty and clemency, as follows: ". . . but above all he must abstain from taking the property of others, for men forget more easily the death of their father than the loss of their patrimony."
22. R. Ardrey. *The Territorial Imperative.* Atheneum, N.Y. 390 p. (1966). Citing of Mr. Ardrey's book may be constructed as agreement with his basic thesis but not necessarily all the "facts" he marshalls to support it. In indirect support one might also cite D. Morris. *The Naked Ape.* McGraw-Hill Book Co., N.Y. 252 p. (1967). i.e., that the veneer of our civilization is, perhaps, thinner than we would like to believe.
23. H. T. Odum. *Limnol. and Oceanogr.* 1(2): 102-117. (1956).
24. H. T. Odum. *Ecol. Monogr.* 27: 55-112. (1957).
25. H. T. Odum, C. M. Hoskin. *Publ. Inst. Marine Sci., Univ. Tex.* 4: 115-133. (1957).
26. H. T. Odum, C. M. Hoskin. *Publ. Inst. Mar. Sci., Univ. Tex.* 5: 16-46. (1958).
27. R. J. Beyers, H. T. Odum. *Limnol. and Oceanogr.* 4(4): 499-502. (1959).
28. R. J. Beyers. *Publ. Inst. Mar. Sci., Univ. Tex.* 9: 19-27. (1963).
29. R. J. Beyers, J. Larimer, H. T. Odum, R. Parker, N. Armstrong. *Publ. Inst. Mar. Sci., Univ. Tex.* 9: 454-489. (1963).
30. J. B. Lackey. *Pub. Health Rep.* 54(18): 740-746. (1939).
31. R. Patrick. *Sewage and Indust. Wastes* 25: 210-214. (1953).
32. R. E. McKinney, A. Gram. *Sewage and Indust. Wastes* 28: 1219-1231. (1956).
33. D. R. Arthur. *New Scientist* 37(589): 625. (March 21, 1968).
34. D. R. Arthur. *New Scientist* 37(589): 625-627. (March 21, 1968).
35. *Time Magazine.* 91(13): 53. (March 29, 1968).

36. Smith, J. E. ed. 'Torrey Canyon' Pollution and Marine Life. A Report by the Plymouth Laboratory of the Marine Biological Association of the United Kingdom. Cambridge Univ. Press. London. 196 p. (1968).
37. K. E. Kamann, U.S.D.I. Grant Nos. WPD-180-01-67, -02-68. (Unpub.).
38. E. Steemann Nielsen. J. Conseil, Conseil Perm. Intern. Exploration Mer 28: 117-140. (1952).
39. J. H. Ryther. Limnol. Oceanog. 1: 61-70. (1956).
40. C. D. McAllister. Limnol. Oceanog. 6: 483-484. (1961).
41. N. J. Antia, C. D. McAllister, T. R. Parsons, K. Stephens and J. D. H. Strickland. Limnol. Oceanog. 8: 166-183. (1963).
42. J. S. Bunt, Nature 207: 1373-1375. (1965).
43. J. H. Ryther and D. W. Menzel. Limnol. Oceanog. 10: 490-491. (1965).
44. A. M. Barnett and J. Hirota. Limnol. Oceanog. 12: 349-352. (1967).
45. J. D. H. Strickland. Measuring the production of marine phytoplankton. Bull. Fisheries Res. Board Can. No. 122. 172 p. (1960).
46. E. Steemann Nielsen. II. Fertility of the Oceans. 7. Productivity definition and measurement. in: The Sea. Vol. 2. Academic Press, New York. (1963).
47. E. Steemann Nielsen and V. K. Hansen. Deep-Sea Res. 5: 222-228. (1959).
48. C. Nalewajko. Limnol. Oceanog. 11: 1-10. (1966).
49. G. E. Fogg, C. Nalewajko and W. D. Watt. Proc. Roy. Soc. London, Ser. B 162: 517-534. (1965).
50. G. E. Fogg. Oceanog. Mar. Biol. Ann. Rev. 4:195-212. (1966).
51. J. A. Hellebust. Limnol. Oceanog. 10: 192-206. (1965).
52. W. D. Watt. Proc. Roy. Soc. London, Ser. B. 164: 521-551.
53. R. Patrick. Proc. Acad. Nat. Sci., Philadelphia 101: 277-341. (1949).
54. R. Patrick. Sewage and Indust. Wastes 22: 926-938. (1953).
55. R. Patrick. Proc. Penna. Acad. Sci. 27: 33-36. (1953).
56. W. M. Snyder. J. Geophys. Res. 67(2): 721-729. (1962).
57. W. M. Ingram, K. M. Mackenthun, A. F. Bartsch. Biological Field Investigative Data for Water Pollution Surveys. WP-13 U.S. Dept. Int., Fed. Water Poll. Cont. Adm. U.S. Govt. Prtg. Off., Wash., D.C. 139 p. (1966).
58. L. E. Keup, W. M. Ingram, K. M. Mackenthun. The Role of Bottom-Dwelling Macrofauna in Water Pollution Investigations. U.S. Publ. Health Serv. Publ. No. 999-WP-38. 23 p. (1966).
59. H. E. Babbitt and E. R. Baumann. Sewage and Sewage Treatment. John Wiley and Sons, Inc., New York. 790 p. (1958).
60. M. L. Rosenzweig. Am. Midland Naturalist 80(2): 299-315. (1968).
61. K. E. F. Watt (ed.). Systems Analysis in Ecology. Academic Press, New York, 276 p. (1966).

62. L. G. Williams. BioScience 18(9): 849. (1968).
63. K. K. Sheaffer, F. J. Little, Jr. Proc. Penna. Acad. Sci. 43: 73-79. (1969).
64. Pennsylvania State Planning Board. Pennsylvania State Water Resources Supplement to the U.S. Army Corps of Engineers "Report for Development of Water Resources in Appalachia." Penna. Dept. Comm. Harrisburg. 161 p. (1968).
65. Pennsylvania State Planning Board. Pennsylvania Appalachian Development Plan. Penna. Dept. Comm. Harrisburg. 555 p. (1968).
66. Beeton, A. M., "Changes in the Environmental Biota of the Great Lakes," in: Eutrophication: Causes, Consequences, Correctives Nat. Acad. Sci., Wash. 661 p. (1969).
67. Time Magazine 97(16): 56. (April 19, 1971). Reporting J. Zavodni and D. Nixon's five month canoe and sampling bottle documentation of 500 cases of industrial water pollution on the Ohio and Monongahela rivers, filing of 362 affidavits and subsequent charging of four companies with 73 violations of the 1899 Refuse Act.
68. Nereus Corporation, Narragansett, R. I. (no longer in operation)
69. InterOcean Systems, Inc., San Diego, Calif. 92110.
70. D. J. Bogue, C. L. Beale. Economic Areas of the United States. Free Press of Glencoe, New York. 1162 p. (1961).
71. L. T. Harms, R. James. Manpower in Pennsylvania. I. Methodological Statement. Commonwealth of Pennsylvania, Department of Community Affairs. Harrisburg. 349 p. (1967).
72. L. T. Harms, R. James. Manpower in Pennsylvania. II. 1940-1963, Projections to 1980. Commonwealth of Pennsylvania, Department of Community Affairs. Harrisburg. 375 p. (1967).
73. U.S. Public Health Service. The Storage and Retrieval of Data for Water Quality Control. Basic Data Branch, Div. of Water Supp. and Pollution Control, U.S.P.H.S., Dept. of Health, Educ. and Welfare. Washington, D.C. P.H.S. Publ. 1263. (1964)
74 Spindletop Research Center, Lexington, Ky. 40505. from: President R. R. Broida and D. H. Spaeth.
75. L. R. Lindgren. Pittsburgh Press. (2 February, 1969).
76. National Oceanographic Data Center. Processing Physical and Chemical Data from Oceanographic Stations. Man. Series Publ. M-2. 105 p. (1962).
77. F. J. Little, Jr. Syst. Zool. 13(4): 191-194. (1964).
78. D. Child, R. T. Oglesby and L. S. Raymond Jr. Land Use Data for the Finger Lakes Region of New York State. Cornell Univ. Water Res. and Mar. Sci. Ctr., Pub. 33. 29 p. (1971).
79. R. T. Oglesby. Personal Communication. (Fall, 1971).

Prof. Frank J. Little, Jr. Ph.D.
Department of Biology,
State University of New York
College at Brockport, New York

# Chapter 29

# Conservation

As the Apollo astronauts returned from the moon they were able to do something that was perhaps even more significant than their observations of our nearest neighbor. They were able to look back at our own blue planet and describe it as an oasis of life in an otherwise dead ocean of black space. This concept of earth itself as a spaceship with its own life support systems, its own limited quantity of resources and its own growing compliment of astronauts, has provided the basis for a new look at the old study of our environment which we call conservation. We all know about conservation. That is the discussion of dated problems to be found in the last chapter of the high school biology text, with pictures of contour plowing and soil erosion. That was the chapter we never studied because the course was too crowded with other more demanding topics. But conservation, along with our other concepts and institutions, has undergone a profound change. As the realization grows in all of us that this globe is all we have to live on, and that its air, water and land are rapidly being polluted by our current way of life, conservation has suddenly become the science of survival.

The word "conservation" was coined near the turn of the nineteenth century by America's first professional forester, Gifford Pinchot. His meaning of the term centered on the development of resources for the widest possible use and his major concern was that the exploitation be carried out as efficiently as possible. While this definition of the word would be disputed by many today, it would be hard to get general agreement on its real meaning. A recent definition by Raymond Dasmann is: "The rational use of the environment to provide a high quality of living for mankind."[1] This presents fresh problems however, as we must decide what uses are rational and what constitutes a "high quality" life.

Any use of natural resources raises conflict since the resource itself may be seen in various ways by various people. In line with Pinchot's thinking, an area that is not being cropped or harvested is being wasted. Use in this sense means economic exploitation, and many people

accept this as the way things should be. The purpose of a herd of deer is to provide sport and meat for humans. If that use is overshadowed by the deer's destructiveness to local crops then there is justification for removing the herd entirely, especially if the land it now occupies can then be used for a shopping center or a new housing development. From this it may be seen that the economic exploitation of a resource may in itself call forth a multiplicity of feelings. The hunter, the farmer and the construction engineer could be expected to express divergent opinions on this problem.

There is, of course, another view on the use of a resource that is essentially non-consumptive. The use implied here is an aesthetic one. As John Muir, naturalist and founder of the Sierra Club, expressed it in 1917: "Everybody needs beauty as well as bread, places to play in and pray in, where Nature may heal and cheer and give strength to body and soul alike."[2] Muir's followers have been called preservationists and reviled as dreamers who cannot see the necessity of things as they are, but as the environment deteriorates and "quality of life" becomes an important consideration, they have become an important force in conservation. Their argument is that the destruction of a resource is an irreversible process and that short-term economic gain is insufficient recompense to balance the long term non-consumptive use of that resource for generations yet unborn.

Both of these positions have some merit and perhaps the best conservation strategy lies somewhere in between. Historically, this course was plotted by some dedicated and little-heeded men like Liberty Hyde Baily, William Vogt, and the current high priest of conservation, Aldo Leopold. Leopold defines conservation as : ". . . a state of harmony between me and land."[3] His land ethic so beautifully expounded in *A Sand County Almanac* (1949), describes his ideas of what harmony should be, and has become a gospel of new conservation thought.

Among these old thoughts, newly delineated, is the concept that man is part of nature. In spite of his technilogical prowess he is still a creature and subject to the cosmic law-order. Because of his technology, however, he has the capacity to destroy his environment. This places a heavy burden of responsibility on his shoulders and makes it of over-riding importance that man understand his place in nature and attain a sense of stewardship toward his environment. This means that he must care for the earth not only because it is essential for mankind that he do so, but also because the earth is in trust to him from all other living things. Conservation becomes a fusion between science and humanism and demands not just the use of a resource for mankind, but an understanding of the place of that resource in the overall scheme of things and its maintenance as a viable part of that scheme. Wilderness, wild animals, free flowing rivers, waving trees, all deserve conservation because of their intrinsic value and the awesome realization that once they are gone they will not come again.

The word *conservation* implies a series of holding actions to maintain things as they are, but this is no longer acceptable. Instead it covers an infinite variety of situations calling for action anywhere from complete preservation to complete management. In any case, action is called for, but only that action that springs from understanding. Too often situations have been worsened because of ill-advised action. Large scale predator poisoning campaigns continue, for instance, in spite of the well documented fact that predators are very important to the ecosystem as a whole.

Many of the ecological principles which are known have been disregarded. Many more remain to be uncovered through further study. All life on earth exists in a thin layer of soil, water and atmosphere that surrounds the globe. This *biosphere* is less than ten miles thick but it covers the earth with remarkable continuity and there are few places where some form of life does not exist. Living things also come in an over-whelming variety of skies, shapes and diversity of form. This diversity is a constant fascination to the mind of man, but it is more than that. Man has discovered that although it is much more efficient to produce food in a monoculture—large areas containing only wheat or, only corn, or only cattle—it creates an ecosystem of great instability. An insect, obscure in the variety of a normal ecosystem, suddenly becomes a pest as it reproduces to keep up with the increase in food supply. A disease minimized in a natural situation, finds epidemic conditions where every individual is susceptible to

it. Extended periods of adverse weather conditions, such as summer drought, affect such simplified ecosystems far more than the naturally diversified systems are affected. Even in natural ecosystems the more diverse and complex tropical communities are more stable and resiliant than the more simplistic ones of the arctic.

Another interesting and important feature of living things that compliments their diversity is their adaptibility. Each creature has its own behavior patterns and its own physical characteristics that enable it to survive and prosper in its own special niche. During the course of evolution the same niche has been filled in widely separated areas by similar adaptations so that we have jaguars in South America and leopards in Africa; deer in the United States and kangaroos in Australia. The isolated continent of Australia shows this evolutionary adaptability in a special way since the pouched marsupials have evolved predators and herbivores along similar lines to the placental mammals of the outside world even though their genetic background is not the same.

An organism's adaptability is its bastion against change, for change is one of the few certainties in the biosphere. If a plant or animal is too specialized—too closely adapted to its own particular way of life—and the environment changes, or is changed by man, the price may be extinction. Less specialized organisms are opportunists and are able to adapt to the change giving them an evolutionary advantage in environments where change is frequent and abrupt.

Change is not always an abrupt affair but more typically has a constant quiet effect on any ecosystem. The phenomenon of succession is the most obvious manifestation of such change. A pond or lake starts to "die" the moment it is formed as silt and plant growth start their slow, but steady action of filling it in. Over a period of time the lake becomes a bog or marsh. As the water-loving plants lend their substance to the soil the wetness becomes less apparant and our former pond becomes a meadow, figure 117. The next stage (a successional stage is called a *sere*) is able to out-compete the last. Thus shrubs and forbs replace the grasses and, in their turn, prepare the area for the coming of trees. Each stage changes the environment in such a way that it makes the area less suitable for the offspring of its own species. This enables a new sere to emerge and the succession to proceed. Eventually a stage is reached where the young trees can grow well in shade of their parents, and can, therefore, replen-

**Figure 117.** Various successional stages in the forest biome.

ish themselves in a more or less sustaining basis. This we have called a climax forest but the term carries too much of the sense of finality. Actually change continues and the effects of windfalls, fire and the activities of such animals as the beaver (and, of course, man) constantly exert their influence on the forest, setting back the clock and starting a new successional progression under new conditions. Since different plants and animals flourish at different levels on the successional gradient it is desirable in the interests of diversity, to have a variety of seres available in any given area. In fact, the edges of *ecotones*, where one stage or ecosystem meets another are especially attractive to a variety of living forms.

Succession and dynamic change are basic to ecology and tie in well with another integral principle: *interrelationship.* Consider for instance a pond with ample sunlight and a good supply of dissolved nutrients. The water is green with algae and single-celled phytoplankton (floating plants) cloud the water. Rocks are carpeted with slimey filamentous algae and water moss, masses of waternet or elodea float in the shallows, while the surface is covered with duckweed or the tiny grains of water meal. Each plant, each green cell is surrounded by simple raw materials. Water is

plentiful, dissolved carbon dioxide and minerals are in abundance and the sun's energy provides the glue for the plants to synthesize food and oxygen. The feast is spread and many are the guests.

Most of them are tiny. Microscopic protozoa and rotifers sweep the green algal cells into their gullets with tirelessly beating cilia. "Waterfleas," huge by comparison but still hardly visible to the naked eye, vie, each in its own special way with other crustaceans, worms and insect larvae for its share of the plant food. Virtual giants to these, the tadpoles, and mayfly nymphs rasp away at the plentiful plants, while certain fish, ducks and even an occasional deer make use of the producer's bounty.

The plant eaters in their turn are the food supply for the second round of consumers. There are many fewer of these voracious creatures for much of the energy supplied by the sun has been dissipated. Even the small portion that has been trapped by the plants is decimated by the transfer to the herbivores. In spite of their numbers, the predators are among the most obvious creatures in the pond. Some, like the diving beetle, are powerful swimmers and chase down their prey. Another is the backswimmer, so-called because he actually does oar his way through the water on his back. His reversed coloration—dark on the ventral side, light on the dorsal—is in defiance of conventional counter-shading. With the help of his folded wings he traps a shimmering bubble of air under his inverted body. This air bubble acts as a "physical gill" exchanging carbon dioxide for oxygen from the surrounding water. He is armed with a sharp, piercing beak but this avails him little if a fish catches him or if a frog makes a meal of him as he surfaces for fresh air. Fish and frogs in their turn fall prey to the kingfisher or the great blue heron stalking patiently in the shallows, or perhaps to the hovering hawk describing lazy circles in the sky, figure 118.

Thus the sun's energy trapped in the plants green cells is passed from one to the next, each step dissipating much of the energy but each creature obtaining enough to use for its own operations. We have called these relationships "food chains" but the image is too simplistic. Even the popular characterization of the "web of life" to denote these complex interactions, is too flimsey a model, for a spider's web is regular and

two dimensional while the interaction of the ecosystem is complicated beyond comprehension and has the effect of binding each organism not only to its neighbor but to all other organisms in the biosphere. The strands may be indirect and difficult to trace but they are there. Man, too, is bound by these obscure and profound connections and one is led to paraphrase John Donne and say "No species is an island complete unto itself."

It is a mistake, therefore, to believe that man, for all his manipulation of the rest of the biosphere is a creature apart from it. Western culture and religion have tended to widen the gulf between man and other forms of life in his own mind, but biologically he is just another primate. One of our greatest tragedies is that we have chosen to believe that we are above nature rather than realizing that we are in reality part of it.

Another misconception that man has developed about himself is the idea that he is a very adaptable animal, and hence can change himself as he changes the environment. For instance, it is predicted that he will "get used to" crowded conditions and polluted air and water. It is true

**Figure 118.** The most pronounced effect of Pesticides are on birds of Prey.

that man can readily adjust to a wide range of different environments, but this is an individual thing. Adaptation, in the biological sense, is a genetic change that makes the new strain physically better able to survive under changed conditions than the old type. In spite of man's proven ability to breed successfully, the long time between generations precludes genetic changes in anything like the short time available. Man's genes are still programmed for a hunting way of life but individuals have learned to survive under his new requirements as a result of one specialization—his brain. It is a mistake to believe that man's adaptability will bail him out as he precipitates new change. He hasn't genetically caught up to the old changes yet. His anatomy and physiology does not fit him any better for urbanization than the Bison on the Plain, figure 119.

Change there will be, however, as man continues to reproduce in numbers that bring him on a collision course with the earth's capacity to support the multitudes. Population grows on the principle of compound interest. Each year the base is increased so that a one percent growth rate will not cause a doubling of the population in 100 years as might be expected but in 70 years. The number of years that it will take for population to double at a given annual rate of increase can easily be found. Simply divide the growth rate into 70. For instance, at the 2% annual rate of increase in world population during the 1960's the population will double in 35 years. The United States with a doubling time of 63 years is not among the fastest growing countries in the world but is surpassed many times over by countries in Asia and South America. It should be noted, however, that the standard of living we enjoy means that a child born here consumes far more in terms of natural resources than a child born in India.

When all the population figures are totaled we are faced with the figure of over 72 million new mouths to feed in the world each year. In addition to the mouths, there is the added demand for energy supplies, for housing, clothing, medicine, waste disposal and, of course, the clamour for the "good life." The question has been asked and must soon be answered: Can we have the "good life" with as many people as we are producing or even with the number we must support now? The answer provided by many competent scien-

**Figure 119.** Will this be the final successional stage of human ecology?

North American Reference Encyclopedia of Ecology and Pollution

tists, sociologists and philosophers is "no, we cannot," and further they say we are doomed to face famine, social unrest and the erosion of human rights and individual freedom as the population spector grows among us.

Is it justified to include this section on human population in a discussion of conservation? It must be included and more. Today's conservationist is an environmentalist. He must be in a position to question many of the heretofore unshakable doctrines of our society. The conservationist is the true revolutionary of the day because it is he who must understand the workings of nature and guide man, his rampant technology, and even his social and religious institutions in new directions acceptable to the natural order. This is a great deal to expect from a dedicated group of "Renaissance men" and a great burden must be placed on education to produce more environmentalists. But time is short and education takes time. The most important thing is to act upon the warnings sounded by the ecologists, the students, the writers, the demographers, the sociologists and all the other modern Cassandra's who have been sounding the alarm. Let us hope that the call is heard and the warning heeded.

## REFERENCES

1. Dasmann, Raymond F., *Environmental Conservation,* New York (1968)
2. Muir, John., *The Mountains of California,* Houghton-Mifflin, New York (1917)
3. Leopold, Aldo., *A Sand Country Almanac,* Oxford University Press, New York (1949)

Prof. Christopher White, Ph.D.
Department of Environmental
Conservation and Resources,
Community College of the Finger Lakes,
Canandaigua, New York.

The following list of governmental agencies responsible for pollution control includes where possible the name of the executive officer and the area of legal responsibility. Some of the more important private and consumer groups are included. Listings are alphabetical by states.

## Environmental Protection Agencies, (State)

W. T. Willis, Director
Alabama Department of Health
645 South McDonough Street
Montgomery, Alabama 36104
(Air)

Aubrey C. Godwin, Director
Alabama Department of Health
Bureau of Environmental Health
Division of Radiological Health
State Office Building
Room 311
Montgomery, Alabama 36104
(Radioactive Materials, Nuclear Power within the state)

Lloyd A. Morley, Chief
Environmental Health Section
Division of Public Health
Department of Health &
Social Services
Pouch H
Juneau, Alaska 99801
(Residential)

Dr. Max C. Brewer, Commissioner
Alaska Department of
Environmental Conservation
419 Sixth Street
Pouch O
Juneau, Alaska 99801
(Air) (Industrial) (Residential)

Association of Soil and Water
Conservation Districts
P.O. Box 128
Peoria, Arizona 85345

Division of Air Pollution Control
4019 N. 33rd Avenue
Phoenix, Arizona 85017

John H. Back, Director
Division of Sanitation
Arizona State Department of Health
2975 West Fairmount Ave.

Phoenix, Ariz. 85017
(Air) (Water) (Residential)
(Pollution of ground (Solid Waste))

S. Ladd Davies
Arkansas—Department of Pollution
Control & Ecology
1100 Harrington
Little Rock, Arkansas 72202
(Air) (Water) (Industrial) (Residential)

Peter Heylyn
Chairman of Board
Ecology Center
2179 Allston Way
Berkeley, Calif. 94704

Fresno Co. Health Dept. Div. of Lab.
515 South Cedar
Fresno, Calif. 93702
(Air) (Water) (Industrial) (Residential)

Robert L. Chass
Los Angeles County Air Pollution
Control District
434 South San Pedro Street
Los Angeles, Calif. 90013
(Air) (Industrial) (Residential)

The Resources Agency
1416 Ninth Street
Sacramento, Calif. 95814

James T. Harrison
Director of Public Health and
Air Pollution Control Officer
Sacramento County Health Dept.
2221 Stockton Blvd.
Sacramento, Calif. 95817
(Air)

Mr. Kerry W. Mulligan, Chairman
State Water Resources
Control Board
Room 1140,
1416 Ninth Street
Sacramento, Calif. 95814
(Water) (Industrial) (Residential)

J. B. Askew, M.D.
Air Pollution Control Service
1600 Pacific Highway
San Diego, Calif. 92101
(Air)

David R. Brower
Friends of the Earth
529 Commercial Street
San Francisco, Calif.

San Francisco Bay Conservation and
Development Commission

507 Polk Street
San Francisco, Calif. 94102

Gerland P. Wood, Ph.D.
Air Pollution Control Div.
Colorado Department of Health
4210 E. 11th Ave.
Denver, Colo. 80220
(Air)

Department of Natural Resources
1845 Sherman
Denver, Colorado 80220

Frank J. Rozich
Water Pollution Control Div.
Colorado Dept. of Health
4210 E. 11th Ave.
Denver, Colorado 80220
(Water)

Department of Agriculture and
Natural Resources
Room 113
State Office Building
Hartford, Conn. 06115

Commissioner Dan W. Luthin
Department of Environmental
Protection
State Office Bldg.
Hartford, Conn.
(Air) (Water) (Industrial) (Residential)

Water Resources Commission
State Office Building
Hartford, Conn. 06115

Secretary Austin N. Heller
Dept. of Natural Resources &
Environmental Control
D Street
Dover, Delaware 19901

Jacob Kreshtool
Delaware Citizens for Clean Air, Inc.
1308 Delaware Ave.
Wilmington, Del. 19806

Edward Debellevue
Environmental Action Group
323 J. Wayne Reitz Union
Gainesville, Fla. 32601

Department of Air
and Water Pollution Control
315 S. Calhoun Street
Tallahassee, Florida 32301

Department of Natural Resources
Larson Building
Gaines Street at Monroe
Tallahassee, Florida

Vincent D. Patton
Florida Department
of Pollution Control
Suite 300, Tallahassee Bank Bldg.
315 S. Calhoun Street
Tallahassee, Fla. 32301
(Air) (Water) (Industrial) (Residential)

Institute of Natural Resources
University of Georgia
203 Forrestry Bldg.
Athens, Georgia 30601

State Soil and Water Conservation
Committee
318 Extension Annex Building
Athens, Georgia 30601

Robert A. Collom, Jr.
Air Quality Control Branch
Georgia Department of Public Health
47 Trinity Ave., S.W.
Atlanta, Ga.

R. S. Howard, Jr.
Georgia Water Quality Control Board
47 Trinity Ave., S.W.
Atlanta, Ga. 30334
(Water) (Industrial) (Residential)

State Department of Public Health
116 Mitchell Street, S.W.
Atlanta, Georgia 30303

State Game and Fish Commission
Trinity-Washington Building
270 Washington Street, S.W.
Atlanta, Georgia 30334

Robert S. Nekomoto
Air Sanitation Branch
1250 Punchbowl Street
Honolulu, Hawaii 96801

Department of Land
and Natural Resources
Box 621
Honolulu, Hawaii 96809

Walter B. Quisenberry, MD.
Director of Health
Hawaii State Department of Health
P.O. Box 3378
Honolulu, Hawaii 96801
(Air) (Water) (Industrial) (Residential)

Water Resources Research Center
University of Hawaii
2525 Correa Road
Honolulu, Hawaii

Alfred J. Eiguren
Air Pollution Control Section
Environmental Improvement Div.
Idahoa Department of Health
Statehouse
Boise, Idlaho 83707
(Air)

Vaugh Anderson
Environmental Improvement Division
Idaho Department of Health
Statehouse
Boise, Idaho 83707
(Air) (Water) (Industrial) (Residential)

Commissioner H. W. Poston
City of Chicago
Department of Environmental
Control
320 N. Clark Street
Chicago, Illinois 60610
(Air) (Water) (Industrial) (Residential)

Mr. E. E. Diddams
McLean County Health Dept.
401 West Virginia Ave.
Normal, Illinois 61761
(Air) (Water) (Industrial) (Residential)

W. L. Blaser
State of Illinois
Environmental Protection Agency
2200 Churchill Rd.
Springfield, Illinois 62706
(Air) (Water) (Industrial) (Residential)

State Natural History Survey Division
179 Natural Resources Building
Urbana, Illinois 61801

Department of Natural Resources
608 State Office Building
Indianapolis, Indiana 46204

Dennis T. Karas
Department of Air Quality Control
4525 Indianapolis Blvd.
East Chicago, Indiana 46312

Joel Johnson
Department of Health
Division of Air Pollution
3600 West 3rd Ave.
Gary, Indiana 46406

Perry E. Miller, Tech. Secy.
Indiana Air Pollution Control
1330 West Michigan Street
Indianapolis, Indiana 46206
(Air) (Water) (Industrial) (Residential)

Lewis Fletcher Scott
Chief Environmental Control Div.
Indianapolis Dept. of Public Works
City-County Bldg.—Room 2401
Indianapolis, Ind. 46204
(Air) (Water) (Industrial) (Residential)

Perry E. Miller, Technical Secy.
Indiana Stream Pollution Control
1330 West Michigan Street
Indianapolis, Indiana 46206
(Air)

Environmental Engineering Service
Iowa State Department of Health
Lucas State Office Building
Des Moines, Iowa 50319

R. J. Schliekelman
Water Pollution Control Div.
Iowa State Dept. of Health
Lucas State Office Bldg.
Des Moines, Iowa 50310
(Water)

Melville W. Gray, P.E.
Division of Environmental Health
Kansas State Dept. of Health
535 Kansas Ave.
Topeka, Kansas 66603
(Air) (Water) (Industrial) (Residential)

Frank P. Partee, Tech. Director
Kentucky Air Pollution Control
Commission
275 E. Main Street
Frankfort, Kentucky 40601

State Department of Health
275 E. Main Street
Frankfort, Kentucky 40601

State Department of Conservation
P.O. Box 44275
Capitol Station
Baton Rouge, Louisiana 70804

Chairman, Clark M. Hoffpauer
La Stream Control Commission
Drawer FC-LSU
E. Baton Rouge, La. 70803
(Water) (Industrial)

Bureau of Environmental Health
Louisiana State Department
of Health
P.O. Box 60630
New Orleans, Louisiana 70160

James R. Renner
President
Ecology Center of Louisiana
P.O. Box 15149
New Orleans, La. 70115

William R. Adams, Jr.
Environmental Improvement
Commission
State House
Augusta, Maine 04330
(Air) (Water) (Industrial) (Residential)

Department of Natural Resources
State Office Building
Annapolis, Maryland 21201

Department of Water Resources
State Office Building
Annapolis, Maryland 21401

Bureau of Air Quality Control
State Department of Health
610 N. Howard Street
Baltimore, Maryland 21201

Howard E. Chaney, Director
Environmental Health Admin.
State Department of Health
and Mental Hygiene
610 North Howard Street
Baltimore, Md. 21201
(Air) (Water) (Residential)

Gilbert T. Joly
Bureau of Air Quality Control
600 Washington Street
Boston, Mass. 02111
(Air)

Department of Natural Resources
Leverett Saltonstall Building
100 Cambridge Street
Boston, Mass. 02202

John C. Collins
Division of Environmental Health
Dept. of Public Health
Room 320, 600 Washington Street
Boston, Mass. 02111
(Air) (Water) (Industrial) (Residential)

John Richter
Enact
146 F. School of Nat. Res.
University of Michigan
Ann Arbor, Michigan 48104

Mr. Morton Sterling
Wayne County Dept. of Health
Air Pollution Control Div.
1311 East Jefferson
Detroit, Michigan 48207
(Air)

Department of Natural Resources
Mason Building
Lansing, Michigan 48926

Ralph W. Purdy, Exec. Secy.
Water Resources Commission
Stevens T. Mason Bldg.
Lansing, Michigan 48926
(Water) (Industrial)

Frederick F. Heisel
Div. of Environmental Health
Minnesota Dept. of Health
717 Delaware S.E.
Minneapolis, Minn. 55440
(Water) (Residential)

Mr. Grant J. Merrit, Director
Minnesota Pollution Control Agency
717 Delaware Street, S.E.
Minneapolis, Minn. 55440
(Air) (Water) (Industrial) (Residential)

Glen Wood, Director
Mississippi Air & Water Pollution
Control Commission
P.O. Box 827
Jackson, Mississippi 39205
(Air) (Water)

Department of Conservation
P.O. Box 180
Jefferson City, Missouri 65101

Missouri Air Conservation
Commission
P.O. Box 1002
112 W. High Street
Jefferson City, Missouri 65101

Jack K. Smith
Missouri Water Pollution Board
P.O. Box 154
Room 102, Capitol Bldg.
Jefferson City, Mo. 65101
(Water)

L. F. Garber, P.E.
Section of Environmental Health
Services Div. of Health of Missouri
Broadway State Office Bldg.
Jefferson City, Missouri 65101
(Residential)

John S. Anderson, M.D.
Dept. of Health & Environmental
Sciences
Cogswell Bldg.
Helena, Montana 59601
(Air) (Water)

Ben Wake
Div. of Environmental Sciences
State Dept. of Health & Env. Science
Helena, Montana 59601
(Air) (Water) (Industrial) (Residential)

Mr. Fred Jolly
Div. of Env. Sanitation
Nebraska Dept. of Health
2nd Floor, Lincoln Bldg.
10th & "O"
State House Station 94757
Lincoln, Neb. 68509
(Air) (Water) (Residential)

State Department of Environmental
Control
411 S. 13th Street
Lincoln, Nebraska 68508

E. G. Gregory, Chief
Bureau of Environmental Health
201 South Fall Street
Carson City, Nevada 89701
(Air) (Water) (Industrial) (Residential)

State Commission of Environmental
Protection
Nye Building
201 South Fall Street
Carson City, Nevada 89701

Forrest H. Bumford
Air Pollution Control Agency
State of New Hampshire
61 S. Spring Street
Concord, N.H. 03301
(Air)

William A. Healy
N. H. Water Supply & Pollution
Control Commission
105 Loudon Rd.
Prescott Park
Box 95
Concord, N.H. 03301
(Water)

Robert E. Herrmann
Hudson Municipal Air Pollution
Commission
532 Summit Ave.
Jersey City, N.J. 07306
(Air)

William A. Munroe
Bureau of Air Pollution Control
Division of Env. Quality
New Jersey Dept. of Env. Protection
P.O. Box 1390
Trenton, N.J. 08625
(Air)

Charles M. Pike
Div. of Water Resources of the
Dept. of Env. Protection
Labor and Industry Bldg.
P.O. Box 1390
Trenton, N.J. 08625
(Water) (Industrial) (Residential)

Larry T. Candill
New Mexico Cons. Coord. Council
Box 142
Albuquerque, New Mexico

Larry J. Gordon, Director
New Mexico Environmental
Improvement Agency
P.O. Box 2348
Sante Fe, New Mexico 87501
(Air) (Water) (Industrial) (Residential)

Alexander Rihm, Jr. P.E.
Div. of Air Resources, N.Y.S. Dept.
of Environmental Conservation
Albany, N.Y. 12201
(Air)

Paul W. Eastman, P.E.
Division of Pure Waters, New York
State Department of Env.
Conservation
50 Wolf Rd.
Albany, N.Y. 12201
(Water) (Industrial) (Residential)

Robert D. Cusumano
Bureau of Air Pollution Control
County Department of Health
240 Old County Rd.
Mineola, N.Y. 11501
(Air)

Acting Commander Fred C. Hart
Dept. of Air Resources—E.P.A.
51 Astor Place
New York, N.Y. 10003
(Air)

Thomas R. Glenn, Jr.
Interstate Sanitation Comm.
10 Columbus Circle, Room 1620
New York, N.Y. 10019
(Air) (Water)

Peter Guala, P.E.
Div. of Env. Sanitation
Onondaga County Health Dept.
300 South Geddes Street
P.O. Box 1325
Syracuse, N.Y. 13201
(Air) (Water) (Industrial) (Residential)

Col. George E. Pickett
Dept. of Natural & Economic
Resources
Office of Water and Air Resources
126 W. Jones Street
Box 27687
Raleigh, N.C. 27611
(Air) (Water) (Industrial) (Residential)

Air Pollution Control Advisory
Council
State Capitol
Bismark, North Dakota 58501

W. Van Heuvelen, Chief
Environmental Health and
Engineering Services
North Dakota State Department
of Health
State Capitol
Bismarck, N.D. 58501
(Air) (Water) (Industrial) (Residential)

Eugene D. Ermene
Air Pollution Control Div.
2400 Beekman Street
Cincinnati, Ohio 45214

James T. Wilburn, Commissioner
Div. of Air Pollution Control
City of Cleveland
2735 Broadway
Cleveland, Ohio 44115
(Air)

Jack W. Wunderle
Engineer-in-Charge
Air Pollution Unit
State of Ohio
1030 King Ave.
Columbus, Ohio 43212
(Air)

John W. Cashman
Director-Designate
Ohio Dept. of Health
450 E. Town Street
P.O. Box 118
Columbus, Ohio 43216
(Air) (Water) (Industrial)

Dr. R. LeRoy Carpenter,
State Commissioner of Health
Loyd Pummill, Deputy Comm. for
Env. Services
Oklahoma State Dept. of Health
3400 North Eastern
Oklahoma City, Oklahoma 73105
(Air) (Water) (Residential)

L. B. Day
Department of Env. Quality
1234 S.W. Morrison Street
Portland, Oregon 97205
(Air) (Water) (Industrial) (Residential)

Peter R. Wells
Citizens Committee for Env. Control
of Eastern Montgomery County, Pa.
P.O. Box 8833
Elkins Park, Pa. 19117

Victor H. Sussman, P.E.
Bureau of Air Quality and
Noise Control
Dept. of Env. Resources
P.O. Box 2351
Harrisburg, Pa. 17105
(Air)

Maurice K. Goddard
Secretary
Dept. of Environmental Resources
P.O. Box 1467
Harrisburg, Pa. 17120
(Air) (Water) (Industrial) (Residential)

Sanitary Water Board
Department of Health
P.O. Box 90
Harrisburg, Pa. 17120

Pennsylvania Environmental
Council, Inc.
1181 Suburban Station Building
Philadelphia, Pa. 19103

Edward F. Wilson
Assistant Health Comm.
Philadelphia Dept. of Public Health
Air Management Services
1701 Arch Street—6th Floor
Philadelphia, Pa. 19103
(Air) (Industrial) (Residential)

Frank B. Clack, V.M.D. Director
Allegheny County Health Dept.
Rondal J. Chleboski, Chief,
Bureau of Air Pollution Control
Allegheny County Health Dept.
Bureau of Air Pollution Control
39th & Penn Ave.
Pittsburgh, Pa. 15224
(Air) (Industrial) (Residential)

Department of Health
335 State Office Building
Providence, Rhode Island 02903

Division of Air Pollution Control
204 Health Building
Davis Street
Providence, Rhode Island 02908

Charleton A. Maine, Chief
Division of Water Supply and
Pollution Control
R.I. Dept. of Health
204 Health Dept. Bldg.
Davis Street
Providence, R. I. 02908

H. J. Webb, Ph.D.
Executive Director
South Carolina Pollution
Control Authority
1321 Lady Street
P.O. Box 11628
Columbia, S.C. 29211
(Air) (Water) (Industrial) (Residential)

Charles E. Carl, Director
Division of Sanitary Engineering
and Environmental Protection
State Department of Health
—Office Bldg. #2
Pierre, S.D. 57501
(Air) (Water) (Industrial) (Residential)

Harold E. Hodges
Division of Air Pollution Control
C2—212 Cordell Hull Bldg.
Nashville, Tenn. 37219
(Air)

Department of Public Health
621 Cordell Hull Bldg.
Nashville, Tennessee 37219

S. Leary Jones, Technical Secretary
Tennessee Water Quality Control
Board
621 Cordell Hull Bldg.
Nashville, Tenn. 37219
(Water)

State Department of Natural
Resources
1100 W. 49th Street
Austin, Texas 78756

Martin C. Wukasch, P.E.
Texas Radiation Control Agency
Div. of Occupation Health of
Addiction Control
Texas Health Dept.
1100 W. 49th Street
Austin, Texas 78756
(Air) (Water) (Industrial)

Hugh C. Yantis, Jr.
Texas Water Quality Board
13246 Capitol Station
Austin, Texas 78711
(Water)

State Department of Natural
Resources
225 State Capitol
Salt Lake City, Utah 84114

State Division of Health
44 Medical Drive
Salt Lake City, Utah 84113

Richard A. Valentinetti,
Air Pollution Control Office
Agency of Env. Conservation
Division of Environmental Protection
Air and Solid Waste Programs
Bailey Bldg., 7 Main Street
Montpelier, Vermont 05602
(Air) (Industrial) (Residential)

A. W. Albert, P.E.
Director Water Supply
and Pollution Control
Agency of Environmental
Conservation
Dept. of Water Resources
Montpelier, Vermont 05602
(Water) (Industrial) (Residential)

State Air Pollution Control Board
Room 1106, Ninth Street
Office Building
Richmond, Virginia 23224

Alfred H. Paessler, Exec. Secy.
State Water Control Board
P.O. Box 11143
Richmond, Va. 23230
(Water)

William R. Meyer, Executive Director
Virginia State Air Pollution
Control Board
Room 1106
9th Street Office Building
Richmond, Virginia 23219
(Air)

John A. Briggs
Washington State Dept. of Ecology
P.O. Box 829
Olympia, Washington 98504
(Air) (Water) (Industrial) (Residential)

Arthur R. Dammkoehler
Air Pollution Control Officer
Puget Sound Air Pollution Control
Agency
410 W. Harrison
Seattle, Washington 98119
(Air)

Roy T. Olson, R.D.
Spokane County Health District
N. 819 Jefferson
Spokane, Washington 99201
(Residential)

Edgar N. Henry, Chief
Dept. of Natural Resources
Div. of Water Resources
1201 Greenbrier Street
Charleston, W. Va. 25311
(Water)

Carl G. Beard, II
West Virginia Air Pollution
Control Commission
1558 Washington Street, East
Charleston, W. Va. 25311
(Air)

Bureau of Air Pollution Control
and Solid Waste Disposal
4610 University Avenue
Madison, Wisconsin 53705

Department of Natural Resources
P.O. Box 450
Madison, Wisconsin 53701

Department of Economic Planning
and Development
210 W. 23rd Street
Cheyenne, Wyoming 82001

Robert E. Sundin
Air Quality Section
Div. of Health & Medical Services
Dept. of Health & Social Services
State Office Bldg.
Cheyenne, Wyoming 82001
(Air)

Lawrence J. Cohen, M.D.
Administrator
Div. of Health and Medical Services
Wyoming Dept. of Health and
Social Services
State Office Bldg.
Cheyenne, Wyoming 82001
(Air) (Water) (Industrial) (Residential)

Wallace L. Ulrich
Region 8 Youth Advisory
Board E. P. A.
Box 3611
University Station
Laramie, Wyoming 82070

**U. S. Territories**
Department of Health
P.O. Box 1442
St. Thomas, Virgin Islands 00801

Department of Health
P.O. Box 9232
San Juan, Puerto Rico 00908

**Environmental Protection
Agencies, (Federal)**

John V. Brink
Bureau of Air & Water Pollution
Control
Dept. of Environmental Services
D.C. Government
Room 310
25 "K" St., N.E.
Washington, D.C. 20002

Department of Environmental
Services
1875 Connecticut Avenue, N.W.
Washington, D.C. 20009

Department of Human Resources
25 K Street, N.E.
Washington, D.C. 20006

Environmental Action
Room 731
1346 Connecticut, N.W.
Washington, D.C. 20036

Federal Power Commission
441 G Street, N.W.
Washington, D.C. 20426

Federal Water Quality Administration
Washington, D.C. 20242

National Air Pollution Control
Administration
5600 Fishers Lane
Rockville, Maryland 20852

National Water Commission
800 N. Quincy Street
Arlington, Va. 22203

U. S. Environmental Protection
Agency
Waterside Mall
401 M Street, S.W.
Washington, D.C. 20460

# Bibliography

**Aaronson, Terri**
"Mystery"
*Environment,* p. 2-10 (May, 1970)

**Aaronson, Terri**
"Out of the Frying Pan"
*Environment,* p. 2-10 (May, 1970)

**Abbywickrama, B. A.**
*Pre-Industrial Man in the Tropical Environment*
London (1964)

**Abelson, Philip H.**
"Costs Versus Benefits of Increased Electric Power"
*Science,* 710: 1159 (1970)

**Abelson, Philip H.**
"Excessive Emotion about Detergents"
*Science,* 169:1033 (1970)

**Abelson, Philip H.**
"Marine Pollution"
*Science,* 171:21 (1971)

**Abelson, Philip H.**
"Methyl Mercury"
*Science,* 169:237 (1970)

**Abelson, Philip H.**
"Pollution by Organic Chemicals"
*Science,* 170:495 (1970)

**Abelson, Philip H.**
"Progress in Abating Air Pollution"
*Science,* 170: (1956)

**Abelson, Philip H.**
"Shortage of Caviar"
*Science,* 168:199 (1970)

**Abrahamson, Dean E.**
Minnesota Committee for Environmental Information
P.O. Box 14207, Univ. Sta.,
Minneapolis, Minn.

**Abrahamson, Dean E.**
*Environmental Cost of Electric Power*
New York (1970)

**Action for Clean Air, Inc.**
Tina Heavrin, 1817 S. 34th St.,
Louisville, Ky.

**Adams, Alexander, B.**
*Eleventh Hour, A Hard Look at Conservation*
New York (1970)

**Adams, Ansel**
"The Eloquent Light"
San Francisco Examiner

**Adams, Ansel and Nancy Newhall**
"This is the American Earth"
San Francisco Examiner

**Adams, Ansel**
"Not Man Apart"
San Francisco Examiner

**Ad Hoc Group for office of Science and Technology, Executive Office of the President**
*Solid Waste Management:*
"A Comprehensive Assessment of Solid Waste Problems, Practices and Needs" New York (1969)

**Air Conservation Committee**
Ann Felber, Chrmn.,
4614 Prospect Ave.,
Cleveland, Ohio

**"Air Pollution"**
*Proceedings of the First European Congress on the Influence of Air Pollution on Plants and Animals.*
Wageningen, Netherlands (1968)

**Alaska Conservation Society**
Robert Weeden, Box 5-192,
College, Alaska

**Alexander, M.**
*Microbial Ecology*
New York (1971)

**Alexander, P.**
*Atomic Radiations and Life*
Baltimore (1965)

**Alexander, Tom**
"Some Burning Questions about Combustion"
*Fortune,* (February, 1970)

**Alee, Warder C.**
*Animal Aggregations*
Chicago: University of Chicago Press (1931)

**Allee, W. C.**
*Cooperation among Animals*
New York (1951)

**Allee, W. C.**
*The Social Life of Animals*
New York (1938)

**Allee, W. C. and Emerson, A. E.**
*The Life Nature Library*
New York (1964-71)

**Allee, W. C., A. E. Emerson, P. Park, T. Park, and K. P. Schmidt**
*Principles of Animal Ecology*
Philadelphia (1949)

**Allee, W. C. and K. Schmidt**
*Ecological Animal Geography*
New York (1951)

**Allen, Durward L.**
*The Life of Prairies and Plains*
(1967)

**Allen, G. M.**
*Extinct and Vanishing Animals of the Western Hemisphere with the Marine Species of all the Oceans*
New York (1942)

**Altman, P. L. and D. S. Dittmer, eds.**
*Environmental Biology* (1966)

**Altmann, S. A. and J.**
*Baboon Ecology*
Chicago: Chicago University Press (1971)

**American Assn. for the Advancement of Science**
*Man, Culture, Animals: The Role of Animals in Human Ecological Adjustments*
Washington (1967)

**American Public Works Assn.**
*Municipal Refuse Disposal*
Chicago: (1970)

**American Public Works Assn.**
*Refuse Collection Practices*
Chicago (1966)

**Amos, William H.**
*The Infinite River*
New York (1970)

**Amos, William H.**
*The Life of the Pond*
(1967)

**Anderson, Dewey**
"Lake Tahoe Then & Now"
*National Parks Magazine*, p. 4-11
(April, 1970)

**Anderson, Edgar**
*Plants, Man and Life*
University of California (1967)

**Anderson, Edwin P.**
*Domestic Water Supply and
Sewage Disposal Guide*
Indianapolis, Ind. (1960)

**Anderson, T. W.**
*An Introduction to Multivariate
Statistical Analysis*
New York (1958)

**Anderson, Walt, ed.**
*Politics and Environment—A
Reader in Ecological Crisis*
Pacific Palisades, Calif.

**Andrewartha, H. G. and L. C. Birch**
*The Distribution and Abundance
of Animals*
University of Chicago Press (1954)

**Angino, E. E.**
"Arsenic in Detergents: Possible
Danger and Pollution Hazard"
*Science*, 168:389-390 (1970)

**Anthrop, Donald F.**
"Environmental Side Effects of
Energy Production"
*Bulletin of the Atomic Scientists*,
p. 39-41 (Oct. 1970)

**Anthrop, Donald F.**
"Environmental Noise Pollution:
A New Threat to Sanity"
*Bulletin of the Atomic Scientists*,
p. 11-16 (May, 1969)

**Abbott, Richard, ed.**
*Readings in the Law of
Environmental Quality*, Vol. III
Montreal (1971)

**American Political Science Review**
"Symposium: Government and
Water Resources," 44:575-649
Sept. 1950

**Archer, Sellers G.**
*Soil Conservation*
Norman: University of Oklahoma
Press (1956)

**Argenbright, L. P.**
"SO₂ from Smelters"
*Environmental Science and
Technology*, 4:554-568 (1970)

**Argenio, Modesto**
"Now They Want to Search for
Oil and Gas," p. 16-18 (1970)
*Lake Erie Sierra Club Bulletin*

**Argonne Universitites Assn. Conference**
"Universities, National
Laboratories, and Man's
Environment"
Springfield, Va. (1969)

**Arizonans in Defense of the
Environment**
Harry Tate, Pres.,
1431 Jen Tilly Lane, Tempe, Ariz.

**Arizonians for a Quality Environment**
Juel Rodack, Chmn.,
P.O. Box 17717, Tucson, Ariz.

**Arkansas Ecology Center**
Pratt Remmel, Jr.,
316 Chester St., Little Rock, Ark.

**Arthur, Don R.**
*Man and His Environment*
New York (1969)

**Artaimonovich, L. A.**
"Controlled Nuclear Fusion:
Energy for the Distant Future"
*Bulletin of the Atomic Scientists*,
p. 47-55 (June, 1970)

**Arvill, Robert**
*Man and Environment, Crisis and
the Strategy of Choice*
Baltimore, Md. (1970)

**Ashley, M.**
*Introduction to Plant Ecology*
New York (1961)

**Atlantic Co. Citizens Council on
Environment Inc.**
B. D. Rehfeld,
137 S. Main St., Pleasantville, N.J.

**Aubert de la Rue, Edgar,
Francois Bourliere and Jean-Paul
Harroy**
*The Tropics* Paris (1957)

**Aunsley, Eric**
"How Air Pollution Alters
Weather"
*New Scientist*, 44:66-67 (1969)

**Avert Man's Extinction Now (AMEN)**
Curt A. Wilberg, Central Wash.
St. College, Ellensburg, Wash.

**Aylesworth, Thomas G.**
*This Vital Air, This Vital Water*
Chicago (1968)

**Bailey, Anthony**
"Noise is a Slow Agent of Death"
*N.Y. Times Magazine*, p. 46
(Nov., 1969)

**Bailey, N. T. J.**
*The Elements of Stochastic
Processes with Applications to the
Natural Sciences*
New York (1964)

**Bailey, N. T. J.**
*Statistical Methods in Biology*
London: English Universities
Press (1959)

**Baker, B. B., W. R. Deebel,
R. D. Geisenderfer**
*Glossary of Oceanographic Terms*
(1966)

**Black, Ralph J. et al.**
*An Interim Report—1968 National
Survey of Community Solid Waste
Practices*
Washington, D.C. (1968)

**Baldwin, Malcolm and
James K. Page, Jr.**
*Law and the Environment*
New York (1971)

**Bardach, John E.**
*Harvest of the Sea*
New York (1968)

**Bardach, John E.**
*Downstream: A Natural History of
the River*
New York (1964)

**Barlow, Boris V.**
"Should We Ban the Book?"
*New Scientist*, 43:434-436 (1969)

**Barnett, Harold J. and
Chandler Morse**
*Scarcity and Growth: The
Economics of Natural Resource
Availability*
Baltimore: Johns Hopkins Univ.
Press (1963)

**Baron, Robert Alex**
"Let Quiet be Public Policy"
*Saturday Review*, p. 66
(Nov. 7, 1970)

**Baron, Robert Alex**
*The Tyranny of Noise*
New York (1970)

**Barrons, K. C.**
"Some Ecological Benefits of
Woody Plant Control with
Herbicides"
*Science*, 165:465-468

**Bartlett, H. H.**
*Fire in Relation to Primitive
Agriculture and Grazing in the
Tropics*
University of Michigan (1955)

**Bartlett, M. S.**
*Stochastic Population Models in Econolgy and Epidemiology*
London (1960)

**Barton, Robert**
*Oceanology Today*
Garden City, New York (1970)

**Bates, D. R., ed.**
*The Earth and Its Atmosphere*
New York (1957)

**Bates, Marston**
*Animal Worlds*
New York (1963)

**Bates, Marston**
*The Forest and the Sea*
New York (1960)

**Bates, Marston**
"The Human Ecosystem"
*Resources and Man*
New York (1968)

**Bates, Marston**
*Man in Nature*
Englewood Cliffs, N.J. (1964)

**Bates, Marston**
*The Prevalence of People*
New York (1955)

**Battan, Louis J.**
*The Unclean Sky. A Meteorologist Looks at Air Pollution*
New York (1966)

**Baver, L. D.**
*Soil Physics*
New York (1940)

**Bazell, Robert J.**
"Water Pollution: Conservationists Criticize New Permit Program"
*Science*, 171:266-268 (1971)

**Beamish, Tony**
*Aldabra Alone*
San Francisco (1970)

**Beaufort, L. F. de**
*Zoogeography of the Land and Inland Waters*
London (1951)

**Bedford, Franklin T.**
*Climates in Miniature*
New York (1955)

**Benarde, Melvin A.**
*Our Precarious Habitat*
New York (1970)

**Bengtsson, Arvid**
*Environmental Planning for Children's Play*
New York (1970)

**Bennett, H. H.**
*Elements of Soil Conservation*
New York (1947)

**Bennett, H. H.**
*Soil Conservation*
London (1939)

**Bennett, Joseph W.**
*Vandals Wild*
Portland, Ore. (1969)

**Beranek, L. L.**
"Noise"
*Scientific American*, p. 66-76
(Dec. 1966)

**Berg, George G.**
*Water Pollution*
New York (1970)

**Berkner, L. V. and L. C. Marshall**
"History of Major Atmospheric Components"
*Proceedings of National Academy of Sciences*, 53:1215-1226 (1965)

**Berkowitz, D. and Squires, A. M.**
*Power Generation and Environmental Change*
Cambridge, Mass.: M.I.T. (1971)

**Berland, Theodore**
*The Fight for Quiet*
Englewood Cliffs, N.J. (1970)

**Berry, Philip and McCloskey, Michael**
"An Analysis: The Public Land Law Review Commission Report"
*Sierra Club Bulletin*, p. 18-20
(Oct. 1970)

**Berry, R. Stephen**
"Perspectives of Polluted Air—1970"
*Bulletin of Atomic Scientists*,
p. 2, 34-41 (1970)

**Bews, J. W.**
*Human Ecology*
London: Oxford University Press
(1935)

**Beyers, R. J.**
"The Microosm Approach to Ecosystem Biology"
*American Biology Teacher*
26:401-498 (1964)

**BICEP (Biological Committee on Ecology and Pollution)**
Richard DelGrosso,
Wayne St. Univ., Dept. of Biology,
Detroit, Mich.

**Black, John**
*The Dominion of Man: The Search for Ecological Responsibility*
New York (1970)

**Black, J. D.**
*The Management and Conservation of Biological Resources*
Philadelphia (1968)

**Bloomfield, L. M. and Gerald F. Fitzgerald**
*Boundary Waters Problems of Canada and the United States*
Toronto (1958)

**Blumenstock, D. I. and C. W. Thornwaite**
"Climate and the World Pattern"
*Climate and Man 1941 Yearbook of Agriculture*
Washington, D.C. (1941)

**Bodenheimer, F. S.**
*Problems of Animal Ecology*
Oxford: Oxford University Press

**Bodenheimer, F. S.**
"Population Problems of Social Insects"
*Biology Review* 12:393-430 (1937)

**Boffey, Phillip M.**
"Energy Crisis: Environmental Issue Exacerbates Power Supply Problem"
*Science* L68:1554-1559 (1970)

**Boffey, Phillip M.**
"Gofman and Tamplin: Harassment Charges against AEC, Livermore"
*Science* 169:838-843 (1970)

**Boffey, Phillip M.**
"Herbicides in Vietnam: AAAS Study Finds Widespread Devastation"
*Science* 171:43-47 (1971)

**Boffey, Phillip M.**
"Hiroshima/Nagasaki. Atomic Bomb Casualty Commission Perserveres in Sensitive Studies"
*Science* 168:679-683 (1970)

**Boffey, Phillip M.**
"Radiation Standards: Are the Right People Making Decisions?"
*Science* 171:780-783 (1971)

**Bohn, Dave**
*Glacier Bay*
San Francisco

**Bolin, Bert**
"The Carbon Cycle"
*Scientific American*, 124-132
(1970)

**Bonner, John Tyler**
*Cells and Societies*
Princeton University (1955)

**Borgstrom, H.**
*Too Many*
New York (1968)

**Bormann, F. H., G. E. Likens and J. S. Eaton**
"Biotic Regulation of Particulate and Solution losses from a forest Ecosystem"
*Bioscience* 19:600-610 (1969)

**Bormann, F. Herbert and Gene E. Likens**
"The Nutrient Cycles of an Ecosystem"
*Scientific America*, p. 92-101 (1970)

**Bornebusch, C. H.**
*The Fauna of Forest Soil*
Copenhagen (1930)

**Boston Area Ecology Action Center**
John McGrame,
925 Massachusetts Ave.,
Cambridge, Mass.

**Boughey, A. S.**
*Ecology of Population*
New York (1968)

**Boughey, A. S.**
*Fundamental Ecology*
Scranton, Pa. (1971).

**Boughey, Arthur S.**
*Man and the Environment*
Riverside, N.J. (1971)

**Bouillene, R.**
"Man, The Destroying Biotype"
*Science* 135:706-712 (1962)

**Boulding, Kenneth E.**
"Ecology and the Environment"
*Trans-Action*, p. 38-44
(March, 1970)

**Bourliere, Francois**
*The Natural History of Mammals*
New York (1954)

**Bovbjerg, R. C.**
"Ecological Isolation and Competative Exclusion in Two Crayfish *(Oaconetes virilisand* and *Orconetes immunis)*
*Ecology*, 51:225-236 (1970)

**Bove, John L. and Stanley Siebenberg**
"Airborne Lead and Carbon Monoxide at 45th St., New York"
*Science*, 167: 986-987 (1970)

**Bowen, Croswell**
"Donora, Pa."
*Atlantic*, p. 27-34 (Nov., 1970)

**Brady, Nyle C., ed.**
Agriculture and the Quality of Our Environment
Washington (1967)

**Bragdon, Clifford R.**
"Noise-a Syndrome of Modern Society"
*Scientist and Citizen*, p. 29-37
(March, 1968)

**Brainerd, J. W.**
*Nature Study for Conservation*
New York (1970)

**Braun, E. Lucy**
*Deciduous Forests of Eastern North America*
Philadelphia (1950)

**Bregman, Jacob I.**
"Membrane Processes Gain Favor for Water Reuse"
*Environmental Science and Technology*, 4:296-306 (1970)

**Breidenbach, Andrew W. and Richard W. Eldredge**
"Research and Development for Better Solid Waste Management"
Bioscience, 19:984-988

**Breland, Osmond P.**
*Animal Life and Lore*
New York (1964)

**Bresler, Jack B.**
*Environments of Man*
Reading, Mass. (1968)

**Bresler, Jack B. (ed.)**
*Human Ecology: Collected Readings*
Reading, Mass. (1966)

**Brierley, John**
*A Natural History of Man*
Cranbury, N.J.: Fairleigh Dickinson Univ. Press (1970)

**Brodeur, Paul**
"The Magic Mineral"
*New Yorker*, p. 117-165
(Oct. 12, 1968)

**Brodine, Virginia**
"Episode 104"
*Environment*, p. 2-27
(Jan.-Feb. 1971)

**Brody, Boruch A. (ed.)**
*Readings in the Philosophy of Science*
Englewood Cliffs, N.J. (1970)

**Broecker, Wallace S.**
"Man's Oxygen Reserves"
*Science* 168:1537-1538 (1970)

**Brookhaven Symposia in Biology**
*Diversity and Stability in Ecological Systems*
Springfield, Va. (1969)

**Brooks, C. E. P.**
*Climate Through the Ages*
London (1949)

**Brooks, Douglas L.**
"A Statement of the Problem: Environmental Quality Control"
*Bioscience*, 17:873-877 (1967)

**Brooks, Paul**
"The Plot to Drown Alaska"
*Atlantic*, p. 53-59 (May 1965)

**Brown, A. W.**
*Insect Control by Chemicals*
New York (1951)

**Brown, Harrison**
*The Challenge of Man's Future*
Chicago (1954)

**Brown, Harrison**
"Human Materials Production as a Process in the Biosphere"
*Scientific American*, 194-208
(Sept. 1970)

**Brown, Harrison S.**
"Population Resources and Technology"
*Bioscience* 18: 31-33 (1968)

**Brown, William**
"The Rape of Black Mesa"
*Sierra Club Bulletin*, p. 14-17
(Aug. 1970)

**Brues, C. T.**
*Insect Dietary*
Cambridge, Mass.: Harvard University Press (1946)

**Brunner, D. R. and Keller, D. J.**
*Sanitary Landfill Design and Operation* (1971)

**Bryerton, Gene**
*Nuclear Dilema*
New York (1970)

**Buchsbaum, Ralph and Mildred Buchsbaum**
*Basic Ecology*
New York (1957)

**Buchsbaum, Ralph and Lorus J. Milne**
*The Lower Animals: Living Invertebrates of the World*
New York (1960)

**Budowski, Gerardo**
"The Quantity-Quality Relationship in Environmental Management"
*Impact*, 20:235-246 (1970)

**"Age and Response to Sonic Booms"**
*Bulletin of Atomic Scientists*
p. 27-28 (May 1970)

**"Man and His Habitat:**
**The Polluted Air"**
*Bulletin of Atomic Scientists*
p. 7-25 (June 1965)

**Burnap, Robert**
Michigan Committee for
Environmental Info.
Environmental Info. Center,
P.O. Box 2281,
Grand Rapids, Mich.

**Burnet, F. McFarlane**
*Ecology and the Appreciation of
Life*
Sydney, Australia (1966)

**Burnet, F. McFarlane**
*Natural History of Infectious
Diseases*
Cambridge University Press (1962)

**Burns, William**
*Noise and Man*
Philadelphia (1969)

**Butcher, Devereaux**
*Exploring Our National Parks and
Monuments*
Boston (1969)

**Buxton, P. A.**
*Animal Life in Deserts*
London (1923)

**Bylinsky, Gene**
"The Limited War on Water
Pollution"
*Fortune*, p. 103 (Feb. 1970)

**Bylinsky, Gene**
"The Long Littered Path to Clean
Air and Water"
*Fortune*, p. 112 (Oct. 1970)

**Bylinsky, Gene**
"Metallic Menaces in the
Environment"
*Fortune*, p. 110 (Jan. 1971)

**Cahn, Robert**
*Will Success Spoil the National
Parks?*
Boston (1968)

**Cain, S. A.**
*Foundations of Plant Geography*
New York (1944)

**Cain, S. A. and G. M. Deo.Castro**
*Manual of Vegetation Analysis*
New York (1959)

**Cairns, John**
"We're in Hot Water"
*Scientist and Citizen*, p. 187-198
(Oct. 1968)

**Calder, Nigel**
*Eden was No Garden*
New York (1967)

**Calder, R.**
*Living with the Atom*
Chicago: Univ. of Chicago Press
(1962)

**Caldwell, Lynton Keith**
*Environment: A Challenge to
Modern Society*
Garden City, N.Y. (1971)

**Calhoun, Alex**
*Inland Fisheries Management*
Sacramento (1966)

**California Air Resources Board**
"Post-1974 Auto Emissions: A
Report from Calif."
*Environmental Science and
Technology* 4:288-294 (1970)

**California Ecology Center**
David Graber, 2179 Alliston Way,
Berkeley, Calif.

**Campaign Against Pollution**
Paul Booth, 611 W. Fullerton,
Chicago, Ill.

**Campbell, R. C.**
*Statistics for Biologists*
Cambridge, Mass.: Cambridge
Univ. Press (England) (1967)

**Canadian Council of Resource
Ministers**
*National Conference on Pollution
and the Environment*
Montreal (1966)

**Cane, B. V.**
*Advance of Life*
London (1966)

**Cannon, Julie**
"Timber Supply Act.
Anatomy of a Battle"
*Sierra Club Bulletin*, p. 8-11
(March 1970)

**Cansdale, G. S.**
*Animals and Man*
London (1952)

**Capital Community Citizens**
Donn J. D'Alessio, Secy.,
1109 Gilbert Rd., Madison, Wis.

**Caras, Roger A.**
*Last Chance on Earth* (1966)

**Carlquist, S.**
*Hawaii, a Natural History*
New York (1970)

**Carpenter, J. R.**
*An Ecological Glossary*
Norman, Okla.: Univ. of
Oklahoma Press (1938)

**Carpenter, J. R.**
"The Biome"
*American Midl. Nat.*, 21:75-91
(1939)

**Carpenter, J. R.**
"The Grassland Biome"
*Ecol. Monogr*, 10:617-684 (1940)

**Carpenter, Kathleen E.**
*Life in Inland Waters with Special
Reference to Animals*
London (1928)

**Carpenter, Richard A.**
"How Congress Focuses on the
Environment"
*Saturday Review*, p. 43
(Aug. 1, 1970)

**Carr, Archie**
"Green Sea Turtles in Peril"
*National Parks Magazine*, p. 19-24
(April 1970)

**Carr, Donald E.**
*The Breath of Life*
New York (1965)

**Carr, Donald E.**
"The Death of Sweet Waters.
The Politics of Pollution"
*Atlantic*, p. 93-106 (May 1966)

**Carson, Rachel**
*The Silent Spring* (1962)

**Carter, Luther J.**
"Air Pollution: Muskie Throws
down the Gauntlet"
*Science*, 169:841 (1970)

**Carter, Luther J.**
"Industrial Minerals: New Study
of How to Avoid a Supply Crisis"
*Science*, 170:147-148 (1970)

**Carter, Luther J.**
"Land Uses: Congress Taking Up
Conflict over Power Plants"
*Science*, 170:718-719 (1970)

**Carter, Luther J.**
"SST: Commercial Race or
Technology Experiment?"
*Science*, 169:352-355 (1970)

**Carter, Luther J.**
"Timber Management:
Improvement Implies New
Land-Use Policies"
*Science*, 170:1387-1390 (1970)

**Carter, Luther J.**
"Water Pollution: Control Program
Lags as Nixon Promises
Cleanup"
*Science*, 167:360-361 (1970)

**Carthy, J. D.**
*Animal Navigation: How Animals Find Their Way About*
New York (1951)

**Cassidy, Harold G.**
"On Incipient Environmental Collapse"
*Bioscience,* 17:878-882

**Caswell, Charles A.**
"Underground Waste Disposal: Concepts and Misconceptions"
*Environmental Science and Technology,* 4:642-647 (1970)

**Caulfield, Patricia**
*Everglades*
San Francisco (1970)

**Cause**
Dr. John Howells,
699 Elmwood Ave., Buffalo, N.Y.

**Center for Human Concerns**
Joseph Phelan,
136 West St., Keene, N.H.

**Center for Law and Social Policy**
20008 Hillyer Pl.,
Washington, D.C. 20009

**Center for the Study of Responsive Law**
1908 Q St., NW,
Washington, D.C. 20009

**Center for Study of Responsive Law Trustee**
Ralph Nader, P.O. Box 19367,
Washington, D.C. 20036

**Chalupnik, James D., ed.**
*Transportation Noises. A Symposium on Acceptability Criteria*
Seattle: Univ. of Washington Press (1970)

**Chacks, Joseph**
*The International Joint Commission between the United States and the Dominion of Canada*
New York: Columbia University Press (1932)

**Chandrasekhar, S., (ed.)**
*Asia's Population Problems*
London (1967)

**Chapman, R. N.**
*Animal Ecology with Special Reference to Insects*
New York (1931)

**Chedd, Graham**
*Sound from Communication to Noise Pollution*
Garden City, N.Y. (1970)

**Chisholm, J. Julian Jr. and Eugene Kaplan**
"Lead Poisoning in Children-Comprehensive Management and Prevention"
*Journal of Pediatrics,* 73:942-950 (1968)

**Charley, Richard J., ed.**
*Water, Earth and Man: A Synthesis of Hydrology, Geomorphology, and Socio-Economic Geography*
London (1969)

**Christensen, J. Roger (Dr.)**
Rochester Committee for Scientific Info.
5236 River Campus Station,
Rochester, N.Y. 14620

**Chow, Tsaihwa, J. and John L. Earl**
"Lead Aerosols in the Atmosphere: Increasing Concentrations"
*Science,* 169:577-580

**Chute, R. M.**
*Environmental Insights*
New York (1971)

**Cipolla, C.**
*Economic History of World Populations*
London (1964)

**Citizens Action Program (CAP)**
Paul R. Booth and
Rev. Leonard Dubi
600 W. Fullerton,
Chicago, Ill. 60614

**Citizens Against Air Pollution**
P. B. Venuto, Pres.,
193 Bangor Ave., San Jose, Calif.

**Citizens Committee Environmental Control, Mont. Co.**
Shirley Merkin, Chmn.,
1108 Rock Creek Dr.,
Wyncote, Pa.

**Citizens Committee for the Hudson Valley**
William Hoppen, P.O. Box 146,
Ardsley on Hudson, N.Y.

**Citizens Council for Cleaner Air of Central Indiana**
Donald L. Kramer,
615 N. Alabama St.,
Indianapolis, Ind.

**Citizens Environment Council**
Dr. Donald Roberts,
855 S. 8th St., Pocatello, Idaho

**Citizens Environmental Council**
C/O Lakeside Nature Center,
5600 Gregory Blvd.,
Kansas City, Mo.

**Citizens for a Better Environment**
Daniel Bowen, Pres.,
Fairchild Hall, Kansas St. Univ.,
Manhattan, Kan. 66502

**Citizens for a Better Environment**
Dr. Robert F. Mueller,
7004 Dolphin Rd., Lanham, Md.

**Citizens for Clean Air, Inc.**
Robert Kafin, Pres,, 502 Park Ave.,
New York, N.Y. 10022

**Citizens for a Clean Environment**
Mrs. Frank Etges,
1297 Sweetwater Dr.,
Cincinnati, Ohio

**Citizens League for Environmental Action Now! (CLEAN)**
Terry Elkins, Chmn.,
Univ. of W. Va., Morgantown, Va.

**Citizens for Environmental Improvements, Inc.**
Dr. Norma Johnson,
P.O. Box 30322, Lincoln, Neb.

**Claiborne, Robert**
*Climate, Man, and History*
New York (1970)

**C.L.E.A.N., Committee to Leave the Environment of America Natural**
Dr. James Williams, Box 103,
Starkville, Miss. 39759

**Clean Air Coordinating Committee**
Richard M. Kates,
1440 W. Washington Blvd.,
Chicago, Ill.

**Clapper, Louis S.**
"Crackdown on Water Pollution"
*National Wildlife,* p. 14-17
(Feb.-Mar. 1970)

**Clark, Earl**
"How Seattle is Beating Water Pollution"
*Harper's,* p. 91-95 (1967)

**Clarke, G. L.**
*Elements of Ecology*
New York (1954)

**Clark, John R.**
"Thermal Pollution and Aquatic Life"
*Scientific American,* p. 18-27
(March 1969)

**Clawson, Marion**
*Land for the Future*
Baltimore: Johns Hopkins Univ. Press (1960)

**Clawson, Marion and Burnell Held**
*The Federal Lands: Their Use and Management*
Baltimore: Johns Hopkins Univ. Press (1957)

**Clean Air Coordinating Committee**
John Kirkwood,
1440 Washington St.,
Chicago, Ill. 60607

**Clean Air for Washington**
J. Porter Relly,
1000 Aurora Ave. N.,
Seattle, Wash. 98109

**Clements, F. E. and V. C. Shelford**
*Bio-ecology*
New York (1939)

**Cloud, Preston, E. Jr.**
"Realities of Mineral Distribution"
*Texas Quarterly*, 11:103-126 (1968)

**Cloudsley-Thompson, J. L.**
*Animal Behavior*
New York (1956)

**Cloudsley-Thompson, J. L.**
*Biology of Deserts*
New York (1954)

**Cloudsley-Thompson, J. L.**
*Microbiology*
London (1967)

**Cloudsley-Thompson, J. L.**
*Rythmic Activity in Animal Physiology and Behavior*
New York (1961)

**Cloudsley-Thompson, J. L.**
*The oology of Tropical Africa*
(1969)

**Coalition for the Environmental St. Louis Region**
Charles A. Schweighauser,
Exec. Dir., 8515 Delmar Blvd.,
St. Louis, Mo.

**Coalition to Tax Pollution**
620 C St., S.E.,
Washington, D.C. 20003

**Cohen, J. E.**
*A Model of Simple Competition*
Cambridge, Mass.:
Harvard University Press (1966)

**Coker, R. E.**
*Streams, Lakes, Ponds*
Univ. of North Carolina Press (1954)

**Coker, R. E.**
*This Great and Wide Sea*
Univ. of North Carolina Press (1947)

**Coker, R. E.**
*This Great and Wide Sea: An Introduction to Oceanography Marine Biology*
New York (1969)

**Cole, LaMont C.**
"Can the World be Saved?"
*New York Times Magazine*
(March 31, 1969)

**Cole, LaMont C.**
"Man's Ecosystem"
*Bioscience* 18:243-248 (1968)

**Cole, LaMont C.**
"Thermal Pollution"
*Bioscience*, 19:989-992 (1969)

**Cole, L. J. and P. C. Nowell**
"Radiation Carcinogenesis: The Sequence of Events"
*Science*, 150:1782-1786 (1965)

**Coleman-Cooke, Jr.**
*The Harvest that Kills*
London (1965)

**Collins, W. B.**
*The Perpetual Forest* (1959)

**Colorado Committee for Public Information**
*Nuclear Explosives in Peacetime*
New York (1970)

**Colwell, R. N.**
"Remote sensing as a Means of Determining ecological Conditions"
*Bioscience*, 17:444-449 (1967)

**Colwell, R. N.**
"Remote Sensing of Natural Resources"
*Scientific American*, p. 54-69 (Jan. 1968)

**Comfort, Alexander**
*The Nature of Human Nature*
New York (1966)

**Committee for Environmental Action**
Mrs. Donna Haines,
7825 Accotink Pl.,
Alexandria, Va.

**Committee for Environmental Info.**
"The Space Available"
*Environment*, p. 2-9 (March 1970)

**Committee for Environmental Preservation**
Ken Hahn, 148 Mackenzie Hall,
Wayne St. Univ., Detroit, Mich.

**Committee on Environmental Quality of the Federal Council of Science and Technology**
*Noise-Sound Without Value*
Washington, D.C. (1968)

**Committee to Leave the Environment of America Natural (CLEAN)**
Boyd Gatlan, Box 643,
Starkville, Miss.

**Commoner, Barry**
"Beyond the Teach-In"
*Saturday Review*, p. 50
(April 4, 1970)

**Commoner, Barry**
*The Closing Circle*
New York (1971)

**Commoner, Barry**
"Lake Erie, Aging or Ill?"
*Scientist and Citizen*, p. 254-265
(Dec. 1968)

**Commoner, Barry**
"Nature Unbalanced. How Man Interferes with Nitrogen Cycle"
*Scientist and Citizen*, p. 9-19
(Jan.-Feb. 1968)

**Commoner, Barry**
"Nature Under Attack"
*Columbia University Forum*,
p. 17-22 (Spring 1968)

**Commoner, Barry**
*Science and Survival*
New York (1966)

**Commoner, Barry**
"Soil and Freshwater: Damaged Global Fabric"
*Environment*, p. 4-11 (April 1970)

**Concerned South Carolinans for Better Environment Inc.**
Box 5844, Columbia, S.C. 29205

**Connell, J. H.**
"The Influence of Interspecific Competition and Other Factors on the distribution of the barnacle *Chthamalus stellatus*"
*Ecology*, 42-710-723 (1961)

**Connel, J. H., Mertz, D. B. and Murdoch, W. W.**
*Readings in Ecology and Ecological Genetics* (1970)

**The Conservation Foundation**
1250 Connecticut Ave. N.W.,
Washington, D.C. 20036

**Conservation 70S**
Loring Lovell, Pres., Ste. 228,
Dorian Bldg., 319 S. Monroe St.,
Tallahassee, Fla. 32301

**Constable, John and Mathew Meselson**
"The Econlogical Impact of Large Scale Defoliation in Vietnam"
*Sierra Club Bulletin* (April 1971)

**Conway, Gordon R.**
"A Consequence of Insecticides"
*Natural History,* p. 46-54
(February 1969)

**Cook, J. G.**
*The Fight for Food*
New York (1957)

**Cooley, Richard**
*Politics and Conservation*
New York (1963)

**Cooper, Charles F. and
William C. Jolly**
*Ecological Effects of Weather,
Modification: A Problem Analysis*
Ann Arbor: Univ. of Michigan
Press (May 1969)

**Cooper, C. F.**
"The Ecology of Fire"
*Scientific American,* p. 150-160
(April 1961)

**Cooper, C. F.**
*The Variable Plot Method of
Estimating Shrub Density*
Chicago (1957)

**Corey, Richard C.**
"Principles and Practices of
Incineration"
New York (1969)

**Corner, E. J. H.**
*The Life of Plants*
New York (1964)

**Cott, H. B.**
*Adaptive Coloration in Animals*
New York: Oxford University Press
(1940)

**Cottam, G. and J. T. Curtis**
"The use of distance measures in
phytosociological sampling"
*Ecology,* 37:451-460 (1956)

**Council on Environmental Issues**
Paul H. Templet, Chmn.,
1025 Carrollton Ave.,
Baton Rouge, La.

**Cox, George W.**
*Readings in Conservation Ecology*
New York (1969)

**Cragg, J. B.**
*Advances in Ecological Research*
New York Vol. I (1962); Vol. II
(1964); Vol. III (1966); Vol. IV
(1967); Vol. V 1968; Vol. VI (1969);
Vol. VII (1971)

**Craig, W. S.**
"Not a Question of Size"
*Environment,* p. 2-5 (June 1970)

**Crowe, Philip Kingsland**
*World Wildlife, The Last Strand*
New York (1970)

**Croxton, F. E.**
*Elementary Statistics with
Applications in Medicine and the
Biological Sciences*
New York (1953)

**Cruickshank, Helen Gere**
*A Paradise of Birds*
New York (1968)

**Crutchfield, James A. and
Guilio Pontecorvo**
*The Pacific Salmon Fisheries,
a Study of Irrational
Conservation*
Baltimore: Johns Hopkins U. Press
(1964)

**Culp, R.**
"Water Reclamation at South
Tahoe".
*Water and Wastes Engineering,*
p. 36-39 (April 1969)

**Curtis, Richard and Elizabeth Hogan**
"The Myth of the Peaceful Atom"
*Natural History,* p. 6-16, 71-76
(March 1969)

**Curtis, Richard and Elizabeth Hogan**
*The Perils of the Peaceful Atom*
New York (1969)

**Dahlsten, Donald L.**
*Pesticides*
New York (1970)

**Dale, M. B.**
"Systems Analysis and Ecology"
*Ecology* 51:2-16 (1970)

**Dales, J. H.**
*Pollution, Property & Prices*
Toronto: Univ. of Toronto Press
(1968)

**Dansereau, Pierre**
*Biogeography*
New York (1957)

**Dansereau, Pierre**
"Challenge for Survival"
New York: Columbia Univ. Press
(1970)

**Darling, F. F. and Dasman, R. F.**
"The Ecosystem View of Human
Society"
*Impact of Science on Society,*
Vol. XIX, No. 2 (April-June 1969)

**Darling, F. F. and J. P. Milton, eds.**
*Future Environments of North
America*
New York (1966)

**Darling, F. Fraser and
Noel D. Eichhorn**
*Man and Nature in the National
Parks*
Washington, D.C. (1969)

**Darlington, P. J. Jr.**
*Zoogeography: The Geographic
Distribution of Animals*
New York (1957)

**Darnell, R. N.**
*Organism and Environment*
San Francisco (1971)

**Dasmann, Raymond F.**
"Conservation in the Antarctic"
*Antarctic Journal* (Jan.-Feb. 1968)

**Dasmann, R. F.**
*The Destruction of California*
New York (1965)

**Dasmann, R. F.**
"A Different Kind of Country"
New York (1968)

**Dasmann, R. F.**
"Environmental Conservation"
New York (1968)

**Dasmann, R. F.**
*The Last Horizon*
New York (1963)

**Daubenmire, Rexford F.**
*Plants and Environment: A
Textbook of Plant Autecology*
(1959)

**David, D. H. S., (ed.)**
*Ecological Studies in Southern
Africa*
New York (1964)

**Delafons, John**
*Land-Use Controls in the
United States*
Cambridge, Mass.: MIT Press
(1969)

**Davies, J. Clarence**
*The Politics of Pollution*
New York (1970)

**DeBach, Paul, (ed.)**
*Biological Control of Insect Pests
and Weeds*
New York (1964)

**de Bell, Garrell, (ed.)**
*The Environmental Handbook*
New York (1970)

**de Bell, G.**
*The Voter's Guide to
Environmental Politics*
New York (1970)

**Deevey, Edward S. Jr.**
"Mineral Cycles"
*Scientific American,* 148-158
(Sept. 1970)

**Deichmann, W. B. and Gerarde, H. W.**
*Toxicology of Drugs and Chemicals*
New York (1969)

**Delaware Citizens for Clean Air, Inc.**
Jacob Kreshtool, Pres.,
1308 Delaware Ave.,
Wilmington, Del.

**Del Moral, R. and R. G. Cates**
"Allelopathic potential of the dominant vegetation of Western Washington"
*Ecology,* 52:1030-1037 (1971)

**Delvey, E. S.**
"The Quality of Environment"
*Topics in the Study of Life,*
372-381 (1971)

**Delwiche, C. C.**
"The Nitrogen Cycle"
*Scientific American,* 136-146
(Sept. 1970)

**De Org, R. E.**
*Chemistry and Uses of Pesticides*
New York (1956)

**Department of Conservation**
Dr. Alfred Eipper, Cornell Univ.,
Ithaca, N.Y.

**Department of Northern Affairs and National Resources**
*Resources for Tomorrow Conference: Backgrounds Papers and Proceedings,* 3 vols.
Ottawa (1961)

**Detwyler, Thomas R.**
*Man's Impact on Environment*
New York (1971)

**Diamond, J.**
"Avifaunal equilibria and Species Turnover Rates on the Channel Island of California"
*Pr. Nat. Acad. Sci.,* (U.S.) 64:57-63
(1969)

**Dice, L. R.**
*The Biotic Provinces of North America*
Ann Arbor: Univ. of Michigan Press (1943)

**Dice, L. R.**
*Natural Communities*
Ann Arbor: Univ. of Michigan Press (1952)

**Dice, L. R.**
*Man's Nature and Nature's Man: The Ecology of Human Communities*
Univ. of Michigan Press (1955)

**Dickerman, Ernest M.**
"The National Park Wilderness Review"
*The Living Wilderness,* p. 40-49
(Spring 1970)

**Dickinson, Robert E.**
*Regional Ecology. The Study of Man's Environment*
New York (1970)

**Dimmick, Robert L. and Ann B. Akers, (eds.)**
*An Introduction to Experimental Aerobiology*
New York (1969)

**Disch, Robert**
*The Ecological Conscience. Values for Survival*
Englewood Cliffs, N.J. (1970)

**Dixon, J. P. and J. P. Lodge**
"Air Conservation Report Reflects National Concern"
*Science,* 148:1060-1066 (1965)

**Dobzhansky, Theodosius**
*Man-kind Evolving*
Yale University Press (1962)

**Dorst, Jean**
"Before Nature Dies"
Boston (1970)

**Dorst, J.**
*The Navigation of Birds*
London (1962)

**Douglas, W. O.**
*A Wilderness Bill of Rights*
Boston (1965)

**Doutt, R. L. and W. W. Kilgore**
*Pest Control*
New York (1967)

**Dowdeswell, W. H.**
*Practical Animal Ecology*
London (1959)

**Dregne, Harold E., (ed.)**
*Arid Lands in Transition*
Washington (1970)

**Drew, Elizabeth**
"Dam Outrage: The Story of the Army Engineers"
*Atlantic,* p. 51-62 (April 1970)

**Dubos, Rene**
"Adapting to Pollution"
*Scientist and Citizen,* p. 1-8
(Jan.-Feb. 1968)

**Dubos, Rene**
*The Dreams of Reason: Science and Utopias*
New York: Columbia Univ. Press
(1961)

**Dubos, Rene**
"The Human Environment in Technological Societies"
*Rockefeller Univ. Review,* p. 2-11
(July-Aug. 1968)

**Dubos, Rene**
*Man Adapting*
New Haven: Yale University Press
(1965)

**Dubos, Rene**
*So Human An Animal*
New York (1968)

**Dyson, James L.**
*The World of Ice*
New York (1962)

**Earth Action Council**
Frank Steen, UCLA, Box 24390,
Los Angeles, Calif.

**ECOS**
Jeannette Luccas, Bd. of Dir.,
Box 4787, Duke Station, Durham,
N.C. 27706

**ECCO (Ecological Coordinating Committee for Okla.)**
Victor L. Jackson,
618 N.E. 15th St., Okla. City, Okla.

**Eckardt, F. E., ed.**
*Functioning of Terrestial Ecosystems at the Primary Production Level*
Paris (1968)

**Eckert, Allen W.**
*Bayou Backwaters*
New York (1968)

**Eckert, Allen W.**
*Wild Season*
New York (1967)

**Eco-Center: Environmental Clearing House**
Sandra H. Cooper, 1424 Pearl St.,
Boulder, Colo. 80302

**Ecology Action Inc.**
Florence Kobernick,
P.O. Box 4661,
Baltimore, Md. 21212

**Ecology Action**
Ted Field, 123 W. Park Ave.,
San Antonio, Tex. 78212

**Ecology Action Educational Institute Inc.**
Clifford C. Humphrey, Box 3895,
Modesto, Calif. 95252

**Ecology Action for Rhode Island Inc.**
Dr. Richard N. Keogh, Pres.,
286 Thayer St., Providence, R.I.

**Ecology Action-Southern Nevada**
Bruce Miller, c/o Dept. of Biology,
Univ. of Nevada, Las Vegas,
Nev. 89109

**Ecology Action-ZPG-San Diego State**
Clay Kemper, 6271 Madeleine St.,
San Diego, Calif. 92115

**Ecology Center of Louisiana, Inc.**
Ross Vincent, Box 15149,
New Orleans, La. 70115

**Ecology Center, Student Activities**
University of Utah, Brian Mason,
Dir., Salt Lake City, Utah

**Ecology Center**
Bill Painter, Dir.,
3256 Prospect St. NW,
Washington, D.C.

**Ecology: The Journal of
Cultural Transformation**
Mary Hunphrey, ed.,
Modesto, Calif. (1969)

**Edwards, C. A.**
"Soil Pollutants and Soil Animals"
*Scientific American*, p. 88-99
(1969)

**Edwards, R. Y. and C. D. Fowle**
*The Concept of Carrying Capacity*
Transcript of 20th No. American
Wildlife Conference (1955)

**Ehlers, V. M. and E. W. Steel**
*Municipal and Rural Sanitation*
(1958)

**Ehrenfeld, David W.**
*Biological Conservation*
New York (1970)

**Ehrensaft, Philip and Amitai Etzioni**
*Anatomies of America*
New York (1969)

**Ehrensvard, G.**
*Life: Origin and Development*
Chicago: University of Chicago
Press (1962)

**Ehrlich, Paul R.**
"The Biological Revolution"
*Center Magazine*, p. 28-49
(Nov. 1969)

**Ehrlich, P. R.**
"Man is the Endangered Species"
*National Wildlife*, p. 38-39
(Apr.-May 1970)

**Ehrlich, P. R.**
*The Population Bomb, Population
Control or Race to Oblivion*
New York (1968)

**Ehrlich, Paul**
"We're Standing on the Edge of
the Earth"
*National Wildlife*, p. 16-17,
(Oct.-Nov. 1970)

**Ehrlich, Paul, et al.**
"The Biological Revolution"
*Center Magazine*, p. 28-49
(Nov. 1969)

**Ehrlich, Paul R. and Anne H.**
"The Food-From-the-Sea Myth"
*Saturday Review*, p. 53-
(April 4, 1970)

**Ehrlich, Paul R. and
Anne H. Ehrlich**
"Population Resources
Environment"
San Francisco (1970)

**Ehrlich, Paul R. and
Harriman, Richard L.**
*How to be a Survivor,* a Plan to
Save Spaceship Earth
New York (1971)

**Ehrlich, Shelton**
"Air Pollution Control Through
New Combustion Processes"
*Environmental Science and
Technology* 4:396-400 (1970)

**Eipper, Alfred W.**
"Pollution Problems, Resource
Policy and the Scientist"
*Science*, 169:11-15 (1970)

**Eipper, Alfred W., C. A. Carlson and
L. S. Hamilton**
"Impact of Nuclear Power Plants
on the Environment"
*The Living Wilderness*, p. 5-12
(Autumn 1970)

**Eiseley, Loren**
*The Invisible Pyramid*
New York (1970)

**Eiseley, Loren**
*The Unexpected Universe*
New York (1969)

**Ekirch, Arthur A.**
*Man and Nature in America*
New York: Columbia University
Press (1963)

**Ellenberg, H.**
*Integrated Experimental Ecology*
New York (1971)

**Elrick, David E.**
"The Land: Its Future-
Endangering Pollutants"
*Impact*, 19:195-207 (1969)

**Elton, C.**
*Animal Ecology*
New York (1927)

**Elton, C.**
*The Ecology of Animals*
New York (1950)

**Elton, C.**
*The Ecology of Invasions by
Animals and Plants*
New York (1958)

**Elton, C.**
*Moles, Mice and Lemmings*
Oxford (1942)

**Engdahl, Richard B.**
*Solid Waste Processing: A State-
of-the-Art Report on Unit
Operations and Processes*
New York (1969)

**Environment**
Tom Stokes, 119 Fifth Ave.,
Rm. 600, New York, N.Y. 10003

**Environment Magazine of the
St. Louis Committee for Public
Information, Inc.**
Sheldon Novick (editor)
St. Louis, Mo. (1958 to the present)

**Environmental Action**
Sam Love, ed.
Washington, D.C. (1970)

**Environmental Action**
c/o King Council,
207 Hampshire House,
Univ. of Mass.
Amherst, Mass. 01002

**Environmental Action**
Weldon Wellingham,
2216 S.E. 180th St., Portland, Ore.

**Environmental Action**
Dennis Knight, Box 3095,
Laramie, Wyo.

**Environment Action Bulletin**
Jerome Goldstein, ed.
Emmaus, Pa. (1970)

**Environmental Action Group**
Allen Sandler, 323 Reitz Union,
Univ. of Fla., Gainesville, Fla.

**Environmental Action Group**
James Ream, Rt. 1, Box 124-A,
Big Spring, Tex. 79720

**Environmental Action Group**
Arthur G. Cleveland, Coord.,
Box 138, Ft. Worth, Tex. 76105

**Environmental Action of Colorado**
Morey Wolfson, Dir.,
1100 14th St., Denver, Colo. 80202

**Environmental Action-Zero Pop. Growth**
Marc Dennis Hiller, Coord.,
4104 Fifth Ave., Schenley Hall,
Pittsburgh, Pa. 15213

**Environmental Action Committee of Mt. Vernon Area**
Doann Haines & John Schelleng,
Co-Chmn.
2000 Janestown Rd.,
Alexandria, Va. 22308

**Environmental Action Council of Memphis**
S. Henry Hall,
2789 Sky Lake Cove,
Memphis, Tenn. 38112

**Environmental Action for Survival—ENACT**
Toby Cooper, Secy., 146 F.
School of Natural Resources,
University of Michigan,
Ann Arbor, Mich. 48104

**Environmental Awareness Society**
Larry Geisman, Box 878,
University of Kentucky,
Lexington, Ky.

**Environmental Awareness Committee**
Patricia G. World,
Green Mountain College,
Poultney, Vt.

**Environmental Conservation Organization**
Tom Dietz, Kent State Univ.,
Kent, Ohio 44240

**Environmental Council of Oak Ridge—ECOR**
Robin Wallace,
105 Monticello Rd.,
Oak Ridge, Tenn. 37839

**Environmental Defense Fund**
1901 N St. NW,
Washington, D.C. 20036

**Environmental Defense Fund**
Roderick A. Cameron, Exec. Dir.,
162 Old Town Rd.,
E. Setauket, N.J.

**Environmental Education Committee Inc.**
David McKain, S.E. Branch,
Univ. Conn., Avery Point
Groton, Conn. 06340

**Environment Group Hawaii**
Jane Proctor/Mark Cockrill,
P.O. Box 1618, Honolulu,
Hawaii 96806

**Environmental Information Center**
Gerry Slater, Pres.,
3207 N. Hackett Ave.,
Milwaukee. Wis.

**Environmental Pollution Panel**
"Restoring the Quality of Our Environment"
Washington (1965)

**Environmental Quality Magazine**
Jeff W. Carter, Ed.
Woodland Hills, Calif. (1970)

**Environmental Science and Technology**
James J. Morgan, ed.
Washington, D.C. (1967)

**Environmental Studies Board**
*Jamaica Bay and Kennedy Airport*
Washington, D.C. (1971)

**Environmental Studies Center**
Robert Bieri, Antioch College,
Yellow Springs, Ohio 45384

**Environmental Teach-In**
H. L. Goodell, Medical Bldg.,
Biology Dept., Univ. of South
Dakota, Vermillion, S.D. 57069

**Epstein, Samuel S.**
"Control of Chemical Pollutant"
*Nature*, 228:816-819 (1970)

**Epstein, Samuel S.**
"NTA"
*Environment*, p. 2-11
(September 1970)

**Erichsen, J. J. R.**
*Fish and River Pollution*
London (1964)

**Erichsen-Brown, J. P.**
"Legal Implications of Boundary Waters Pollution"
*Buffalo Law Review* 17:65-69
(Fall 1967)

**Errington, Paul L.**
*Of Predation and Life*
New York (1967)

**Esposito, John C.**
*Vanishing Air*
New York (1970)

**Evans, David M. and Albert Bradford**
"Under the Rug"
*Environment*, p. 3-13, 31
(Oct. 1969)

**Evans, F. C.**
"Relative Abundance of Species and the Pyramid of Numbers"
*Ecology* 31:631-632 (1950)

**Evans, James O.**
"The Soil as a Resource Renovator"
*Environmental Science and Technology* 4:732-735 (1970)

**Evenari, Michael**
"The Land: Ecological Farming"
*Impact*, 19:209-216 (1969)

**Everyman's Way**
Greg Cailliet, (Ed.)
Santa Barbara, Calif.
(1968 to present)

**Ewald, William R.**
*Environment for Man: The Next Fifty Years*
Bloomington, Ind.: Indiana Univ.
Press (1967)

**Fairbrother, Nan**
*New Lives, New Landscapes. Planning for the 21st Century*
New York (1970)

**Family Planning Perspectives**
Richard Lincoln, (Ed.)
New York (1969)

**Farb, Peter and the Editors of Life**
*Ecology*
New York (1963)

**Farb, Peter (ed.)**
*Living Earth*
New York (1959)

**Farvar, M. Taghai and John P. Milton, eds.**
*The Careless Technology*
New York (1971)

**Farvar, M. Taghai**
"The Unforseen International Ecologic Boomerang"
*Natural Hisory*, p. 41-72
(Feb. 1969)

**Featherstone, Joseph**
"The Silent Epidemic"
*New Republic*, p. 13-14
(Nov. 8, 1969)

**Feiss, Julian W.**
"Minerals"
*Scientific American*, 209:128-136
(1963)

**Fenner, J. and F. N. Ratcliffe**
*Myxomatosis*
Cambridge, Mass.: Cambridge
Univ. Press (1965)

**Ferkiss, V. C.**
"Toward the Creation of Technological Man"
*Technological Man: The Myth and the Reality*
New York (1969)

**Fermi, Laura**
"Cars and Air Pollution"
*Bulletin of the Atomic Scientists*,
p. 35-37 (Oct. 1969)

**Ferry, W. H.**
"The Unanswerable Questions"
*Center Magazine*, p. 2-7
(July 1969)

**Fertig, Fred**
"Child of Nature. The American
Indian as an Ecologist"
*Sierra Club Bulletin*, p. 4-7
(Aug. 1970)

**Fishbein, Lawrence, W. Gary Flamm
and Hans L. Falk**
*Chemical Mutagens,
Environmental Effects on
Biological Systems*
New York (1970)

**Fisher, James, Noel Simon and
Jack Vincent**
*Wildlife in Danger*
New York (1969)

**Fisher, R. A. and F. Yates**
*Statistical Tables for Biological
and Medical Research*
Edinburgh (1948)

**Fittkau, E. J.**
*Biogeography and Ecology in
South America*, Vol. I & II
New York (1968-69)

**Flaherty, D. L. and C. B. Huffaker**
"Biological Control of Pacific
Mites and Willamette Mites in San
Joaquin Valley Vineyards"
*Hilgardia* 40:267-330 (1970)

**Florkin, M. and Schoffeniels, E.**
*Molecular Approaches to Ecology*
New York (1969)

**Fonselius, Stig H.**
"Stagnant Sea"
*Environment*, p. 2-11, 40-48
(July-Aug. 1970)

**Forbes, R. J.**
*The Conquest of Nature,
Technology and its Consequences*
New York (1968)

**Forbes, R. J. and Dijksterhuis, E. J.**
*A History of Science and
Technology—18th and 19th
Century* (2 vols.)
London (1963)

**Forbes, Stephen A.**
"The Lake as a Microcosm"
*Natural History Survey Bulletin
No. 15*
Urbana, Ill. (Reprint of 1925)

**Ford, E. B.**
*Ecological Genetics*
London (1964)

**Ford Foundation**
"Ecology: The New Great Chain
of Being"
*Natural History*, p. 8-16, 60-69
(Dec. 1968)

**Foreman, H.**
*Nuclear Power and the Public*
Minneapolis: Univ. of Minnesota
Press (1970)

**Fosberg, F. R. (ed.)**
*Man's Place in the Island
Ecosystem*
Honolulu (1963)

**Fox, Irving K.**
"The Use of Standards in
Achieving Appropriate Levels of
Tolerance"
*National Academy of Sciences*,
67:877-886 (1970)

**Frank, Bernard and Anthony Netboy**
*Water, Land, and People*
New York (1950)

**Fraser, D.**
*People Problems*
Bloomington, Ind.:
Indiana Univ. Press (1971)

**Frazer, R.**
*The Habitable Earth*
London (1964)

**Freeman, A., Myrick III and
Leonard Ross**
"Cleaning up Foul Waters"
*New Republic*, p. 13-16
(June 20, 1970)

**Friedlander, C. P.**
*Heathland Ecology*
London (1960)

**Friedlander, Michael W. and
Joseph Klarmann**
"How Many Children?"
*Environment*, p. 2-13 (Dec. 1969)

**Friends of the Earth**
David R. Bower, 529 Commercial
St., San Francisco, Calif.

**Fuller, R. Buckminster**
*Operating Manual for Spaceship
Earth* (1969)

**Fuller, R. Buckminster**
*Utopia or Oblivion*
New York (1969)

**Gabrielson, Ira N.**
*Pest Control and Wildlife
Relationships: A Symposium*
Washington, D.C. (1961)

**Gabrielson, I. N.**
*Wildlife Conservation*
New York (1959)

**Gallup Organization**
*The U.S. Public Considers Its
Environment*
Washington, D.C. (1969)

**Galton, Arthur W.**
"Plants, People and Politics"
*Bioscience*, 20:405-410 (1970)

**Garlick, J. P. and Keay, R. W. J.**
*Human Ecology in the Tropics*
New York (1970)

**Gates, D. M.**
*Energy Exchange in the Biosphere*
(1962)

**Gates, D. M.**
"Energy, Plants and Ecology"
*Ecology* 46:1-13 (1965)

**Gates, D. M.**
"Microclimatology"
*Topics in the Study of Life*
Philadelphia (1971)

**Gates, Paul W.**
*History of Public Land Law
Development*
Washington (1968)

**Gause, G. F.**
*The Struggle for Existence*
Baltimore (1934)

**Gaussens, Jacques and Robert Bonnet**
"The Applications of Nuclear
Energy"
*Impact*, 17:75-100 (1967)

**Geiger, R.**
*The Climate Near the Ground*
Cambridge, Mass.: Harvard
Univ. Press (1957)

**Georgia Conservangcy Inc.**
Walter M. Mitchell, Chmn.,
3376 Peachtree Rd., Atlanta, Ga.

**Gerking, S. D.**
*Biological Systems*
Philadelphia (1969)

**Giles, R., ed.**
*Wildlife Managament Techniques*
Washington, D.C. (1969)

**Gillette, Robert**
"The Economics of Lead
Poisoning"
*Sierra Club Bulletin*, p. 14-17
(Sept. 1970)

**Gilpin, A.**
*Control of Air Pollution*
London (1963)

**Gilpin, Alan**
*Dictionary of Fuel Technology*
New York (1969)

**Glass, David C.**
*Biology and Behavior:*
*Environmental Influences*
New York: Rockefeller Univ. Press
(1968)

**Glover, K. M., K. R. Hardy,**
**T. G. Konrad, W. N. Sullivan and**
**A. S. Michaels**
"Radar Observations of Insects in
Free Flight"
*Science*, 154:967-972 (1966)

**Godfrey, Arthur**
*Environmental Reader,*
*A Compilation*
New York (1970)

**Gofman, John W. and Arthur Tamplin**
"Radiations: The Invisible
Casualties"
*Environment*, p. 12-19, 49
(April 1970)

**Goldman, Marshall I., ed.**
*Controlling Polution: The*
*Economics of a Cleaner America*
Englewood Cliffs, N.J. (1967)

**Goldman, Marshall I.**
"The Convergence of
Environmental Disruption"
*Science*, 170:37-42 (1970)

**Goldring, Irene P.**
"Pulmonary Hemmorage in
Hamsters after Exposure to
Proteolytic Enzymes of Bacillus
subtilis"
*Science*, 170:73-74 (1970)

**Goldsmith, J. R. and S. A. Landaw**
"Carbon Monoxide and
Human Health"
*Science*, 162:1352-1359 (1969)

**Goldsmith, Maurice**
"Crisis in Aspen"
*Bulletin of the Atomic Scientists*,
p. 18-30, 45 (Nov. 1970)

**Goldstein, Jerome**
*Garbage as You Like It*
Emmaus, Pa. (1969)

**Golley, F. B. and H. K. Beuchner**
*A Practical Guide to the Study of*
*the Productivity of Large*
*Herbivores*
Oxford (1968)

**Golley, F. B., McGinnis, J. T.,**
**Clements, R. G., Child, G. I. and**
**Duever, M. J.**
"The Structure of Tropical Forests
in Panama and Columbia"
*Bioscience*, 19:693-696 (1969)

**Gomer, Robert**
"The Tyranny of Progress"
*Bulletin of the Atomic Scientists*,
p. 4-8 (1968)

**Good, Ronald**
*Geography of Flowering Plants*,
2nd ed.,
New York (1953)

**Goodall, D. W.**
"Statistical Plant Ecology"
*Anun. Review Ecology and*
*Systematics*, Vol. 1:99-124 (1970)

**Goodman, Gordon T.**
*Ecology and the Industrial Society*
New York (1965)

**Gowett, J.**
*Life in Ponds*
New York (1970)

**Gotaas, Harold B.**
"Outwitting the Patient Assassin:
The Human Use of Lake Pollution"
*Bulletin of the Atomic Scientists*,
p. 8-10 (May 1969)

**Gough, William C. and**
**Bernard J. Eastlund**
"The Prospects of Fusion Power"
*Scientific American*, p. 50-64
(Feb. 1971)

**Graham, Edward H.**
*Natural Principles of Land Use*
New York: Oxford Univ. Press
(1944)

**Graham, E. H.**
*Water for America*
New York: Oxford Univ. Press
(1965)

**Graham, Frank**
*Disaster by Default:*
*Politics and Water Pollution*
New York (1966)

**Graham, Frank Jr.**
"Pesticides"
*Atlantic* 22-25 (September 1970)

**Graham, Frank Jr.**
"Pesticides, Politics and the
Public"
*Audubon Magazine*, 69:54-62
(1967)

**Graham, Frank Jr.**
*Since Silent Spring*
Boston (1970)

**Grava, Sigurd**
*Urban Planning Aspects of Water*
*Pollution Control*
New York: Columbia Univ. Press
(1969)

**Gray, Peter**
*The Dictionary of the Biological*
*Sciences*
Philadelphia (1967)

**Greater Alliance to Stop Pollution**
**(G.A.S.P.)**
Dr. Dan Prince, Pres.,
900 S. 18th St., Birmingham, Ala.

**Green, I.**
*Water, Our Most Valuable*
*Resource*
New York (1958)

**Green, L. Jr.**
"Energy Needs Versus
Environmental Pollution.
A Reconilation?"
*Science*, 156:1448-1450 (1967)

**Greene, Wade**
"What Happened to the Attempts
to Clean up the Majestic,
the Polluted Hudson?"
*New York Times Magazine*, p. 27
(May 10, 1970)

**Greig-Smith, P.**
*Quantative Plant Ecology* 2nd ed.
London (1964)

**Griffin, William L.**
"A History of the Canadian-
United States Boundary Waters
Treaty of 1909"
*University of Detroit Law Journal*,
37:76-96 (Oct. 1959)

**Grinstead, Robert B.**
"The New Resource"
*Environment*, p. 2-17 (Dec. 1970)

**Grinstead, Robert B.**
"No Deposit, No Return"
*Environment*, p. 17-23 (Nov. 1969)

**Grossinger, R.**
*Solar Journal: Oecological*
*Sections*
Los Angeles (1970)

**Grossman, Mary Louise, S. Grossman**
**and J. J. Hamlet**
*Our Vanishing Wilderness*
New York (1969)

**Grossman, Sally**
*The Struggle for Life in the*
*Animal World*
New York (1967)

**Gruchow, Nancy**
"Detergents: Side Effects of the
Washday Miracles"
*Science*, 167:151 (1970)

**Grzimek, Bernhard**
"The Last Great Herds of Africa"
*Natural History*, p. 8-21
(January 1961)

**Grzimek, Bernard and Michael Grzimek**
*Serengeti Shall Not Die*
London (1961)

**Guggisberg, C. A. W.**
*Man and Wildlife*
New York (1970)

**Gulland, J. A.**
*Manual of Sampling and Statistical Methods of Fishery Biology*, Part 1
Rome (1966)

**Gullion, Edmund A., ed.**
*Uses of the Seas*
Engelwood Cliffs, N.J. (1968)

**Gunter, Peter A.**
"Mental Inertia and Environmental Decay: The End of and Era"
*Living Wilderness*, p. 3-7
(Spring 1970)

**Guthrie, John A. and George R. Armstrong**
*Western Forest Industry: An Economic Outlook*
Baltimore: Johns Hopkins Univ. Press (1961)

**Guyol, N. C.**
*The World Electric Power Industry*
Berkeley, Calif.: University of Calif. Press (1969)

**Haagen-Smit, A. J.**
"The Control of Air Pollution"
*Scientific American*, p. 24-31
(Jan. 1964)

**Haagen-Smit, A. J.**
"A Lesson from the Smog Capital of the World"
*Academy of Sciences*, 67:887-897
(1970)

**Haagen-Smit, A. J.**
"Man and His Home"
*The Living Wilderness*, p. 38-46
(Summer 1970)

**Haddow, A., ed.**
*Biological Hazards of Atomic Energy*
Oxford University Press (1952)

**Haden-Guest S., Wright, T. K. and Teclaff, E. M.**
*A World Geography of Forest Resources*
New York (1956)

**Haines, Francis**
*The Buffalo*
New York (1970)

**Halcomb, Robert W.**
"Power Generation: The Next 30 Years"
*Science*, 167:159-160 (1970)

**Hall, Warren A. and John A. Dracup**
*Water Resources Systems Engineering*
New York (1970)

**Halliday, W. R. and A. M. Woodbury**
"Protection of Rainbow Bridge National Monument"
*Science*, 133:1572-1583 (1961)

**Hammond, Allen L.**
"Mercury in the Environment, Natural and Human Factors"
*Science*, 171:788-789 (1971)

**Hammond, P. B.**
*Lead Poisoning an Old Problem with a New Dimension*
New York (1969)

**Handler, P., ed.**
*Biology and the Future of Man*
Oxford University Press (1970)

**Hanks, Thrift G.**
*Solid Waste/Disease Relationships*
Public Health Service Publ. #999-UIH-6
Washington (1967)

**Hanson, Earl D.**
*Animal Diversity* (1961)

**Hanson, H. C.**
*Dictionary of Ecology*
London (1962)

**Hanson, Herbert C. and Eathan D. Churchill**
*The Plant Community* (1961)

**Hardenbergh, W. A. and E. B. Rodie**
*Water Supply and Waste Disposal*
Scranton, Pa. (1966)

**Hardin, Garrett**
"Ghost of Authority"
*Perspectives in Biology and Medicine*, p. 289-297 (1966)

**Hardin, G.**
"Nobody ever Dies of Overpopulation"
*Science* 171:52 (1971)

**Hardin, G.**
*Population, Evolution and Birth Control*, 2nd Ed.
San Francisco (1969)

**Hardy, Alister C.**
*Great Waters*
New York (1968)

**Hardy, A. C.**
*The Open Sea*
London (1956)

**Hare, F. K.**
"Climatic and Zonal Divisions of the Boreal Forest Formation of Eastern Canada"
*Geographical Review*, 40:615-635 (1950)

**Harrison, H. L.**
"Systems Studies of DDT Transport"
*Science*, 170:503-508 (1970)

**Hart, Gary Wanert**
"Creative Federalism—Recent Trials in Regional Water Resources"
*University of Colorado Law Review*, 39:29-47 (Fall 1966)

**Hart, H. C.**
*The Dark Missouri*
Madison, Wis.: Univ. of Wisconsin Press (1957)

**Hartesveldt, Richard J.**
"Fire Ecology of the Giant Sequoias"
*Natural History*, p. 12-19 (Dec. 1964)

**Hartley, Harold and Others**
"Energy for the World's Technology"
*New Scientist*, p. 1-24 (Nov. 1969)

**Hassler, A. D.**
"Cultural Eutrophication is Reversible"
*Bioscience*, 19:425-431 (1969)

**Hasler, Arthur D. and Warren J. Wisby**
"Discrimination of Stream Odors by Fishes and Its Relation to Parent Strean Behavior"
*American Naturalist*, 85:223-238 (1951)

**Haviland, Maud D.**
*Forest, Steppe and Tundra*
Cambridge, Mass.: Cambridge Univ. Press (1926)

**Haw, R. C.**
*The Conservation of Natural Resources*
London (1959)

**Hay, John**
*In Defense of Nature*
Boston (1969)

**Hazen, W. E. (ed.)**
*Readings in Population and Community Ecology*
Philadelphia (1964)

**Healy, H. H.**
"The Denver Earthquakes"
*Science*, 161:1301-1310 (1968)

**Heape, W.**
*Emigration, Migration and
Nomadism*
London (1931)

**Hedgpeth, Joel**
"The Oceans: World Sump"
*Environment*, p. 40-47 (1970)

**Hedgpeth, J. W.**
*Treatise on Marine Ecology and
Paleoecology* (1957)

**Helfrich, Harold W. Jr., ed.**
*The Environmental Crisis,
Agenda for Survival*
New Haven: Yale Univ. Press
(1970)

**Helfrich, H. W.**
*Agenda for Survival*
New Haven: Yale Univ. Press
(1971)

**Helfrich, H. W.**
*The Environmental Crisis*
New Haven: Yale Univ. Press.
(1970)

**Helfrich, Harold W., ed.**
*The Environmental Crisis: Man's
Struggle to Live with Himself*
New Haven: Yale Univ. Press
(1970)

**Henderson, L. J.**
*The Fitness of the Environment*
New York (1913)

**Henkin, Harmon**
"Side Effects Report of a
Conference on Ecological Aspects
of International Development"
*Environment*, p. 28-35, 48
(Jan.-Feb. 1969)

**Hennigan, Robert D.**
"Water Pollution"
*Bioscience*, 19:976-978 (1969)

**Herfindahl, O. C.**
*What is Conservation?*
Washington, D.C. (1961)

**Herfindahl Orris C. and Allen V. Keese**
*Quality of the Environment*
Baltimore: Johns Hopkins Univ.
Press (1965)

**Herrero, Stephen**
"Human Injury Inflicted by
Grizzly Bears"
*Science*, 170: 593-598 (1970)

**Hesse, Richard, W. C. Allee
and K. P. Schmidt**
*Ecological Animal Geography*
New York (1951)

**Hibbard, W. R. Jr.**
"Mineral Resources:
Challenge or Threat?"
*Science*, 160:143-149 (1968)

**Hickey, J. J., ed.**
*Peregrine Falcon Populations:
Their Biology and Decline*
Madison, Wis.: University of
Wisconsin Press (1968)

**Hickey, Joseph J. and
Donald W. Anderson**
"Chlorinated Hydrocarbons and
Eggshell Changes in Raptorial and
Fish-Eating Birds"
*Science*, 162:271-273

**Hilgard, W. E.**
*Soils, Their Growth, Formation,
Properties, Composition and
Relations to Climate and Plant
Growth in the Humid and
Arid Regions*
New York (1960)

**Hill, Gladwin**
"The Great and Dirty Lakes"
*Saturday Review*, p. 32
(Oct. 23, 1965)

**Hill, R. and C. P. Whittingham**
*Photosynthesis*
New York (1955)

**Hillaby, John**
"Primate Overkill"
*New Scientist*, 40:93 (1968)

**Hillaby, John**
"Seals, Scientists and Stalemate"
*New Scientist*, 37:140-143 (1968)

**Hillaby, John**
"The Wind, the White Man and
the Caribou"
*New Scientist*, 38:222-224 (1968)

**Hirsch, S. Carl**
*The Living Community: A Venture
into Ecology* (1966)

**Hockey, R. and C. Holloway**
"Noise and Efficiency"
*New Scientist*, 42:244-248 (1969)

**Hocking, Bryan**
*Biology or Oblivion*
Cambridge, Mass. (1965)

**Hogerton, J. F.**
"The Arrival of Nuclear Power"
*Scientific American*, p. 21-31
(Feb. 1968)

**Hogg, Tony**
"The Passing of the Eternal
Infernal Internal Combustion
Engine"
*Esquire*, p. 80-92, 242
(October 1970)

**Hohenemser, Kurt H.**
"Onward and Upward"
*Environment*, p. 22-27 (May 1970)

**Hohenemser, Kurt H.**
"The Supersonic Transport"
*Scientist and Citizen*, p. 1-10
(April 1966)

**Holcomb, Robert W.**
"Insect Control: Alternatives to
the Use of Pesticides"
*Science*, 168:456-458 (1970)

**Holcomb, Robert W.**
"Radiation Risk: A Scientific
Problem"
*Science*, 167:853-855 (1970)

**Holcomb, Robert W.**
"Waste-Water Treatment:
The Tide is Turning"
*Science*, 169:457-459 (1970)

**Holdgate, N. W.**
*Antarctic Ecology*, 2 Vols.
New York (1970)

**Holme, N. A. and MacIntyre, A. D., ed.**
*Methods for the Study of Marine
Benthos*
Philadelphia (1971)

**Hope, Jack**
"The King Besieged"
*Natural History*, p. 52-56, 72-82
(Nov. 1968)

**Hope, Jack**
"Prosperity and the National
Parks"
*Natural History*, p. 6-23
(February 1968)

**Hornbein, Thomas**
"Everest: The Wild Ridge"
San Francisco (1970)

**Howard, Walter E.**
"The Population Crisis is
Here Now"
*Bioscience*, Vol. 19, No. 9
(Sept. 1969)

**Howells, Gwyneth Parry**
"Water Quality in Industrial Areas.
Profile of a River"
*Science and Technology*, 4:26-35
(1970)

**Hubbert, M. K.**
*Energy Resources*
Washington, D.C. (1962)

**Hubschman, Jerry H.**
"Lake Erie Pollution Abatement.
Then What?"
*Science*, 171:536-540 (1971)

**Hungerford, H. R.**
*Ecology*
Chicago (1971)

**Hunt, Eldrige, G.**
"Biological Magnification of Pesticides"
London (1966)

**Huntington, E.**
*Civilization and Climate*
New Haven: Yale Univ. Press (1924)

**Hurtig, H. and C. R. Harris**
*The Pollution Reader*
Montreal (1968)

**Hutchings, M. M. and Caver, M.**
*Man's Dominion*
New York (1970)

**Hutchinson, G. Evelyn**
"The Biosphere"
*Scientific American*, 44-53 (September 1970)

**Hutchinson, G. Evelyn**
*The Ecological Theatre and the Evolutionary Play*
New Haven: Yale Univ. Press (1965)

**Huxley, Julian, A. C. Hardy and E. B. Ford**
*Evolution as a Process*
London (1958)

**Hyams, E.**
*Soil and Civilization*
London (1952)

**Hynes, H. B. N.**
*The Biology of Polluted Waters*
Liverpool Univ. Press (1960)

**Hynes, H. B. N.**
*The Ecology of Running Waters*
Univ. of Toronto Press (1970)

**Idaho Environmental Council**
Jerry Jayne, Pres., P.O. Box 3371, Univ. Sta., Moscow, Id.

**Iglauer, Edith**
"The Ambient Air"
*New Yorker*, (April 13, 1968)

**Iltis, Hugh H.**
"Criteria for an Optimum Human Environment"
*Bulletin of the Atomic Scientists*, p. 2-6 (Jan. 1970)

**Iltis, Hugh H.**
"Man First? Man Last?"
*Bioscience*, 20.820 (1970)

**Inove, E.**
*The $CO_2$-Concenetration Profile within Crop Canopies and its Significance for the Productivity of Plant Communities*
Paris (1968)

**International Joint Commission Canada and the United States**
*Pollution of Lake Erie, Lake Ontario and the International Section of the St. Lawrence River*
Washington, D.C. (1970)

**International Union for Conservation of Nature and Natural Resources**
2000 "P" St., NW, Washington, D.C. 20006

**Iowa Confederation of Environmental Organizations**
Dr. David L. Trauger, P.O. Box 1147, Univ. Station, Ames, Iowa

**Ise, John**
"Our National Park Policy"
Baltimore: Johns Hopkins Univ. Press (1961)

**Issue**
Mr. Hank Hassell, P.O. Box 728, Cedar City, Utah 84720

**Izaak Walton League of America**
1326 Waukegon Rd., Glenview, Ill. 60025

**Jacobs, G. V. and Whyte, R. O.**
*The Rape of the Earth, A World Survey of Soil Erosion*
London (1939)

**Jackson, Henry M.**
"Public Policy and Environmental Administration"
*Bioscience*, 17:883-885 (1967)

**Jackson, James P.**
"Death of a River"
*National Parks Magazine*, p. 19-22 (1970)

**Jackson, Nora and Philip Penn**
*A Dictionary of Natural Resources and Their Principal Uses*
New York (1969)

**Jacoby, Neil H.**
"The Environmental Crisis"
*The Center Magazine*, p. 36-48 (Nov.-Dec. 1970)

**Jaguaribe, H.**
"World Order Rationality and Socio-Economic Developments"
(Spring 1966)

**Jamison, Andrew**
*The Steampowered Automobile*
Bloomington: Indiana Univ. Press (1970)

**Jarrett, H.**
*Environmental Quality in a Growing Economy*
Baltimore: Johns Hopkins Univ. Press (1966)

**Jarrett, Henry**
*Perspectives on Conservation: Essays on America's Natural Resources*
Baltimore: Johns Hopkins Univ. Press (1958)

**Jenkins, Samuel H.**
"British Water Pollution Control"
*Environmental Science and Technology*, 4:204-209 (1970)

**Jenkins, Samuel H., ed.**
*Advances in Water Pollution Research*
New York (1969)

**Jennings, Burgess H. and Murphy, John R.**
*Interactions of Man and His Environment*
New York (1966)

**Jensen, Pennfield**
"A Student Manifesto on the Environment"
*National History*, p. 20-22 (April 1970)

**Johnson, Cecil E., (ed.)**
*Eco-crisis*
New York (1970)

**Johnson, Gerald W.**
"Plowshare at the Crossroads"
*Bulletin of the Atomic Scientists*, p. 83-91 (June 1970)

**Johnson, Huey D. (ed.)**
*No Deposit, No Return*
Reading, Mass. (1970)

**Johnson, P. L.**
"Remote Sensing as an Ecological Tool"
*Symposium on Ecoolgy of Subarctic Regions*
Helsinki (1966)

**Johnson, P. L.**
*Remote Sensing in Ecology*
Athens: Univ. of Georgia Press (1969)

**Johnston, Douglas M.**
*The International Law of Fisheries*
New Haven: Yale Univ. Press (1965)

**Johnston, Nancy and Retta**
*Central Park Country*
San Francisco, Calif.

**Jones, Arthur W.**
*Introduction to Parasitology*
Philadelphia (1967)

**Jones, J. R. Erichsen**
*Fish and River Pollution*
London (1964)

**Jordan, P. A.**
"Ecology, Conservation and
Human Behavior"
*Bioscience*, 18:1023-1029 (1968)

**Journal of Water Pollution Control**
Bob G. Rogers, Ed.,
Washington, D.C.
(1928 to the present)

**Judson, Sheldon**
"Erosion of the Land—Or What's
Happening to our Continents?"
*American Scientist*, 56:356-374
(1968)

**Jungk, R. and Galtung, J.**
*Mankind 2000*
London (1969)

**Kahn, Herman and Wiener, Anthony J.**
*The Year 2000:* A Framework for
Speculation on the Next
Thirty-Three Years
New York (1967)

**Kaplan, Abraham, ed.**
*Individuality and the New Society*
Washington: Univ. of Washington
Press (1970)

**Kardos, Louis T.**
"A New Prospect"
*Environment*, p. 10-21, 27
(March 1970)

**Katz, Alfred H. and
Felton, Jean Spencer**
*Health and Community*
New York (1965)

**Kaufman, Herbert**
*The Forest Ranger*
Baltimore: Johns Hopkins Univ.
Press (1960)

**Kauffman, Richard**
*Gentle Wilderness: The Sierra
Nevada*
San Francisco, Calif. (1969)

**Keast, A., ed.**
*Biogeography and Ecology in
Australia*
New York (1959)

**Keeling, Charles D.**
"Is Carbon Dioxide from Fossil
Fuel Changing Man's
Environment?"
*Proceedings of the American
Philosophical Society* 114:10-17
(1970)

**Keen, B. A.**
*The Physical Properties of the Soil*
New York (1931)

**Kellogg, C. E.**
*Development and Significance of
the Great Soil Groups of the
United States*
Boston (1936)

**Kempthorne, O., T. A. Bancroft,
J. W. Gowen, and J. L. Lush, eds.**
*Statistics and Mathematics in
Biology*
Ames, Iowa: Iowa State College
Press (1954)

**Kendrew, W. G.**
*The Climates of the Continents*
London: Oxford Univ. Press (1937)

**Kendeigh, S. Charles**
*Animal Ecology*
Englewood Cliffs, N.J. (1961)

**Kershaw, K. A.**
*Quantative and Dynamic Ecology*
New York (1964)

**Kethley, John**
"Population Regulation in Quill
Mites"
*Ecology*, 52:1113-1118 (1971)

**Keyfitz, N.**
"Population Density and the Style
of Social Life"
*Bioscience*, 16:868-873 (1966)

**Kilburn, Paul D.**
"Endangered Relic Trees"
*Natural History*, p. 56-63
(Dec. 1961)

**Kilgore, Bruce**
"Restoring Fire to the Sequoias"
*National Parks and Conservation
Magazine*, p. 16-22 (October, 1970)

**Kimball, Thomas L.**
"One-Third of the U.S. is up for
Grabs"
*National Wildlife*, p. 38-40
(Dec.-Jan. 1971)

**King, Cuchlaine**
*Beaches and Coasts*
London (1959)

**King-Hele, Desmond**
*The End of the 20th Century*
London (1970)

**Klausner, Samuel Z.**
*On Man and His Environment*
San Francisco (1971)

**Klein, L.**
*Aspects of River Pollution*
New York (1957)

**Klein, Louis**
*River Pollution*
Vol. 1—Chemical Analysis,
Vol. 2—Causes and Effects,
Vol. 3—Control
New York (1959-66)

**Klopfer, P. H.**
*Behavioral Aspects of Ecology*
Englewood Cliffs, N.J. (1962)

**Kneese, Ayres, D'Arge**
*Economics and the Environment*
Baltimore (1970)

**Kneese, Allen V., and Blair T. Bower**
*Managing Water Quality:
Economics, Technology, and
Institutions*
Baltimore: Johns Hopkins Univ.
Press (1968)

**Kneese, Allen V.**
*Water Pollution: Economic
Aspects and Research Needs*
Baltimore: Johns Hopkins Univ.
Press (1962)

**Kohn, Sherwood Davidson**
"Warning: the Green Slime is
Here"
*New York Times Magazine*, p. 26
(March 22, 1970)

**Konkle, Ward W.**
*Providing Quality Environment in
Our Communities*
Washington, D.C. (1968)

**Konrad, T. G.**
"Radar as a Tool in Meterology,
Entomology and Orinthology"
*Remote Sensing of the
Environment*
Ann Arbor: Univ. of Michigan
Press (1968)

**Kormondy, Edward J.**
*Concepts of Ecology*
Englewood Cliffs, N.J. (1969)

**Kormondy, Edward J.**
*Readings in Ecology* (1965)

**Krenkel, Peter A. and Frank L. Parker**
*Biological Aspects of Thermal
Pollution*
Nashville, Tenn.: Vanderbilt
Univ. Press (1969)

**Krogh, A.**
*Osmotic Regulation in Aquatic Animals*
New York: Cambridge Univ. Press (1939)

**Krutch, Joseph Wood**
*The Voice of the Desert:*
*A Naturalists' Interpretation*
New York (1965)

**Kryter, Karl D.**
*The Effects of Noise on Man*
New York (1970)

**Kryter, Karl D.**
"Psychological Reactions to Aircraft Noise"
*Science*, 151:1346-2355 (1966)

**Kryter, Karl D.**
"Sonic Booms from Supersonic Transport"
*Science*, 163:359-367 (1969)

**Lack, D.**
*Ecological Isolation in Birds*
Cambridge, Mass.: Harvard Press (1971)

**Lack, D.**
*Darwin's Finches*
London: Cambridge Univ. Press (1947)

**Lack, D.**
*Natural Regulation of Animal Numbers*
London (1954)

**Lafayette Environmental Action Federation**
David Page, Pres., P.O. Box 2103, W. Lafayette, Ind. 47906

**Lang, A. S. and R. M. Soberman**
*Urban Rail Transit: Its Economics and Technology*
Cambridge, Mass. (1964)

**Langbein, W. B. and W. G. Hoyt**
*Water Facts for the Nation's Future*
New York (1959)

**Landsberg, Hans H. and Leonard L. Fischman, and Joseph L. Fisher**
*Resources in America's Future*
Johns Hopkins Press (1963)

**Landsberg, Helmut E.**
"Metropolitan Air Layers and Pollution"
*Garden Journal*, p. 54-57 (March-April 1969)

**Lansford, Henry**
"The Superivilized Weather and Sky Show Brought to You by the Producers of Contrails, Dust and $CO_2$"
*Natural History*, p. 92-97, 112-113 (1970)

**Lapp, Ralph E.**
"Power-Hungry America: Where Will We Get the Energy?"
*New Republic*, p. 17-21 (July 11, 1970)

**Lauff, George H. (ed.)**
*Estuaries*
Washington (1967)

**Lave, Lester B. and Eugene P. Seskin**
"Air Pollution and Human Health"
*Science*, 169:723-733 (1970)

**Law School, Environmental Committee,**
Rm. 105, Univ. of North Dakota, Grand Forks, N.D.

**Laycock, George**
*The Aliens Animals: The Story of Imported Wildlife*
New York (1966)

**Laycock, George**
"The Diligent Destroyers.
A Critical Look at the Industries and Agencies that are Permanenly Defacing the American Landscape"
New York (1970)

**Leadley, Brown, A.**
*Ecology of Fresh Water*
Cambridge, Mass.: Harvard Press (1971)

**League of Conservation Voters**
620 C St., SE, Washington, D.C. 20003

**Lear, John**
"Clean Power from Inside the Earth"
*Saturday Review*, p. 53 (December 5, 1970)

**Lear, John**
"Green Light for the Smogless Car"
*Saturday Review*, p. 81-86 (December 6, 1969)

**Lear, John**
"A Progress Report on Smogless Motoring"
*Saturday Review*, p. 44 (August 1, 1970)

**Lear, John**
"Teaching in the Big School"
*Saturday Review*, p. 63 (January 2, 1971)

**Leavitt, Helen**
*Superhighway-Superhoax*
New York (1970)

**Lee, D. H. K., and Minard, D., eds.**
*Physiology, Environment and Man*
New York (1970)

**Lee, R. B. and I. DeVore**
*Man the Hunter*
Chicago (1969)

**Leff, David N.**
"Familiar Story"
*Environment*, p. 11-14 (May 1970)

**Legge, R. F. and D. Dingeldein**
"We Hung Phosphates without a Fair Trial"
*Canadian Research and Development* (March 1970)

**Leigh-Pemberton, John**
*Vanishing Wild Animals of the World*
New York (1968)

**Leinward, G.**
*Air and Water Pollution*
New York (1969)

**Leipzig, Arthur**
"Old Africa's People of the Village"
*Natural History*, 10-19 (April 1964)

**Leithe, W.**
*The Analysis of Air Pollution*
Ann Arbor: Univ. of Michigan (1970)

**Lemon, E. R.**
"Aerodynamic Studies of $CO_2$ Exchange between the Atmosphere and the Plant"
*Harvesting the Sun:*
*Photosynthesis in Plant Life*
New York (1967)

**Lemon, E. R.**
*Introduction to Experimental Ecology*
New York (1967)

**Leopold, A.**
*Game Management*
New York (1933)

**Leopold, A.**
*A Sand County Almanac*
New York: Oxford Univ. Press (1966)

**Leopold, Luna B.**
"Landscape Esthetics"
*Naturali History*, p. 36-45 (October 1969)

**Lessing, Lawrence**
"New Ways to Move Power with Less Pollution"
*Fortune,* p. 78 (November 1970)

**Lessing, Lawrence**
"Power from the Earth's Own Heat"
*Fortune,* (June 1969)

**Lewis, T. and Taylor, L. R.**
*Introduction to Experimental Ecology*
New York (1967)

**Leydet, Francois and James D. Rose**
*The Last Redwoods and the Parkland of Redwood Creek*
San Francisco, Calif. (1969)

**Leydet, Francois**
*Time and the River Flowing: Grand Canyon*
San Francisco, Calif. (1970)

**Lilienthal, David E.**
"300,000,000 Americans Would be Wrong"
*New York Times Magazine,* p. 25 (Jan. 9, 1966)

**Lillard, Richard G.**
"Eden in Jeopardy"
New York (1966)

**Lipsett, Charles H.**
*Industrial Wastes and Salvage. Conservation and Utilization*
New York (1963)

**Lipton, Ron M.**
*Terracide, America's Destruction of Her Living Environment*
Boston (1970)

**Lisk, Donald J.**
"The Analysis of Pesticide Residues: New Problems and Methods"
*Science,* 170:589-593 (1970)

**Lisk, Donald J.**
"Detention and Measurement of Pesticide Residues"
*Science,* 154:93-98 (1966)

**Little, Charles E.**
*Challenge of the Land*
New York (1968)

**Little, Charles E.**
*Challenge of the Land*
New York (1969)

**Livingston, D. A.**
"Biochemicals Cycles"
*Topics in the Study of Life* (1971)

**Livingston, Dennis**
"Pollution Control: An International Perspective"
*Scientist and Citizen,* p. 172-182 (Sept. 1968)

**Living Wilderness**
Richard C. Olson, Ed.
Washington, D.C. (1935)

**Litis, Hugh H., et al**
"Criteria for an Optimum Human Environment"
*Bulletin of the Atomic Scientists,* p. 2-6 (Jan. 1970)

**Litis, Hugh H.**
"Man First? Man Last? The Paradox of Human Ecology"
*Bioscience,* 20:820 (1970)

**Lofrath, Foran**
"Pesticides and Catastrophe"
*New Scientist,* 40:567-568 (1968)

**Lotka, A. J.**
*Elements of Physical Biology*
Baltimore (1925)

**Lotspeich, Frederick B.**
"Water Pollution in Alaska: Present and Future"
*Science,* 166:1239-1245 (1969)

**Love, Glen A. and Love, Rhoda M.**
*Ecological Crisis. Readings for Survival*
New York (1970)

**Love, R. Merton**
"The Rangelands of the Western United States"
*Scientific American,* p. 88-96 (Feb. 1970)

**Love, S., ed.**
*Earth Tool iKt*
Washington, D.C. (1971)

**Lovering, T. S.**
"New Fuel Mineral Resources in the Next Century"
*Texas Quarterly,* 11:127-147 (1968)

**Lowdermilk, W. C.**
"The Reclamation of a Man-Made Desert"
*Scientific American,* p. 54-63 (March 1960)

**Lukacs, John**
*The Passing of the Modern Age*
New York (1970)

**Lund, J. W. G. and J. F. Talling**
"Botanical Limnological Methods with Special Reference to Algae"
*Botanical Review,* 23:489-583 (1957)

**Lutz, H. J.**
"Forest Ecosystems: Their Maintenance, Amelioration and Deterioration"
*Journal of Forestry,* 61:563-569 (1963)

**Lutz, H. J. and R. F. Chandler, Jr.**
*Forest Soils*
New York (1946)

**Lyle, David**
"The Human Race Has, Maybe, Thirty-Five Years Left"
*Esquire,* p. 116-118, 176-183 (Sept. 1967)

**Lyles, Charles H.**
*Fishery Statistics of the United States*
Washington, D.C. (1968)

**Lynch, P. and Chandler, R.**
*National Environmental Test*
New York (1970)

**MacArthur, Robert and Joseph Connell**
*The Biology of Populations* (1966)

**MacFadyen, A.**
*Animal Ecology*
New York (1957)

**McAtee, W. L.**
"Census of Four Square Feet"
*Science,* 26:447-449 (1907)

**McCabe, Louis C.**
"Can One State's Code Solve Another State's Problems?"
*Environmental Science and Technology,* 4:210-213 (1970)

**McCaull, Julian**
"Who Owns the Water?"
*Environment,* p. 30-39 (October 1970)

**McCloskey, Maxine E.**
*Wilderness*
San Francisco (1970)

**McClung, Robert M.**
*Lost Wild America*
New York (1969)

**McCormick, Jack**
*The Life of the Forest*
New York (1966)

**McCormick, Jack**
*The Living Forest*
New York (1959)

**McCue, Gerald and Ewald, William R.**
*Creating the Human Environment*
Urbana: Univ. of Illinois Press (1970)

**McCullough, David G.**
"The Lonely War a Good Angry
Man"
*American Heritage*, p. 97-113
(Dec. 1969)

**McDermott, W.**
"Air Pollution and Public Health"
*Scientific American*, p. 49-57
(October 1961)

**McDivitt, James F.**
*Minerals and Men: An Exploration
of the World of Minerals and its
Effect on the World We Live In.*
Baltimore: Johns Hopkins Univ.
Press (1965)

**McDonald, Rita**
*Guide to Literature on
Environmental Sciences*
Washington, D.C. (1971)

**McDougal, Myres and William Burke**
*Public Order of the Oceans*
New Haven: Yale Univ. Press
(1962)

**McGinnies, William G.,
Bram J. Goldman and
Patricia Paylore, (eds.)**
*Deserts of the World*
Tucson: Univ. of Arizona Press
(1969)

**McHale, John**
"The Ecological Context"
New York (1970)

**McHale, John**
*The Future of the Future*
New York (1969)

**McHarg, Ian L.**
*Design with Nature*
New York (1969)

**McHenry, R. and Van Doren, C. (eds.)**
*A Documentary History of
Conservation in America*
New York (1972)

**McHugh, J. L.**
"The United States and
World Whale Resources"
*Bioscience*, 19:1075-1078 (1969)

**McKay, Stewart**
*Bio-medical Telemetry: Sensing
and Transmitting Biological
Information from Animals and
Men*, 2nd ed.
New York (1970)

**McLean, Louis A.**
"Pesticides and the Environment"
*Bioscience*, 17:613-617 (1967)

**McLean, R. C. and Cook, W. R.**
*Practical Field Ecology*
London (1950)

**McMillan, Ian**
*Man and the California Condor*
New York (1968)

**McVay, Scott**
"Can Leviathan Long Endure so
Wide a Chase?"
*Natural History*, p. 36-40, 68-72
(Jan. 1971)

**McVay, Scott**
"The Last of the Great Whales"
*Scientific American*, p. 13-21
(August 1966)

**Macinko, John**
"The Tailpipe Problem"
*Environment*, p. 6-13 (1970)

**Magan, T. T.**
*Freshwater Ecology*
London (1963)

**Magan, T. T. and Worthington, E. B.**
"Life in Lakes and Rivers"
*The New Naturalist*, No. 15
London (1951)

**Malin, James C.**
*The Grasslands of North America:
Prolegomena to Its History*
Lawrence, Kan. (1965)

**Malthus, Thomas R.**
*An Essay on the Principles of
Population*
Homewood, Ill. (1963)

**Margalef, D. R.**
"Information Theory in Ecology"
*General Systems* 3:36-71 (1958)

**Margalef, R.**
*Perspectives in Ecological Theory*
Chicago: Univ. of Chicago Press
(1968)

**Margolin, Malcolm**
"Desert Under the Trees"
*National Parks and Conservation
Magazine*, p. 8-13
(December 1970)

**Margolis, Jon**
"Our Country 'Tis of Thee,
Land of Ecology"
*Esquire*, p. 124 (March 1970)

**Marine, Gene**
*America the Raped*
New York (1969)

**Marsh, George P.**
*The Earth as Modified by Human
Action*
New York (1874), reprinted (1971)

**Marsh, George P.**
*Man and Nature: Or, Physical
Geography as Modified by Human
Action*
New York (1874), reprinted (1971)

**Marshall, Nelson**
"Ecology: The Social Science and
Resource Use"
*Bioscience*, 18:765-766 (1968)

**Martell, E. A.**
"Fire Damage"
*Environment*, p. 14-21 (May 1970)

**Martin, John Stuart**
"Rebirth of the Shad"
*Atlantic*, p. 90-93 (June 1965)

**Martin, Paul S.**
"Pleistocene Overkill"
*Natural History*, p. 32-28
(Dec. 1967)

**Martin, R. S.**
"Ecology in the Fourth
Demension"
*Topics in the Study of Life*
New York (1971)

**Marx, Wesley**
*The Frail Ocean*
New York (1967)

**Matson, Floyd W.**
*The Broken Image: Man,
Science and Society*
Garden City, N.Y. (1966)

**Matthews, W .H., ed.**
*Man's Impact on the Climate*
Cambridge, Mass.: M.I.T. Press
(1971)

**Matthews, W. H. (ed.)**
*Man's Impact on the Global
Environment*
Cambridge, Mass.: M.I.T. Press
(1970)

**Matthieson, Peter**
"Waiting for the Last Whales"
*Esquire*, p. 64-65, 124-126
(Feb. 1971)

**Matthieson, Peter**
*Wildlife in America*
New York (1959)

**Mayer, Lawrence A.**
"Why the U.S. is in an
'Energy Crisis' "
*Fortune*, p. 75 (Nov. 1970)

**Mayr, Ernst**
*Animal Species and Evolution*
Harvard Univ. Press (1963)

**Mayr, E.**
"Ecological Factors in Speciation"
*Evolution*, 1:263-288 (1947)

**Mead, Margaret, (ed.)**
*Cultural Patterns and Technical Change*
New York (1961)

**Means, Richard L.**
*The Ethical Imperative: The Crisis in American Values*
Garden City, N.Y. (1969)

**Mech, L. David**
*The Wolf*
Garden City, N.Y. (1970)

**Mellanby, Kenneth**
*Pesticides and Pollution*
London (1969)

**Mercklin, John M.**
"It's Time to Turn Down All that Noise"
*Fortune*, p. 130-133, 188-195
(Oct. 1969)

**Mero, J. L.**
*The Mineral Resources of the Sea*
New York (1965)

**Merriman, Daniel**
"The Calefaction of a River"
*Scientific American*, p. 42-52
(May 1970)

**Metress, James F.**
"Bibliography, Physical Anthropology"
*Bioscience*, 20:252-255 (1970)

**Metro Clean Air Committee**
Mrs. Sandi Knudson, Dir.,
1829 Portland Ave.,
Minneapolis, Minn.

**Metro Washington Coalition for Clean Air**
John S. Winder, Exec. Dir.,
1714 Mass. Ave., N.W.,
Washington, D.C.

**Meyer, J. R., J. F. Kain, M. Wohl**
*The Urban Transportation Problem*
Cambridge, Mass. (1965)

**Meyerhoff, H. A.**
"Mineral Raw Materials in the National Economy"
*Science*, 135:510-516 (1962)

**Michael, Donald N.**
*The Unprepared Society*
New York (1968)

**Michener, James A.**
*The Quality of Life*
Philadelphia (1970)

**Mihursky, J. A.**
"On Possible Constructive Use of Thermal Additions to Estuaries"
*Bioscience*, 17:698-702 (1967)

**Mikesell, M. W.**
"Deforestation in Northern Morrocco"
*Science*, 132:441-448 (1960)

**Miles, D. A. and H. B.**
*Freshwater Ecology*
London (1967)

**Mills, G. Alex, Harry R. Johnson and Harry Perry**
"Fuels Management in an Environmental Age"
*Environmental Science and Technology*, 5:30-38 (1971)

**Miller, Henry**
*Rich Man, Poor Man*
New York (1971)

**Miller, Morton W. and George G. Berg, eds.**
*Chemical Fallout: Current Research on Persistent Pesticides*
Springfield, Ill. (1969)

**Miller, R. S., and Others**
*Man and His Environment: The Ecological Limits of Optimism*
New Haven: Yale Univ. Press
(1970)

**Miller, Stanton**
"Water, Pollution in the States"
*Environmental Science and Technology*, 5:120-125 (1971)

**Milne, Lorus J. and Margery J.**
*The Biotic World and Man*
New York (1958)

**Milne, Lorus J. and Margery Milne**
*The Balance of Nature*
New York (1960)

**Milne, Lorus J. and Margery J. Milne**
*Paths Across the Earth*
New York (1958)

**Milne, L. and Milne, M.**
*The Nature of Life*
New York (1971)

**Milne, Lorus J. and Margery Milne**
*Patterns of Survival*
New York (1967)

**Milne, Lorus J. and Margery J. Milne**
*The Mating Instinct*
New York (1954)

**Milner, C. and R. E. Hughes**
*Methods for Estimating the Primary Production of Grasslands, Arid Land and Dwarf Shrublands*
Oxford, England (1968)

**Minnesota Environmental Control Citizens Assn.**
Paul H. Engstrom, Pres.,
26 Exchange St., St. Paul, Minn.

**Mitchell, G. E. and S. E. Thrower**
*Sanitation, Drainage and Water Supply*
London (1961)

**Mitchell, John G.**
"The Bitter Struggle for a National Park"
*American Heritage*, p. 97-198
(April 1970)

**Moats, Sheila and William A. Moats**
"Toward a Safer Use of Pesticides"
*Bioscience*, 20:459-464

**Momaday, N. Scott**
"An American Land Ethic"
*Sierra Club Bulletin*, p. 8-11
(Feb. 1970)

**Moment, G. B.**
"Bears: The Need for New Sanity in Wildlife Conservation"
*Bioscience*, 18:1105-1108 (1968)

**Moncrief, Lewis W.**
"The Cultural Basis for Our Environmental Crisis"
*Science*, 170:508-512 (1970)

**Montague, Peter and Katherine**
"Mercury: How Much are we Eating?"
*Saturday Review*, p. 50
(Feb. 6, 1971)

**Montana Environmental Council**
Don Aldrich, 410 Woodworth,
Missoula, Mont.

**Montana Environmental Task Force**
Gordon Whirry, Box 14,
Montana St. Univ., Bozeman, Mont.

**Monterey Committee for Environmental Info.**
Peter Resag, Ph.D., 980 Fremont,
Monterey, Calif.

**Montrerth, J. L.**
"Measurement and Interpretation of Carbon Dioxide Fluxes in the Field"
*Netherland Journal of Agricultural Science*, 10.334-346 (1962)

**Moore, H. B.**
*Marine Ecology*
New York (1958)

**Moorman, James**
"DDT—The Ban that Isn't"
*Sierra Club Bulletin*, 9-11
(January 1970)

**Moran, William E.**
"A Sourcebook on Population"
*Population Bulletin*, 25:No. 5.
(1969)

**Morgan, George B.**
"Air Pollution Surveillance
Systems"
*Science*, 170:289-296 (1970)

**Morgan, Karl Z.**
"Never Do Harm"
*Environment*, p. 28-30
(January-February 1971)

**Morgan, K. Z.**
"Permissible Exposure to
Ionizing Radiation"
*Science*, 139:565-571 (1963)

**Mossman, Archie S.**
"Environmental Crisis and the
Wildlife Ecologist"
*Bioscience*, 20:813-814 (1970)

**The Mother Earth News**
John Shuttleworth, (ed.)
P.O. Box 38, Madison, Ohio

**Mowbray, A. O.**
*Road to Ruin*
Philadelphia (1969)

**Mumford, Lewis**
*The Myth of the Machine:*
Technics and Human Development
New York (1967)

**Mumford, Lewis**
*The Myth of the Machine*, Vol. II.
The Pentagon of Power
New York (1970)

**Mungc, Ray**
"If Mr. Thoreau Calls, Tell Him
I've Left the Country"
*Atlantic*, p. 72-86 (May 1970)

**Munn, R. E.**
*Biometerological Methods*
New York (1970)

**Murdoch, William and Connell, Joseph**
"All About Ecology"
*The Center Magazine*, p. 56-63
(Jan. 1970)

**Murphy, Robert**
*Wild Sanctuaries: Our National
Wildlife Refuges. A Heritage
Restored*
New York (1968)

**Murray, D. L. and Bond, J. A.**
*An Experience with Populations*
Reading, Mass. (1971)

**Murray, J. and J. Hjort**
*The Depths of the Ocean*
New York (1912)

**Nadler, Allen C. (Dr.)**
Monterey Bay Committee for
Environmental Info.
800 Fremont, Monterey, Calif.
93940

**Nadler, Allen A.**
*Air Pollution*
New York (1970)

**Nash, R.**
*The American Environment:
Readings in the History of
Conservation*
Menlo Park, Calif. (1972)

**Nash, Roderick**
*Wilderness and the American
Mind*
New Haven: Yale Univ. Press
(1967)

**Nashville Committee for
Scientific Info.**
Dept. of Chemistry, Vanderbilt
Univ., Nashville, Tenn.

**National Academy of Engineering—
National Academy of Sciences**
*Policies for Solid Waste
Management* (1970)

**National Academy of Sciences**
"Committee on Resources and
Man of the Div. of Earth Sciences"

**Resources and Man**
San Francisco (1969)

**National Academy of Sciences**
*Effects of Chronic Exposure to
Low Levels of Carbon Monoxide
on Human Health, Behavior and
Performance*
Washington, D.C. (1969)

**National Academy of Sciences**
*Eutrophication: Causes,
Consequences, Correctives*
Washington, D.C. (1969)

**National Academy of Sciences**
*The Great Alaska Earthquake
of 1964*
Washington, D.C. (1970)

**National Academy of Sciences**
*Jamaica Bay and Kennedy
Airport: A Multidisiplinary
Environmental Study*
Washington, D.C. (1971)

**National Academy of Sciences**
*The Life Sciences*
Washington, D.C. (1970)

**National Academy of Sciences**
*Resources and Man*
San Francisco (1969)

**National Academy of Sciences**
"Symposium on Aids and Threats
to Society from Technology"
*Proc. of the National Academy of
Sciences*, 67:857-915 (1970)

**National Academy of Sciences,
Committee on Pollution**
*Waste Management and Control*
Washington, D.C. (1966)

**National Air Pollution
Control Administration**
*Control Techniques for Particulate
Air Pollutants*
Washington, D.C. (1969)

**National Audubon Society**
Dr. Elvis Stahr, 950 Third Ave.,
New York, N.Y. 10022

**National Geographic Society Staff**
*America's Wonderlands:
The Senic National Parks and
Monuments of the United States*
Washington, D.C. (1966)

**National Parks and
Conservation Magazine**
Eugenia H. Connally, (ed.)
Washington, D.C. (1919)

**National Recreation and Park Assn.**
1700 Pennsylvania Avenue,
Washington, D.C. 20006

**National Recreation and Park Assn.**
1601 N. Kent St.,
Rosslyn, Va. 22209

**National Reporter**
Rob Sauer, (ed.)
Los Altos, Calif. (1969)

**National Parks Assn.**
1701 18th St., NW,
Washington, D.C. 20009

**National Research Council**
*Land Use and Wildlife Resources*
Washington, D.C. (1970)

**National Science Teachers Assn.**
*Science Looks at Itself*
New York (1970)

**National Tuberculosis and
Respiratory Disease Assn.**
*Air Pollution Primer*
New York (1969)

**National Wildlife Federation**
Dr. James H. Schaeffer, Pres.,
1412 16th St., NW,
Washington, D.C.

**National Wildlife Federation**
"1970 National EQ Index"
*National Wildlife*, p. 25-40
(October-November 1970)

**Nature Conservancy**
1522 "K" St., NW,
Washington, D.C.

**Nature Conservancy News**
Arlington, Va. (1950)

**Natural Resources Council of Maine**
Dr. Robert Mohlar,
20 Willow St., Augusta, Me.

**Neal, E.**
*Woodland Ecology*
London (1958)

**Nearing, Helen and Scott**
*Living the Good Life,*
How to Live Sanely and Simply in
a Troubled World
New York (1970)

**Nef, John U.**
"The Search for Civilization"
*Center Magazine,* p. 2-6
(May 1969)

**Nelson, Bryce**
"Colorado Environmentalists:
Scientists Battle AEC And Army"
*Science,* 168:1324-1328 (1970)

**Nelson, J. G. and R. C. Scace**
*The Canadian National Parks:
Today and Tomorrow*
Alberta, Canada:
University of Calgary (1969)

**Netboy, Anthony**
*The Atlantic Salmon:
A Vanishing Species*
Boston (1968)

**Netchert, Bruce C.**
"The Economic Impact of Electric
Vehicles: A Scenario"
*Bulletin of the Atomic Scientists,*
p. 29-35 (May 1970)

**Netchert, Bruce C.**
*The Future Supply of Oil and Gas*
Baltimore: Johns Hopkins Press
(1958)

**Newbigin, M. I.**
*Plant and Animal Geography*
New York (1948)

**Newbould, P. S.**
*Methods for Estimating the
Primary Production of Forests*
Oxford (1967)

**Newell, N. D.**
"Crisis in the History of Life"
*Scientific American,* (Feb. 1963)

**Newell, Reginald E.**
"The Global Circulation of
Atmospheric Pollutants"
*Scientific American,* p. 32-42
(Jan. 1971)

**New Mexico Conservation
Coordinating Council**
Cliff Crawford, Pres.,
433 Maple, N.E.,
Albuquerque, N.M.

**Nicholson, Max**
*The Environmental Revolution*
New York **(1970)**

**Nicol, E.**
"The Ecology of a Salt Marsh"
*Journal of Marine Biological Assn.
United Kingdom,* 20:203-261
(1935)

**Nikolsky, G. V.**
*Ecology of Fishes*
New York (1963)

**Northern Calif. Committee for
Environmental Info.**
Dr. Donald Dahlsten,
P.O. Box 761,
Berkeley, Calif. 94701

**Not Man Apart**
John Muir Inst., San Francisco
(1970)

**Novick, Sheldon**
*The Careless Atom*
Boston (1969)

**Novick, Sheldon**
"Earthquake and Nuclear Power"
*Scientist and Citizen,* p. 230-241
(Nov. 1968)

**Novick, Sheldon**
"Earthquake at Giza"
*Environment,* p. 2-15
(January-February 1970)

**Novick, Sheldon and Others**
"A New Pollution Problem"
*Environment,* p. 2-23, 43-44
(May 1969)

**Nuessle, Virginia D. and
Robert W. Halcomb**
"Will SST Pollute the
Stratosphere?"
*Science,* 168:1562 (1970)

**Oberle, Mark W.**
"Endangered Species: Congress
Curbs International Trade in
Rare Animals"
*Science,* 167:152-154 (1970)

**Odum, Eugene P.**
"The Concept of the Biome as
Applied to the Distribution of
North American Birds"
*Wilson Bulletin,* 57:191-201 (1945)

**Odum, E. P.**
*Ecology,*
New York (1963)

**Odum, E. P.**
*Fundamentals of Ecology,* 3rd. ed.
Philadelphia (1971)

**Odum, E. P.**
"The Strategy of Ecosystem
Development"
*Science,* 164:262-270 (1969)

**Odum, Eugene**
"The Role of Tidal Marshes in
Estuarine Production"
*New York State Conservationist*
(June-July 1961)

**Odum, Howard T.**
*Environment, Power and Society*
New York (1971)

**Ohio Students Environmental Council**
Environmental Studies Center,
Antioch Union, Antioch College,
Yellow Springs, Ohio 45387

**Ohio Valley Environmental Council**
Ben Weber, Pres.,
224 N. 5th St., Box 399,
Steubenville, Ohio

**Oklahoma Coalition for Clean Air**
Lois Blanche, P.O. Box 53303,
Oklahoma City, Okla.

**Olsen, Jack**
*Night of the Grizzlies*
New York (1970)

**Olson, T. A. and F. J. Burgess, (eds.)**
*Pollution and Marine Ecology*
New York (1967)

**Oort, Abraham H.**
"The Energy Cycle of the Earth"
*Scientific American,* p. 54-63
(1970)

**Oosting, Henry J.**
*Plant Communities*
San Francisco (1948)

**Oosting, H. J.**
*The Study of Plant Communities*
(1956)

**Organization for Environmental
Quality**
Michael Baranski, Chmn.,
Box 5536, College St.,
Raleigh, N.C.

**Orleans, Leo and
Suttmeier, Richard P.**
"The Mao Ethic and
Environmental Quality"
*Science,* 170:1173-1176 (1970)

**Orloci, L.**
"Geometric Models in Ecology:
The Theory and Application of
Some Ordination Methods"
*Journal of Ecology,* 54:193-215
(1966)

**Orwell, George**
*1984*
New York (1949)
(reprinted in 1962)

**Osborn, Fairfield**
*Our Plundered Planet*
Boston (1948)

**Owen, F. D.**
*Animal Ecology in Tropical Africa*
Edinburgh and London (1966)

**Owen, Wilfred**
*Cities in the Motor Age*
New York (1959)

**Pack, D. H.**
"Meteorology of Air Pollution"
*Science*, 146:1119-1128 (1964)

**Paddock, W. and P.**
"The Explosion in Humans"
*Topics in the Study of Life* (1971)

**Paddock, William and Paul**
*Hungry Nations*
Boston (1964)

**Paine, R. T.**
"A Short-Tern Experimental
Investigation of Resource
Partitioning in a New Zealand
Rocky Intertidal Habitat"
*Ecology*, 52:1096-1106 (1971)

**Panofsky, Hans A.**
"Air Pollution Meteorology"
*American Scientist*, 57:269-285
(1959)

**Paradise, Scott**
"The Vandal Ideology"
*Nation*, 29, 720-732 (Dec. 1969)

**Park, Charles F.**
*Affluence in Jeopardy: Minerals
and the Political Economy*
San Francisco (1968)

**Park, Robert E.**
*Human Communities: The City
and Human Ecology*
Glencoe, Ill. (1952)

**Parker, Carl E.**
"Mercury—Major New
Environmental Problem"
*The Conversationist*, p. 6-9
(Aug.-Sept. 1970)

**Patrick, R.**
"The Structure of Diatom
Communities in Similar
Ecological Conditions".
*American Naturalists*, 102:173-183
(1968)

**Patten, B. C., ed.**
*Systems Analysis and Simulation
in Ecology*
New York
Vol. I (1971) Vol. II (1972)

**Pattison, E. Scott**
"Arsenic and Water Pollution
Hazard"
*Science*, 170:870-872 (1970)

**Pauling, Linus**
"Genetic and Somatic Effects of
High-Energy Radiation"
*Bulletin of the Atomic Scientists*,
p. 3-5 (Sept. 1970)

**Paulson, Glenn (President)**
New York Scientists' Committee
for Public Info.
30 E. 68th St., New York,
N.Y. 10021

**Peakall, David B.**
"Pesticides and the Reproduction
of Birds"
*Scientific American*, 72-78
(April 1970)

**Pearl, R.**
*The Biology of Population Growth*
New York (1925)

**Pearl, R.**
"The Growth of Populations"
*Quarterly Review of Biology*,
2:532-548 (1927)

**Pearse, A. S.**
*The Emigration of Animals from
the Sea*
Dryden, N.Y. (1950)

**Pecora, W. T.**
"Science and the Quality of
Our Environment"
*Bulletin of the Atomic Scientists*,
p. 20-23 (Oct. 1970)

**Pennak, R. W.**
*Fresh-Water Invertebrates of the
United States*
New York (1935)

**Perin, Constance**
*With Man in Mind*
Cambridge, Mass.: M.I.T. Press
(1970)

**Perry T.**
*Our Polluted World*
New York (1967)

**Peter, Walter G. III**
"Controlled Fusion: A Multifaceted
Approach to Solving
Environmental Degradation"
*Bioscience*, 20:717-719 (1970)

**Peter, Walter G. III**
"New York Blight: A Case Study"
*Bioscience*, 20:617-619, 669-671
(1970)

**Peter, Walter G. III and
Fling, Karen J.**
"Conference on Environment:
A Quest for Quality"
*Bioscience*, 20:563-564

**Peterson, Malcolm L.**
"Environmental Contamination
from Nuclear Reactors"
*Scientist and Citizen*, p. 1-11
(Nov. 1965)

**Philip, Duke of Edinburgh and
James Fisher**
*Wildlife Crisis*
New York (1970)

**Phillips, T.**
*Agriculture and Ecology in Africa*
London (1959)

**Phillipson, John**
*Ecological Energetics*
New York (1966)

**Phillipson, J. (ed.)**
*Methods of Study in Soil Ecology*
New York (1970)

**Phillipson, J.**
*Methods of Study in Quantative
Soil Ecology*
Philadelphia (1971)

**Pielou, E. C.**
*An Introduction to Mathematical
Ecology*
New York (1969)

**Pielou, E. C.**
"A Single Mechanism to Account
for Regular, Random and
Aggregated Populations"
*Journal of Ecology*, 48:575-584
(1960)

**Pielou, E. C.**
*The Use of Information Theory in
the Study of the Diversity of
Biological Populations*
New York, (1967)

**Pinchot, G.**
*The Fight for Conservation*
New York (1910)

**Pitts, James N. Jr. and
Metcalf, Robert L.**
*Advances in Environmental
Sciences*
Vol. I, Vol. II
New York (1969), (1971)

**Plainfield Area Coalition for the Environment**
Jasmine King, Pres.,
1454 Maplewood Terr.,
Plainfield, N.J.

**Planned Parenthood—World Population Chief**
John C. Robbins, Exec. Off.,
810 Seventh Ave., New York, N.Y.

**Platt, John R.**
"Diversity"
*Science,* 154:1132-1139 (1966)

**Platt, John R.**
*Perception and Crange*
Ann Arbor:
Univ. of Michigan Press (1970)

**Platt, John R.**
"The Step to Man"
*Science,* 149:607-613 (1965)

**Platt, John R.**
"What We Must Do."
*Science,* 166:1115-1121 (1969)

**Platt, R. B. and J. Griffiths**
*Environmental Measurement and Interpretation*
New York (1964)

**Polak, F. L.**
*The Images of the Future*
New York (1961)

**Pollution Control Committee**
Anita Pericolosi, Co-Founder,
P.O. Box 271, W. Hartford, Conn.

**Pond, Alonzo W.**
*The Desert World* (1962)

**Porter, Eliot**
*Baja California*
San Francisco, Calif. (1969)

**Porter, Eliot**
*Galapagos: The Flow of Wildness*
San Francisco, Calif. (1968)

**Porter, Eliot**
*In the Wilderness is the Preservation of the World*
San Francisco, Calif. (1967)

**Porter, Eliot**
*The Place No One Knew: Glen Canyon on The Colorado*
San Francisco, Calif. (1970)

**Porter, Eliot**
*Summer Island: Penobscot Country*
San Francisco, Calif. (1969)

**Porter, Richard D. and Stanley N. Wiemeyer**
"Dieldrin and DDT, The Effects on Sparrow Hawk Eggshells and Reproduction"
*Science,* 165:199-200 (1969)

**Portmann, Adolf**
*Animal Camouflage*
Univ. of Michigan: Ann Arbor (1959)

**Portmann, Adolf**
*Animals as Social Beings*
Univ. of Michigan: Ann Arbor (1961)

**Potts, A. M.**
"Toward an Esthetics of Science"
*Perspectives in Biology and Medicine,* 9:137-141 (1965)

**Potter, Frank M. Jr.**
"Everyone Wants to Save the Environment But No One Knows Quite What to Do."
*The Center Magazine,* p. 34-40 (March 1970)

**Potter, V. R.**
"Purpose and Function of the University"
*Science,* 167:1590-1593 (1970)

**Poulton, E. C.**
*Environment and Human Efficiency*
Springfield, Ill. (1970)

**Powledge, Fred**
"Life in the Campsites"
*Esquire,* p. 101-103, 165-169
Walden, Ill. (1970)

**Pratt, Christopher J.**
"Sulfur"
*Scientific American,* p. 62-72 (May 1970)

**Princeton Ecology Action**
Larry Campbell,
1 Murray Dodge Hall,
Princeton Univ., Princeton, N.J.

**Pringle, L.**
*In A Beaver Valley*
New York (1970)

**The Progressive**
"The Crisis of Survival"
Glenview, Ill. (1970)

**Protect Your Environment**
Nancy Siegler, 40 Highland Ave.,
Rowayton, Conn.

**Pryde, Philip R.**
"Victors are not Judged"
*Environment,* p. 30-39 (Nov. 1970)

**P.U.R.E.**
Bill Ellis, 409 E. Greenwood,
Bend, Ore. 97701

**Pyle, L.**
*The Pill and Birth Regulation*
London (1964)

**Quality Environment Council**
Dr. Larry C. Holcomb,
Box 7025, Omaha, Neb.

**Raleigh, ECOS**
Bob Ambrose, Box 5536,
College Sta., Raleigh, N.C.

**Ramparts Magazine**
"Ecology Special"
(May 1970)

**Ramparts Magazine**
"Eco-Catastrophe"
New York (1970)

**Randolph, Theron G.**
*Susceptibility to the Chemical Human Ecology and Environment*
Springfield, Ill. (1962)

**Ranson, S. L. and Thomas, M.**
"Crassulacean Acid Metabolism"
*Annual Review of Plant Physiology*
11: (1960)

**Rapoport, Roger**
"Los Angeles Has a Cough"
*Esquire,* p. 83-85, 32 (1970)

**Raskin, Edith**
*The Pyramid of Living Things*
New York (1967)

**Rayner, A. C. (ed.)**
*Land against the Sea*
Washington, D.C. (1964)

**Redfield, A. C. and E. S. Deevey**
"Temporal Sequences and Biotic Successions"
U.S. Naval Institute, Annapolis (1952)

**Reich, Charles A.**
*The Greening of America*
New York (1970)

**Reichle, David E., ed.**
*Analysis of Temperate Forest Ecosystems*
New York (1970)

**Reid, George K.**
*Ecology of Inland Waters anl Estuaries* (1961)

**Reid, L.**
*The Sociology of Nature*
London (1962)

**Reinig, William C., ed.**
*Environmental Surveillance in the Vicinity of Nuclear Facilities*
Springfield, Ill. (1970)

**Reno County Environmental Action**
Mrs. Betty Davis, 6601 Monroe, Hutchinson, Kan. 67501

**Resources and Man (A Scientific American Reader)**
San Francisco (1969)

**Revelle, Roger and Landsberg, Hans L.**
*America's Changing Environment*
Boston (1970)

**Richards, Paul W.**
*The Tropical Rain Forest*
Cambridge Univ. Press (1952)

**Ricketts, Edward F. and Jack Calvin**
*Between Pacific Tides*
Stanford Univ. Press (1962)

**Ridgeway, James**
*The Politics of Ecology*
New York (1970)

**Rienow, Robert**
*Man Against His Environment*
New York (1970)

**Rienow, Robert and Rienow, Leona T.**
*Moment in the Sun*
New York (1967)

**Rillo, T. J.**
*Ecology*
Elgin, Ill. (1972)

**Ritchie-Calder, Lord**
"Mortgaging the Old Homestead"
*Foreign Affairs*, 48:207-220 (1970)

**Ritchie-Calder, Lord**
"Polluting the Environment"
*Center Magazine*, p. 7-12
(May 1969)

**Ritterbush, Philip C.**
"The Biological Muse"
*Natural History*, p. 26-31
(October 1968)

**Rivera-Cordero, Antonio**
"The Nuclear Industry and Air Pollution"
*Environmental Science and Technology*, 4:392-395 (1970)

**Roberts, Keith**
"Federal Power Commission"
*Sierra Club Bulletin*, p. 9-11
(August 1970)

**Roberts, Walter Orr**
"Man on a Changing Earth"
*American Scientist*, 59:16-19
(1971)

**Roberts, Walter Orr**
*A View of Century 21*
Claremont, Calif.:
Claremont Colleges (1969)

**Rochester Committee for Scientific Info.**
Dr. Robert E. Lee, Box 5236, River Campus Station, Rochester, N.Y.

**Rock, J.**
*The Time Has Come*
New York (1963)

**Rockefeller, Nelson A.**
"Our Environment Can Be Saved"
Garden City, N.Y. (1970)

**Rogers, G. W., ed.**
*Change in Alaska*
Seattle, Wash.:
Univ. of Washington Press (1970)

**Rolan, Robert G.**
"Must We Choose Between Men in Space and Mankind on Earth?"
*Bioscience*, 20:797-806 (1970)

**Roosevelt, Nicholas**
*Conservation: Now or Never*
New York (1970)

**Rorison, I. H.**
*Ecological Aspects of Mineral Nutrition in Plants*
Philadelphia (1969)

**Rose, Sanford**
"The Economics of Environmental Quality"
*Fortune*, p. 120 (Feb. 1970)

**Rosen, Walter G.**
"The Environmental Crisis: Through a Glass Darkly"
*Bioscience*, 20:1209-1211, 1216
(1970)

**Rosene, Walter**
*The Bobwhite Quail. Its Life and Management*
New Brunswick, N.J.:
Rutgers Univ. Press (1969)

**Roslansky, J. D.**
*Genetics and the Future of Man*
New York (1966)

**Ross, H. H.**
*A Synthesis of Evolutionary Theory*
Englewood Cliffs, N.J. (1962)

**Ross, MacDonald, M. (ed.)**
*The World Wildlife Guide*
New York (1971)

**Roueche, Berton**
*What's Left:*
*Reports on a Diminishing America*
Boston (1968)

**Roughton, R. O.**
"A Review of Literature on Dendrochronology and Age Determination of Woody Plants"
*Technical Bulletin No. 15*,
Denver (1962)

**Rudd, Robert L.**
*Pesticides and the Living Landscape*
Madison, Wis.:
Univ. of Wisconsin Press (1964)

**Russell, Bertrand**
*Has Man a Future?*
New York (1958)

**Sagan, Leonard A.**
"Infant Mortality Controversy: Sternglass and His Critics"
*Bulletin of the Atomic Scientists*,
p. 26-32
(October 1969)

**Salisbury, E.**
"Weeds and Allies"
*New Naturalist*, No. 43
London (1961)

**Sanford, David**
"Bad Riddance: Turning Trash into Cash"
*New Republic*, p. 14-16
(Dec. 5, 1970)

**Sanky, J.**
*Chalkland Ecology*
London (1966)

**Santa Barbara Underseas Foundation**
David Doerner, Pres.,
P.O. Box 4815,
Santa Barbara, Calif.

**Sargent, Frederick**
"Adaptive Strategy for Air Pollution"
*Bioscience*, 17:691-697 (1967)

**Savage, Henry J.**
*Lost Heritage*
New York (1970)

**S.A.V.E. (Student Assn. of Voters for Ecology)**
Tom Moser, ASC Office CWSC, Ellensburg, Wash. 98926

**Sax, Joseph L.**
*Defending the Environment*
New York (1971)

**Schaller, George B.**
*The Deer and the Tiger:*
*A Study of Wildlife*
Univ. of Chicago (1967)

**Schaefer, Vincent J.**
"Auto Exhaust, Pollution and
Weather Patterns"
*Bulletin of the Atomic Scientists,*
p. 31-33 (Oct. 1970)

**Schaefer, Vincent J.**
"Some Effects of Air Pollution on
Our Environment"
*Bioscience,* 19:896-897 (1969)

**Schaefer, Vincent J.**
"The Threat of the Unseen"
*Saturday Review,* p. 55
(Feb. 6, 1971)

**Schaeffer, John R.**
"Reviving the Great Lakes"
*Saturday Review,* p. 62
(Nov. 7, 1970)

**Scheffer, Victor B.**
*The Year of the Seal*
New York (1970)

**Scheffer, Victor B.**
*The Year of the Whale*
New York, 1969)

**Schmidt-Nielsen, Knut**
*Desert Animals: Physiological*
*Problems of Heat and Water*
(1964)

**Schmitt, Peter J.**
*Back to Nature: The Arcadian*
*Myth in Urban America*
Oxford, England:
Oxford Univ. Press (1969)

**Schnell, R.**
*Introduction à la Phytogeographie*
*des Pays Tropically*
2 Vols, Paris (1971)

**Schrader, Gene**
"Atomic Doubletalk"
*The Center Magazine,* p. 29-52
(February 1971)

**Schrag, Peter**
"Life on a Dying Lake"
*Saturday Review,* p. 19-21, 55-56
(Sept. 1969)

**Schrimshaw, Nevin S.**
"Food"
*Scientific American,* 209 (3):72-80
(Sept. 1963)

**Schroeder, H. A.**
*Pollution, Profits and Progress*
(1971)

**Schulman, Edmund**
*Dendrochlimatic Changes in*
*Seimarid America*
Univ. of Arizona Press (1956)

**Schultze, Charles L.,**
**Hamilton, Edward K. and Schick, Allen**
*Setting National Priorities*
Washington, D.C. (1970)

**Schurcliff, William A.**
*SST And Sonic Boom Handbook*
New York (1970)

**Schurr, Sam H.**
*Energy in the American*
*Economy 1850–1975*
Baltimore: Johns Hopkins
Univ. Press (1960)

**Schwartz, William, (ed.)**
*Voices for the Wilderness*
New York (1969)

**Science Info. Committee of the**
**Staten Island Inst. of Arts & Sciences**
Arts & Sciences
Terence Benbow,
75 Stuyvesant Pl.,
Staten Island, N.Y.

**Scientist and Citizen**
(April 1968) and (October 1968),
p. 199-205

**Scientist and Citizen**
"Sonic Boom in Relation to Man"
p. 223-229 (Nov. 1968)

**Scientists Inst. of Public Info.**
Dr. Philip Sickevitz, 30 E. 68th St.,
New York, N.Y.

**Scott, John Paul**
*Animal Behavior*
Univ. of Chicago (1958)

**Seaborg, Glenn T.**
"An International Challenge"
*Bulletin of the Atomic Scientists,*
p. 5-7 (Nov. 1970)

**Seaborg, Glenn T.**
"Our Nuclear Future—1995"
*Bulletin of the Atomic Scientists,*
p. 7-14 (June 1970)

**Sears, Paul B.**
"Beyond the Forest"
*American Scientist,* 55:338-346
(1967)

**Sears, Paul B.**
*Deserts on the March*
Univ. of Oklahoma Press (1937)

**Sears, Paul B.**
*Lands Beyond the Forest*
New York (1969)

**Sears, Paul B.**
*The Living Landscape*
New York (1966)

**Segerberg, O.**
*Where Have All the Flowers,*
*Fishes, Birds, Trees, Water and*
*Air Gone?*
New York (1971)

**Seidenberg, Roderick**
*Posthistoric Man*
Chapel Hill:
Univ. of North Carolina Press
(1950)

**Selikoff, I. J.**
"Asbestos:
The Hazards of Asbestos Dust"
*Environment,* p. 2-7 (March 1969)

**Selsam, Millicent E.**
*How Animals Tell Time*
New York (1967)

**Selye, Hans**
"Mercury Poisoning:
Prevention by Spronolactone"
*Science,* 169:775-776 (1970)

**Shaplen, Robert**
*Toward the Well-Being of*
*Mankind*
New York (1964)

**Shea, Kevin P.**
"Dead Stream"
*Environment,* p. 12-15
(July-August 1970)

**Shelford, Victor E.**
*Animal Communities in*
*Temperate America*
Chicago: Univ. of Chicago Press
(1913)

**Shelford, V. E.**
*The Ecology of North America*
Univ. of Illinois (1963)

**Shelford, V. E.**
*Laboratory and Field Ecology*
Baltimore (1929)

**Shelford, V. E.**
"Basic Principles of the
Classification of Communities and
Habitats and the Use of Terms"
*Ecology,* 13:105-120 (1932)

**Shen-Miller, J.**
"Notes After a Conference on
Great Lakes Research"
*Bioscience,* 20:1294-1296 (1970)

**Shepard, Paul and McKinley, Daniel**
*The Subversive Science, Essays*
*Toward an Ecology of Man*
New York (1969)

**Shepard, Paul**
"What Ever Happened to Human Ecology?"
*Bioscience,* 17:891-894, 911 (1967)

**Shepard, Paul and McKinley, Daniel**
"Environ-Mental"
Boston (1971)

**Shils, Edward**
"Daydreams and Nightmares: Reflections on the Criticism of Mass Culture"
*Sewanee Review,* 65:587-608 (1957)

**Sierra Club**
Raymond J. Sherwin, Pres.,
1050 Mills Tower,
San Francisco, Calif.

**Sierra Club**
*Ecotactics*
San Francisco, Calif (1970)

**Simms, D. Harper**
The Soil Conservation Service
New York (1970)

**Simon, Arthur**
"Battle of Beaufort: Conservation Collides with the Jobless"
New Republic, p. 11-15 (May 23, 1970)

**Simon, H.**
*The Science of the Artificial*
Cambridge, Mass.: M.I.T. Press (1969)

**Simon, Noel and Paul Geroudet**
*Last Survivors. The Natural History of Animals in Danger of Extinction*
New York (1970)

**Simpson, George Gaylord**
*This View of Life:*
*The World of an Evolutionist*
New York (1964)

**Simpson, G. G., A. Roe and R. C. Lewontin**
*Quantitative Zoology* (Revised ed.)
New York (1960)

**Singer, S. Fred**
"The Energy Revolution: Population Growth and Environmental Change"
*Bioscience,* 21:163 (1971)

**Singer, S. Fred (ed.)**
*Global Effects of Environmental Pollution*
New York (1970)

**Singer, S. Fred**
"Human Energy Production as a Process in the Biosphere"
*Scientific American,* 174-190 (Sept. 1970)

**Singer, S. Fred**
"Will the World Come to a Horrible End?"
*Science,* 170:125 (1970)

**Skinner, Brian J.**
*Earth Resources*
Englewood, N.J. (1969)

**Sladechova, Alena**
"Limnological Investigation Method for the Periphyton (aufwuchs) Community"
*Botanical Review,* 28:286-350 (1962)

**Slater, L. E., ed.**
"Biotelemetry"
*Bioscience,* 15:81-120 (1965)

**Slobodkin, Lawrence B.**
"Aspects of the Future of Ecology"
*Bioscience,* 18:16-23 (1968)

**Slobodkin, L. B.**
*Growth and Regulation in Animal Populations*
New York (1961)

**Small, William E.**
*Third Pollution*
New York (1971)

**(1940)**
**Smith, Grahame, J. C.**
"The Ecologist at Bay"
*Saturday Review,* p. 68 (Jan. 2, 1971)

**Smith, Guy-Harold**
"Conservation of Natural Resources"
New York (1971)

**Smith, J. M.**
*The Theory of Evolution*
Middlesex, England (1958)

**Smith, M. Brewster**
*Social Psychology and Human Values*
Chicago (1970)

**Smith, N. G.**
"The Advantage of Being Parasitized"
*Nature,* 219:690-694 (1968)

**Smith, T. E., ed.**
'Torrey Canyon' Pollution and Marine Life
Cambridge, Univ. Press (1968)

**Smith, R. L.**
*Ecology and Field Biology*
New York (1966)

**Smith, R. L.**
*The Ecology of Man*
New York (1972)

**Snedecor, G.**
*Statistical Methods Applied to Experiments in Agriculture and Biology*
Ames, Iowa:
Iowa State College Press (1956)

**Snow, C. P.**
"The Two Cultures and the Scientific Revolution"
New York:
Cambridge Univ. Press (1959)

**Snow, Joel A.**
"Radioactivity from "Clean" Nuclear Power"
*Scientist and Citizen,* p. 97-101 (May 1968)

**Society for Environmental Stabilization**
Delbert S. Plante,
P.O. Box 252,
Fayetteville, Ark. 72701

**Soil Conservation Society of America**
*Soil and America's Future*
Ankeny, Iowa (1967)

**Sokal, R. R. and F. J. Rohlt**
*Biometry, The Principles and Practice of Statistics in Biological Research*
San Francisco (1969)

**Solow, R. N.**
"The Economists Approach to Pollution and Its Control"
*Science,* 173(3996):498-503 (August 6, 1971)

**Sondheimer, E. and Simone, J. B. (eds.)**
*Chemical Ecology*
New York (1970)

**Sorg. Thomas J. and Hickman, H. L. Jr.**
*Sanitary Landfill Facts* (1970)

**South Maine Environmental Action Committee**
Cliff Goodall,
Univ. of Maine Law School,
68 High St., Portland, Me.

**Southwood, T. R. E.**
*Ecological Methods*
London (1966)

**Spilhaus, Athelstan**
"Technology, Living Cities and Human Environment"
*American Scientist,* 57:24-36 (1969)

**Spofford, Walter O. Jr.**
"Closing the Gap in Waste Management"
*Environmental Science and Technology,* 4:1108-1114 (1970)

**Sproull, Wayne T.**
*Air Pollution and its Control*
New York (1970)

**Spurr, Stephen H.**
*Wilderness Management:* The Horace M. Albright Conservation Lectureship Berkeley, Calif.: Univ. of Calif. School of Forestry (1966)

**Squires, Arthur M.**
"Clean Power from Coal"
*Science,* 169:821-828 (1970)

**Stamp, L. D.**
*The Land of Britain, Its Use and Misuse*
London (1962)

**Stamp, L. Dudley**
*Land for Tomorrow, Our Developing World*
Bloomington, Ind.: Indiana Univ. Press (1969)

**Stamp, L. D.**
"Man and the Land"
*New Naturalist,* No. 31
London (1955)

**Stanford Committee for Environmental Info.**
Dr. John W. Farquar, Stanford Univ. Medical Center, Stanford, Calif.

**Starr, Roger**
"Garbage, or, Can We Ever Get Away From It All?"
*Horizon,* p. 48-51 (Winter 1969)

**Starr, Roger**
"This is the Way the World Ends"
*American Heritage,* p. 95-101 (Oct. 1970)

**Stear, James R.**
*Municipal Incineration, a Review of the Literature*
(1971)

**Steinhaus, E. A.**
*Insect Pathology*
London (1963)

**Stent, Gunther S.**
*The Coming of the Golden Age,* A View of the End of Progress
Garden City, N.Y. (1970)

**Stern, Arthur C., ed.**
*Air Pollution Control*
New York (1968)

**Sternglass, Ernest**
"Has Nuclear Testing Caused Infant Deaths?"
*New Scientist,* 43:178-181 (1969)

**Stewart, George Rippey**
*Not So Rich as You Think*
Boston (1968)

**Still, Henry**
*In Quest of Quiet, Meeting the Menace of Noise Pollution*
Harrisburg, Pa. (1970)

**Stock, R. W.**
"Saving the World the Ecologist's Way"
*New York Times Magazine,* (October 5, 1969)

**Stone, E. C. and R. B. Vasey**
"Preservation of Coast Redwood Alluvial Flats"
*Science,* 150:157-161 (1968)

**Stone, E. C.**
"Preserving Vegetation in Parks and Wilderness"
*Science,* 150:1261-1267 (1965)

**Storer, John H.**
*Man in the Web of Life*
New York (1968)

**Strauss, Werner, (ed.)**
*Air Pollution Control, Part I*
New York (1970)

**Street, Philip**
*Vanishing Animals: Preserving Nature's Rarities*
New York (1963)

**Students for Environmental Concerns**
Dick Pastor;
Univ. of Ill. YMCA,

**Students for Environmental Controls**
Don Yon, 1001 S. Wright St., Champaign, Ill. 61820

**Study of Critical Environmental Problems**
"The Williamstown Study"
*Bulletin of the Atomic Scientists,* p. 24-30 (Oct. 1970)

**Sullivan, James B. and Albert J. Fritsch**
"Getting the Lead Out"
*New Republic,* p. 9-10 (Nov. 21, 1970)

**Summerhayes, V. S.**
"The Effect of Coles (Microtus Agrestis) on Vegetation"
*Journal of Ecology,* 29:14-48 (1941)

**Sutton, Ann and Myron Sutton**
*New Worlds for Wildlife*
Chicago (1970)

**Sutton, Horace**
"The Erosion of Eden"
*Saturday Review,* p. 58 (June 6, 1970)

**Sutton, Horace**
"Is the SST Really Necessary?"
*Saturday Review,* p. 14 (August 15, 1970)

**Sverdrup, H. M., M. W. Johnson and R. H. Fleming**
*The Oceans: Their Physics, Chtmistry and General Biology*
Englewood Cliffs, N.J. (1942)

**Swan, L. A.**
*Beneficial Insects*
New York and London (1964)

**Swatek, P.**
*The User's Guide to the Protection of the Environment*
New York (1970)

**Swinnerton, J. W.**
"The Ocean: A Natural Source of Carbon Monoxide"
*Science,* 167: 984-986 (1970)

**Swisher, R. D.**
"Detergent Enzymes, Biodegradation and Environmental Acceptability"
*Bioscience,* 19:1093-1094 (1969)

**Szent-Gyorgyi, Albert**
*The Crazy Ape*
New York (1970)

**Tamplin, Arthur R., and John W. Gofman**
*Population Control Through Nuclear Pollution*
Chicago (1970)

**Tamplin, Arthur R. and John W. Gofman**
"The Radiation Effects Controversy"
*Bulletin of the Atomic Scientists,* p. 2, 5-8 (Sept. 1970)

**Tarring, R. C.**
"Chemistry and Society, IV: Progress in the Chemistry of Detergents and its Social Consequences"
*Impact,* 16:277-296 (1966)

**Tax, S. (ed.)**
Evolution After Darwin (2 Vols.)
Chicago: Univ. of Chicago Press
(1960)

**Taylor, G. R.**
The Biological Time Bomb
New York (1968)

**Taylor, Gordon Rattray**
The Doomsday Book, Can the
World Survive?
New York (1970)

**Taylor, Ron**
"Subdividing the Wilderness"
Sierra Club Bulletin, p. 4-9
(Jan. 1971)

**Taylor, Wallis**
"Our World, AD 2000"
"Our World, A.D. 2000"
New Scientist, 43:146-147 (1969)

**Ternes, Alan P.**
"The State of the Species"
Natural History, p. 43-74
(Jan. 1970)

**Terry, Mark**
"Hello Ants, Goodbye Aristotle"
Natural History, p. 6-23
(Jan. 1971)

**Terry, Mark**
Teaching for Survival
New York (1971)

**Texas Wesleyan Environmental
Action Group**
Dr. Arthur G. Cleveland,
Box 138, Texas Wesleyan College,
Ft. Worth, Tex. 76105

**Theobold, Robert (ed.)**
Social Policies for America in the
Seventies: Nine Divergent Views
Garden City, N.Y. (1969)

**Thomas, Lowell**
"The Painful Lessons of the
Cedars of Lebanon"
National Wildlife, p. 50-55
(Dec.-Jan. 1970)

**Thomas, William L.**
Man's Role in Changing the Face
of the Earth
Chicago: Univ. of Chicago Press
(1970)

**Thompson, Theos J. and
William R. Bibb**
"Response to Gofman and
Tamplin: The AEC Position"
Bulletin of the Atomic Scientists,
p. 9-12, 48 (1970)

**Thompson, W. F.**
"Fishing Treaties and Salmon of
the North Pacific"
Science, 150:1786-1789 (1965)

**Thomson, George**
The Foreseeable Future
Cambridge:
Cambridge Univ. Press (1965)

**Tilden, Freeman**
National Parks
New York (1968)

**Tinker, Jon**
"Marrying Wildlife to Forestry"
New Scientist, 42:518-520 (1969)

**Tinbergen, N.**
Social Behavior in Animals with
Social Behaviour in Animals with
Special Reference to Vertebrates
New York (1953)

**Toffler, Alvin**
Future Shock
New York (1970)

**Tomasevich, Jozo**
International Agreements for
Conservation of Marine Resources
London: Oxford Univ. Press (1943)

**Topp, Robert W.**
"Interoceanic Sea-Level Canal:
Effects on the Fish Faunas"
Science, 165:1324-1327 (1969)

**Toward the Conquest of Hunger**
New York (1967)

**Trever, Lloyd**
"A Water Resource Policy for
Canada"
Canadian Geographic Journal,
73:2-17 (July 1966)

**Trefethen, James B.**
"The Return of the White-Tailed
Deer"
American Heritage, p. 97-103
(Feb. 1970)

**Tuan, Yi-Fu**
"Our Treatment of the
Environment in Idea and
Actuality"
American Scientist, 58:244-249
(1970)

**Tucker, A.**
The Toxic Metals
New York (1972)

**Udall, Stewart L.**
"The Ecology of Man and the
Land Ethic"
Natural History, p. 32-41
(June-July 1965)

**Udall, S. L.**
1976 Agenda for Tomorrow
New York (1968)

**Udall, S. L.**
The Quiet Crisis
New York (1964)

**Udvardy, Miklos**
Dynamic Zoogeography: With
Special Reference to Land
Animals
New York (1969)

**Ulfstrand, S.**
"Ecology in Semi-Arid East Africa.
A Selection of Recent Ecological
References"
Bulletins from the Swedish
Research Committee, II,
Stockholm (1971)

**UNESCO**
Soil Biology:
A Review of Research
Paris (1969)

**United Nations**
Environmental Contamination by
Radioactive Materials
Vienna (1969)

**United Nations**
Report of the U.N. Scientific
Committee on the Effects of
Atomic Radiations
New York (1970)

**U.S. Congress**
Changing National Priorities
Washington, D.C. (1970)

**U.S. Congress**
Designation of Wilderness Areas
Washington, D.C. (1970)

**U.S. Council on Environmental Quality**
Environmental Quality
Washington, D.C. (1970)

**U.S. Dept. of Agriculture, Yearbooks**
Soils and Man (1938); Climate and
Man (1941); Grass (1948);
Trees (1949; Water (1955)
Washington, D.C.

**U.S. Dept. of Agriculture Forest
Services**
Timber Resources for America's
Future
Washington, D.C. (1958)

**U.S. Dept. of Agriculture Forest
Service**
Timber Trends in the United
States
Washington, D.C. (1965)

**U.S. Dept. of Agriculture**
*Yearbook for 1958: Land*
Washington, D.C. (1958)

**U.S. Dept. of Health, Education and Welfare. Natl. Air Pollution Control Admin.**
*Air Quality Criteria for Particulate Matter*
Washington, D.C. (1969)

**U.S. Dept. of the Interior**
*Fisheries Statistics of the United States*
Washington, D.C. (Yearly Issue)

**U.S. Dept. of the Interior**
*Public Land Statistics 1968*
Washington, D.C.
(1969) and yearly

**U.S. Dept. of the Interior**
*United States Petroleum Through 1980*
Washington, D.C. (1968)

**U.S. Dept. of the Interior**
*Waterfowl for Tomorrow*
Washington, D.C. (1964)

**U.S. Dept. of the Interior**
*Waterfowl Status Report 1969*
Washington, D.C.
(1970 and yearly)

**U.S. 89th Congress, Committee on Govt. Operations**
*Environmental Hazards Coordination*
Washington, D.C. (1966)

**U.S. Federal Committee on Research National Areas**
*A Dictionary of Research Natural Areas on Federal Lands of the United States of America*
Washington, D.C. (1968)

**U.S. Forest Service**
*Range Research Methods: A Symposium*
Washington, D.C. (1962)

**U.S. Forest Service**
*Techniques and Methods of Measuring Understory Vegetation, a Symposium*
Washington, D.C. (1958)

**U.S. House of Representatives**
*Environmental Quality Education Act of 1970*
Washington, D.C. (1970)

**U.S. House of Representatives**
*Fisheries and Wildlife Conservation, Committee on Merchant Marine and Fisheries*
Washington, D.C. (1969)

**U.S. House of Representatives**
*Hearings before the Subcommittee on Science, Research, and Development of the Committee on Science and Astronautics*
Washington, D.C. (1968)

**U.S. House of Representatives**
*Water Pollution Control and Abatement*
Washington, D.C. (1964)

**U.S. House of Representatives**
*Public Lands of the Committee on Interior and Insular Affairs*
Washington, D.C. (1962)

**U.S. House of Representatives**
*Hearings before the Committee on Public Works, Federal Water Pollution Control Act Amendments*
Washington, D.C. (1969)

**U.S. House of Representatives**
*Phosphate in Detergents and the Eutrophication of America's Waters*
Washington, D.C. (1970)

**U.S. National Park Service**
*National Parks and Landmarks*
Washington, D.C. (1970)

**U.S. National Park Service**
*Parks for America*
Washington, D.C. (1964)

**U.S. Public Health Service**
*Carbon Monoxide. A Bibliography with Abstracts*
Washington, D.C. (1966)

**U.S. Public Health Service**
*Comprehensive Studies of Solid Waste Management*
Washington, D.C. (1970)

**U.S. Public Health Service**
*Handbook of Air Pollution*
Washington, D.C. (1968)

**U.S. Public Health Service**
*Air Pollution Publications. A Selected Bibliography with Abstracts 1966–1968*
Washington, D.C. (1969)

**U.S. Public Health Service**
*Air Quality Criteria for Carbon Monoxide*
Washington, D.C. (1970)

**U.S. Public Health Service**
*Air Quality Criteria for Hydrocarbons*
Washington, D.C. (1970)

**U.S. Public Health Service**
*Air Quality Criteria for Photochemical Oxidants*
Washington, D.C. (1970)

**U.S. Public Health Service**
*Guide to Research in Air Pollution*
Washington, D.C. (1970)

**U.S. Public Health Service**
*Hydrocarbons and Air Pollution: An Annotated Bibliography*
Washington, D.C. (1970)

**U.S. Public Health Service**
*NAPCA Abstract Bulletin*
Raleigh, N.C. (1970)

**U.S. Public Health Service**
*Nitrogen Oxides: An Annotated Bibliography*
Washington, D.C. (1970)

**U.S. Public Health Service**
*Proceedings of a Symposium on Human Ecology*
Washington, D.C. (1969)

**U.S. Public Health Service**
*Solid Waste/Disease. Relationships. A Literature Survey*
Washington, D.C. (1968)

**U.S. Public Health Service**
*Solid Waste Management: Abstracts and Excerpts from the Literature*
Washington, D.C. (1970)

**U.S. Public Health Service**
*Sulfur Oxides and Other Sulfur Compounds*
Washington, D.C. (1965)

**U.S. Public Land Law Review Commission**
*Digest of Public Land Laws*
Washington, D.C. (1968)

**U.S. Senate**
*Air Pollution—1970. Hearings before the Subcommittee on Air Pollution of the Committee on Public Works*
Washington, D.C. (1971)
*Air Pollution—1970. Hearings before the Subcommittee on Air Pollution of the Committee on Public Works*

**U.S. Senate**
*Disposal of Junked and Abandoned Motor Vehicles*
Washington, D.C. (1970)

**U.S. Senate**
*Effects of Mercury on Man and the Environment*
Washington, D.C. (1970)

**U.S. Senate**
*Environmental Protection Act of
1970*
Washington, D.C. (1970)

**U.S. Senate**
*Hearings before the Committee on
Commerce, Radiation Control for
Health and Safety Act of 1967*
Washington, D.C. (1968)

**U.S. Senate**
*Thermal Pollution*
Washington, D.C. (1968–1969)

**U.S. Senate**
*Timber Management Policies*
Washington, D.C. (1969)

**U.S. Senate**
*Water Pollution in the National
Capital Region*
Washington, D.C. (1970)

**U.S. White House**
*Office of Science and Technology
—Electric Power and the
Environment (Position Paper*
Washington, D.C. (1970)

**U.S. Senate**
*Hearings before the Committee
on Interior and Insular Affairs,
North Cascades—Olympic
National Park*
Washington, D.C. (1966)

**U.S. Senate**
*Hearings before the Subcommittee
on Air and Water Pollution of the
Committee of Public Works*
Washington, D.C. (1969)

**U.S. Senate**
*Hearings before the Subcommittee
on Air and Water Pollution of the
Committee on Public Works*
Washington, D.C. (1970)

**U.S. Senate**
*Lead-based Paint Poisoning*
Washington, D.C. (1970)

**U.S. Senate**
*Resource Recovery Act of 1969.
Hearings before the Subcommittee
on Air and Water Pollution of the
Committee on Public Works*
Washington, D.C. (1969–1970)

**Vandevere, Judson E. and
James A. Mattison**
"Sea Otters"
*Sierra Club Bulletin*, p. 12-15
(October 1970)

**Van Doren, Charles**
*The Idea of Progress*
New York (1967)

**Van Duuren, B. L.**
"Is Cancer Airborne?"
*Scientist and Citizen*, p. 1-6
(Jan. 1966)

**Van Dyne, G. M., ed.**
*The Ecosystem Concept in
Natural Resource Management*
New York (1969)

**Van Dyne, G. M.**
*Ecosystems, Systems Ecology,
and Systems Ecologists*
(1966)

**Van Oye, P. and Van Miegham, J., eds.**
*Biogeography and Ecology in
Antarctica* (1965)
(1965)

**Velz, Clarence J.**
*Applied Stream Sanitation*
New York (1970)

**Venuto, Peter D.**
Citizens Against Air Pollution,
193 Bangor Ave., San Jose, Calif.

**Vernburg, F. J. & W. B.**
*The Animal and the Environment*
New York (1970)

**Volpe, E. P.**
*Understanding Evolution*, 2nd ed.
Dubuque, Iowa (1970)

**von Eckardt, Woel**
"People, Yes: Cars, No."
*Saturday Review*, p. 62
(October 3, 1970)

**Wadsworth, R. N.**
*The Measurement of
Environmental Factors in
Terrestial Ecology*
Philadelphia (1968)

**Waelder, Robert**
*Progress and Revolution*
New York: Intl. Universities Press
(1967)

**Wagar, J. Alan**
"Growth versus the
Quality of Life"
*Science*, 168:1179-1184 (1970)

**Wagner, Philip**
*The Human Use of the Earth*
New York (1960)

**Wagner, Richard W.**
*Environment and Man*
New York (1971)

**Walford, L. A.**
*Living Resources of the Sea*
New York (1958)

**Wallace, Bruce and Adrian M.**
*Adaption*, 2nd ed.
New York (1964)

**Wallen, I. Eugene**
"Preserving the Quality of the
Oceans"
*Bioscience*, 20:533 (1970)

**Ward, M. A., ed.**
*Man and His Environment*, Vol. 1
New York (1970)

**Warming, J. E. B.**
*Oecology of Plants*
Oxford (1909)

**Warp, Max**
"How to Think about the
Environment"
*Fortune*, p. 98 (Feb. 1970)

**Warren, C. E.**
*Biology and Water Pollution
Control*
Philadelphia (1971)

**Warren, C. E. and G. E. Davis**
"Laboratory Stream Research:
Objectives, Possibilities, and
Constraints"
*Annual Review of Ecology and
Systematics*, Vol. 2 p. 111-144
(1971)

**Warshofsky, Fred**
*The Control of Life/
The 21st Century*
New York (1969)

**Washington Ecology Center**
William G. Painter,
2000 "P" St., NW,
Washington, D.C.

**WASTE**
John Freeman, Dept. of Biology,
Winthrop College, Rock Hill, S.C.

**Water Pollution Control Federation**
Joseph F. Lagnese, Jr.,
3900 Wisconsin Ave.,
Washington, D.C.

**Watkins, Lucy Scott**
"Conservation—For Whom?"
*The Living Wilderness*, p. 51-52
(Autumn 1970)

**Watson, A.**
*Animal Populations—in Relation
to Their Food Resources*
Philadelphia (1970)

**Watt, K. E. F.**
*Ecology and Resource Management*
New York (1968)

**Watt, K. E. F. (ed.)**
*Systems Analysis in Ecology*
New York (1966)

**Wayman, Cooper H.**
"A Hard Look at Soft Detergents"
*Bulletin of the Atomic Scientists,*
p. 22-26 (April 1965)

**Ways, Max**
"O Say Can Your See? The Crisis in Our National Perception"
*Fortune,* (October 1968)

**Weaver, J. E.**
*North American Prairie*
Lincoln, Neb. (1954)

**Weaver, J. E. and F. E. Clements**
*Plant Ecology*
New York (1938)

**Weaver, Lynn E. (ed.)**
*Education for the Peaceful Use of Nuclear Explosives*
Tucson:
University of Arizona Press (1970)

**Webb, S. D.**
"Extinction—Organization Equilibria in Late Cenozoic Land Mammals of North America"
*Evolution* 23:688-702 (1969)

**Wegner, A.**
*The Origin of Continents and Oceans*
London (1924)

**Weidenbaum, Murray L.**
"How to Buy a Cleaner Environment"
*Bulletin of the Atomic Scientists,*
p. 19-21 (Nov. 1970)

**Weinberg, Alvin M.**
"Nuclear Energy and the Environment"
*Bulletin of the Atomic Scientists,*
p. 69-74 (June 1970)

**Weinberg, Alvin M. and Philip Hammond**
"Limits to the Use of Energy"
*American Scientist,* 58: 412418
(1970)

**Weins, Herold J.**
*Atoll Environment and Ecology*
Yale University Press (1962)

**Weisz, Paul B. (ed.)**
*The Contemporary Scene:*
Readings on Human Nature, Race, Behavior, Society and Environment
New York (1970)

**Welch, P. S.**
*Limnological Methods*
New York (1948)

**Welch, P. S.**
*Limnology*
New York (1952)

**Welford, G. A. and W. R. Collins, Jr.**
"Fallout in New York City during 1958"
*Science,* 131:1711-1715 (1960)

**Welty, Joel Carl**
*The Life of Birds* (1963)

**Wenkam, Robert**
*Kauai, and the Park County of Hawaii*
San Francisco, Calif (1969)

**Went, F. W.**
"The Ecology of Desert Plants"
*Scientific American,* 192:68-75
(April 1955)

**Western Montana Scientists' Committee for Public Info.**
Dr. M. J. Law, Chmn.,
1650 Madeline Ave.,
Missoula, Mont.

**Wheeler, W. M.**
*Social Life among the Insects*
New York (1923)

**Whitaker, Bob**
"Primitive Areas—A New Designation Under BLM"
*The Living Wilderness,* p. 12-14
(Autumn 1969)

**White, G. F.**
*Strategies of American Water Management*
Ann Arbor:
Univ. of Michigan Press (1969)

**White, L. Jr.**
"The Historical Roots of Ecologic Crisis"
*Science,* 155(3767:1203-1207
(1967)

**White, L. Jr.**
*Machina ex Deo*
Cambridge: M.I.T. Press (1968)

**Whitehead, A. N.**
*Science and the Modern World*
New York (1948)

**Whittaker, Robert H.**
*Communities and Ecosystems*
New York (1970)

**Wiggins, Dorothy and Loren I.**
"A Winter Journey to Point Barrow, with Summer Pictures of the Tundra Vegetation"
*Asa Gray Bulletin,* 2:83-92 (1953)

**Wilber, Charles G.**
*Biological Aspects of Water Pollution*
Springfield, Ill. (1969)

**Wiley, John P. Jr.**
"Space: A Barrier to the Species"
*Natural History,* p. 70-73
(January 1970)

**Williams, C. B.**
*Patterns in the Balance of Nature*
New York (1964)

**Williams, C. M.**
"Third-Generation Pesticides"
*Scientific American,* p. 13-17
(1967)

**Williamson, Francis S. L.**
"The SST, the Environment and the Credibility Gap"
*Bioscience,* 20:995 (1970)

**Williams, G. C.**
*Adaption and Natural Selection*
Princeton Univ. Press (1966)

**Wilson, B. R.**
*Environment, The University and the Welfare of Man*
Philadelphia (1969)

**Wilson, B. R. (ed.)**
*Environmental Problems*
Philadelphia (1971)

**Wilson, Billy Ray**
*Environmental Problems: Pesticides, Thermal Pollution and Environmental Synergisms*
Philadelphia (1968)

**Wilson, C. D.**
*Inadvertant Climate Modification Report of the Study of Man's Impact on Climate*
Cambridge, Mass.: M.I.T. Press
(1971)

**Wilson, Carroll L. and Matthews, William H.**
*Man's Impact on the Global Environment*
Cambridge, Mass.: M.I.T. Press
(1970)

**Wilson, D. P.**
*Life of the Shore and Shallow Sea*
London (1951)

**Wilson, E. O. and Bossert, W. H.**
*A Primer of Population Biology*
Stamford, Conn. (1971)

**Wilson, H. A. Jr.**
"Sonic Boom"
*Scientific American*, p. 36-43
(January 1962)

**Wolf, Leonard**
"Cleaning up the Merrimack"
*Bulletin of the Atomic Scientists*,
p. 16-22 (April 1965)

**Wolman, A.**
"Air Pollution:
Time for Appraisal"
*Science*, 159:1437-1440 (1968)

**Wolozin, Harold, ed.**
*The Economics of Air Pollution*
New York (1966)

**Wolstenholme, G., ed.**
*Man and His Future*
Toronto (1963)

**Wong, H. W. and Vessel, M. F.**
*Pond Life*
Reading, Mass. (1970)

**Wood, David L., Robert M. Silverstein and Minoru Kakajima, (eds.)**
*Control of Insect Behavior by Natural Products*
New York (1970)

**Woodbury, A. M.**
*General Ecology*
Philadelphia (1954)

**Woodwell, George M.**
"The Ecological Effects of Radiation"
*Scientific American*, p. 40-49
(June 1963)

**Woodwell, G. M.**
"Effects of Pollution on the Structure and Physiology of Ecosystem"
*Science*, 168:429-433 (1970)

**Woodwell, George M.**
"The Energy Cycle of the Biosphere"
*Scientific American*, p. 64-74
(Sept. 1970)

**Woodwell, George M.**
"Radiation and the Patterns of Nature"
*Science*, 156:461-470 (1967)

**Woodwell, George M.**
"Radioactivity and Fallout: The Model Pollution"
*Garden Journal*, p. 100-104
(August 1968)

**Woodwell, George M.**
"Radioactivity and Fallout: The Model Pollution"
*Bioscience*, 19:884-887 (1969)

**Woodwell, George M.**
"Toxic Substance and Ecological Cycles"
*Scientific American*, p. 24-31
(March 1967)

**World Health Organization**
*Problems in Community Waste Management*
New York (1969)

**World Health Organization**
*Research into Environmental Pollution*
New York (1968)

**World Wildlife Fund**
Dr. Ira N. Gabrielson,
910 17th St., NW,
Washington, D.C.

**Wormer, J. Van**
*The World of the American Elk*
Philadelphia (1969)

**Wright, Richard T.**
"Responsibility for the Ecological Crisis"
*Bioscience*, 20:851-853 (1970)

**Wrigley, E. A.**
*Population and History*
New York (1969)

**Wurtz, C. B.**
"Thermal Pollution"
*Environmental Problems*
Philadelphia (1968)

**Wyant, William K. Jr.**
"The Aspnall Report. This Land is Whose Land?"
*New Republic*, p. 10-11
(July 11, 1970)

**Wylie, Philip**
*The Magic Animal*
Garden City, N.Y. (1968)

**Wnne-Edwards, W. C.**
*Animal Dispersion in Relation to Social Behavior*
Edinburgh and London (1962)

**Wyoming Environmental Group**
Dan Tolin, 1754 S. Conwell St.,
Casper, Wyo.

**Yapp, W. B.**
*The Effects of Pollution on Living Material*
London (1959)

**Yerges, Lyle F.**
*Sound, Noise, and Vibration Control*
New York (1969)

**Yonge, C. M.**
*The Sea Shore*
London (1949)

**Young, Stanley P.**
*The Last of the Loners*
New York (1970)

**Zero Population Growth Inc.**
Mr. Harold Seielstad,
4080 Fabian Way,
Palo Alto, Calif.

**Z. P. G. Honolulu**
Sheila Babbie, P.O. Box 11127,
Honolulu, Hawaii 96822

**Zeuner, Frederick E.**
*A History of Domesticated Animals*
New York (1962)

**Ziswiler, Vinzenz**
*Extinct and Vanishing Animals.
A Biology of Extinction and Survival*
New York (1967)

**Zurhorst, Charles**
*The Conservation Fraud*
New York (1970)

**Zwerdling, Daniel**
"And Now, Mercury"
*New Republic*, p. 17-18
(August 1, 1970)

# Index